Other Titles in This Series

Volume

4 **William F. Langford and Wayne Nagata, Editors**
Normal forms and homoclinic chaos
1995

3 **Anthony Bloch, Editor**
Hamiltonian and gradient flows, algorithms and controls
1994

2 **K. A. Morris, Editor**
Control of flexible structures
1993

1 **Michael J. Enos, Editor**
Dynamics and control of mechanical systems: The falling cat and related problems
1993

Normal Forms and Homoclinic Chaos

Fields Institute Communications

The Fields Institute for Research in Mathematical Sciences

Normal Forms and Homoclinic Chaos

William F. Langford
Wayne Nagata
Editors

American Mathematical Society
Providence, Rhode Island

The Fields Institute
for Research in Mathematical Sciences

The Fields Institute is named in honour of the Canadian mathematician John Charles Fields (1863–1932). Fields was a remarkable man who received many honours for his scientific work, including election to the Royal Society of Canada in 1909 and to the Royal Society of London in 1913. Among other accomplishments in the service of the international mathematics community, Fields was responsible for establishing the world's most prestigious prize for mathematics research—the Fields Medal.

The Fields Institute for Research in Mathematical Sciences is supported by grants from the Ontario Ministry of Education and Training and the Natural Sciences and Engineering Research Council of Canada. The Institute is sponsored by McMaster University, the University of Toronto, and the University of Waterloo and has affiliated universities in Ontario and across Canada.

1991 *Mathematics Subject Classification.* Primary 58F36, 58F13.

Library of Congress Cataloging-in-Publication Data
Normal forms and homoclinic chaos / William F. Langford, Wayne Nagata, editors.
p. cm.—(Fields Institute communications / The Fields Institute for Research in Mathematical Sciences, ISSN 1069-5265; v. 4)
"Proceedings of the Workshop on Normal Forms and Homoclinic Chaos, held at the Fields Institute for Research in Mathematical Sciences, November 13 to 16, 1992, in Waterloo, Canada"—Pref.
Includes bibliographical references.
ISBN 0-8218-0326-3
1. Differentiable dynamical systems—Congresses. 2. Chaotic behavior in systems—Congresses. 3. Normal forms (Mathematics)—Congresses. I. Langford, W. F. (William F.). II. Nagata, Wayne. III. Workshop on Normal Forms and Homoclinic Chaos (1992 : Fields Institute for Research in Mathematical Sciences). IV. Series.
QA614.8.N67 1995
515′.352—dc20 95-20344
CIP

Printed in the United States of America.
∞ The paper used in this book is acid-free and falls within the guidelines established to ensure permanence and durability.
Information on copying and reprinting can be found in the back of this volume.
♻ Printed on recycled paper.
This publication was prepared by the Fields Institute using AMS-TEX and AMS-LATEX, the American Mathematical Society's TEX macro systems.
10 9 8 7 6 5 4 3 2 1 00 99 98 97 96 95

Contents

Preface

This volume contains the proceedings of the Workshop on Normal Forms and Homoclinic Chaos, held at The Fields Institute for Research in Mathematical Sciences, November 13 to 16, 1992, in Waterloo, Canada, as a part of The Fields Institute Program Year on Dynamical Systems and Bifurcation Theory. The Workshop bridged the local and global analysis of dynamical systems, with emphasis on normal forms and the homoclinic phenomena which may arise in normal forms. Both dissipative and conservative, and both discrete and continuous systems were considered, including perturbed systems which cross these boundaries. Specific topics explored in the Workshop included: the effects of symmetry on normal forms and on the persistence of homoclinic orbits, exchange of stability, symmetry-breaking, resonances, integrability, local chaos, the effects of truncation of the normal form tail, efficient symbolic computation of normal forms, numerical computation of orbits in phase space, roundoff errors and shadowing. These topics were motivated by, and have application to, systems in physics, biology, engineering and other fields.

The Workshop assembled outstanding researchers as well as students from diverse backgrounds and geographical locations, to share ideas and to work together with longer term participants at the Institute. Some of the workshop visitors stayed at the Institute for periods preceding or following the Workshop, for extended discussions and collaborations. In all, there were 32 presentations at the Workshop, of which 14 are represented by papers in this volume.

The method of normal forms has its roots in the work of Poincaré [1892] and Birkhoff [1927]. It is a powerful technique for analyzing dynamical systems near nonhyperbolic equilibrium points, that is, points where the linearization has one or more eigenvalues with zero real part. This is the only place where local behaviour may become complicated (as follows from the Rectification Theorem and the Hartman-Grobman Theorem), and therefore "interesting" from both the mathematical and applied viewpoints. The method applies generally to parametrized families of differential equations on $\mathbb{R}^n$

$$\frac{dy}{dt} = f(y, \mu), \quad y \in \mathbb{R}^n, \ \alpha \in \mathbb{R}^p \tag{1}$$

where y is the state variable, μ is the parameter (both may be vectors) and t is an independent real variable usually referred to as "time". The parametrized vector field $f(y, \mu)$ is assumed to be smooth. In the construction of a normal form, a sequence of canonical changes of coordinates is performed, to successively "simplify" the Taylor polynomials, of increasing degree, in the series expansion of f in terms of y at the singular point. Classical work on normal forms often focussed on questions of convergence. In fact, convergence rarely holds for normal forms of practical interest. Modern work, as in this volume, deals with polynomial truncations of normal form expansions and their relationship to the flow generated by the original vector field.

The normal form for a given vector field is not unique, but it depends on the criteria chosen to simplify the Taylor polynomials. One particularly astute choice is always possible; one may eliminate all terms except those which commute with the flow generated by the adjoint of the linearization of the vector field at the equilibrium point (Belitskii [1981], Cushman and Sanders [1986], Elphick et al. [1987]). This choice has two advantages: it eliminates the nonuniqueness in the normal form in a consistent, rational manner, and, more importantly, it introduces an explicit equivariance (or symmetry) into the normal form which facilitates the analysis and may not have been apparent in the original system.

The method of normal forms gains power when coupled with the Center Manifold Theorem, which allows one to separate the (well-understood) hyperbolic behavior from the more interesting nonhyperbolic behavior in the neighborhood of an equilibrium. A major difficulty in the study of equation (1) is that its dimension may be large; it may even be infinite-dimensional, as for example in problems of fluid dynamics (Chossat and Iooss [1994]) or reaction-diffusion problems (Scott [1991]). However, often one is interested in changes in the qualitative behavior of (1). Such changes can occur locally only at values of μ for which (1) is nonhyperbolic; these are called bifurcation points. For example, the Hopf bifurcation occurs when a pair of complex-conjugate simple eigenvalues crosses the imaginary axis as μ varies; then, subject to certain generic transversality conditions, a periodic solution is created or destroyed around the equilibrium, see e.g. Golubitsky and Langford [1981]. Typically for (1), only a small number of eigenvalues cross the imaginary axis at one time; the corresponding eigenspaces (or modes) are finite dimensional. In this situation, the Center Manifold Theorem guarantees the existence of a manifold, called the center manifold, which is invariant for the flow of (1) and of the same dimension as, and tangent to, the sum of the (generalized) eigenspaces corresponding to those eigenvalues with zero real part. If the remaining eigenvalues have negative real part, then the flow off of this manifold is characterized by exponential decay towards the manifold. Thus, all the interesting behavior occurs on this low-dimensional center manifold, and it is desirable to reduce (1) to an equation for the flow on the center manifold. Fortunately for applications, there is an algorithm to calculate the center manifold and this reduced equation to any order, see Iooss and Adelmeyer [1992], Wiggins [1990]. The reduction to a center manifold has become so standard in recent years that most of the papers in this volume proceed from the assumption that it has been performed.

Let us now assume that (1) has been reduced to an equation on a center manifold. Then, in spite of the low dimensionality of (1), the analysis is still difficult, even if one is interested only in qualitative behavior. Because of the assumed nonhyperbolicity, the linear part no longer determines the local dynamics as it does in the hyperbolic case (Hartman-Grobman Theorem). The problem is intrinsically nonlinear and the best one can hope for is that the dynamics on the center manifold is determined by a finite degree truncation of the Taylor series, i.e. by a Taylor polynomial of (1). The number of terms in the Taylor polynomial grows rapidly with the degree, and quickly becomes unmanageable, even in dimension 2 or 3. The method of normal forms allows one to dramatically reduce the number of terms in these polynomials, without changing the qualitative behavior of solutions.

Often it is possible to give a complete description of the dynamics defined by this polynomial normal form (as in the case of classical Hopf bifurcation); several papers in this volume do precisely this. Of course, it still remains to show that the dynamics found for the truncated polynomial normal form persists when the higher order remainder or "tail" is restored. In some cases, this follows easily by transversality arguments (using the Implicit Function Theorem); in other cases the tail changes the qualitative dynamics and sometimes there remain open questions regarding the effects of the tail.

Symmetry is an important sub-theme of this volume. First, there is the normal form symmetry mentioned above, whereby the truncated normal form commutes with the flow generated by the adjoint of the linear part. It is this symmetry which characterizes the terms which remain in the normal form. In addition, the original system (1) may possess spatial or other symmetries, which can be expressed as equivariance with respect to the action of a group Γ, i.e.

$$f(\gamma y, \mu) = \gamma f(y, \mu) \tag{2}$$

for all $\gamma \in \Gamma$ and for all (y, μ). In this case, the reduction to the center manifold and further reduction to the normal form may be performed in such a way as to preserve the symmetry; more precisely, the normal form also satisfies the equivariance equation (2), where Γ is now understood to be an appropriate representation of the original symmetry group, on the center manifold. This additional symmetry has the effect of knocking out additional terms in the normal form and simplifying the analysis. Thus, symmetry is a double-edged sword: although symmetry can force multiple eigenvalues and increase the dimension of the center manifold, thus complicating the analysis, it also dramatically cuts down the number of terms in the normal form. The net effect is that new and rich structures with interesting dynamics, previously difficult to understand, become amenable to analysis. A classic example which provides a fascinating variety of bifurcations is the Taylor-Couette problem, see Golubitsky and Langford [1988] or Chossat and Iooss [1994]. Several new examples are presented in this volume.

The second main theme of the Workshop centered around homoclinic orbits; that is, solutions of (1) which have an equilibrium point (or limit cycle) as both α-limit and ω-limit sets. (The case where the α-limit and ω-limit sets are distinct points (or limit cycles) is properly called a heteroclinic orbit, but is included in this discussion without distinction.) Homoclinic orbits are not generic in (1); typically a homoclinic orbit will disappear under a small perturbation, for example due to a change in μ. Furthermore, they are generally difficult to calculate, or even to detect, due to their nonlocal nature. Yet, homoclinic orbits are of great interest, due in part to their important role in the onset of chaos, see Newhouse [1980], Wiggins [1990]. One context in which homoclinic orbits have been found to occur, and can even be calculated, is in the perturbations (or unfoldings) of a differential equation in normal form. Thus, paradoxically, the "global dynamics" of a homoclinic orbit can be found by "local analysis", in the neighborhood of a nonhyperbolic equilibrium point. In fact, to date, this has been the most effective means to study homoclinic behavior in real physical systems. Therefore, the linking of normal forms with homoclinic orbits and chaos in the theme of the Workshop and in this volume is seen to be natural and fruitful. Recently, it was observed that the addition of a third ingredient, namely symmetry, to this recipe can lead to a new phenomenon: normal forms of symmetric systems can have homoclinic orbits which are structurally stable; that is, which persist under small perturbations which respect the symmetry (see Field [1980], Armbruster et al. [1988], Campbell and Holmes [1992]). These structurally stable homoclinic orbits (or cycles of heteroclinic orbits) were subjects of discussion at the Workshop, and continue to be investigated intensively.

All that has been said above regarding the differential equation (1) and its flow, normal form and homoclinic orbits, holds equally well for *maps* and in particular diffeomorphisms; in fact chaos occurs even more readily for maps than for flows. Periodic orbits of (1) lead naturally to the definition of Poincaré maps, which are investigated in several papers of this volume. Similarly, the above comments apply also to systems which are Hamiltonian. The additional structure assumed by a Hamiltonian system at once facilitates the analysis (e.g. by providing first integrals) and leads to complex dynamics, as seen in Marsden [1992] as well as this volume.

The Workshop represented by this volume was part of the Program Year on Dynamical Systems and Bifurcation Theory, held at The Fields Institute for Research in Mathematical Sciences, from September 1992 to August 1993. The Program Committee consisted of J. Chadam, L. Glass, W. Langford, J. Marsden and W. Shadwick. The main areas of emphasis were: bifurcations with symmetry; low dimensional dynamics including normal forms and invariant manifolds; geometric methods from Hamiltonian mechanics including higher dimensional Melnikov methods and Lagrangian reduction; applications to many disciplines, especially to problems involving pattern formation in partial differential equations. There were seven workshops held during the Program Year, including this one; proceedings of all the workshops are being published.

The Fields Institute for Research in Mathematical Sciences is supported by the Ontario Ministry of Education and Training and by the Natural Sciences and Engineering Research Council of Canada, and also by the sponsoring and affiliated universities. The Fields Institute has three main mandates: to promote leading edge research in the mathematical sciences, to increase opportunities for graduate and postdoctoral training in Ontario and Canada and to foster interactions between university based research and users of mathematics in industry. In addition, the Institute serves as a resource for mathematics education at all levels.

References

Armbruster, D., Guckenheimer, J. and Holmes, P. [1988] *Heteroclinic cycles and modulated travelling waves in systems with $O(2)$ symmetry.* Physica **29D**, 257-282.

Belitskii, G. [1981] *Normal forms relative to a filtering action of a group.* Trans. Moscow Math. Soc. **40**, 1-39.

Birkhoff, G.D. [1927] *Dynamical Systems.* AMS Publ., Providence RI.

Campbell, A.S. and Holmes, P. [1992] *Heteroclinic cycles and modulated traveling waves in a system with D_4 symmetry.* Physica **59D**, 52-78.

Chossat, P. and Iooss, G. [1994] *The Couette-Taylor Problem.* Springer-Verlag, New York.

Cushman, R. and Sanders, J. [1986] *Nilpotent normal forms and representation theory of $sl(2, \mathbb{R})$.* Multiparameter Bifurcation Theory, ed M. Golubitsky and J. Guckenheimer, Contemp. Math. **56**, 31-51, AMS, Providence RI.

Elphick, C., Tirapegui, E., Brachet, M.E., Coullet, P. and Iooss, G. [1987] *A simple global characterization for normal forms of singular vector fields.* Physica **29D**, 95-127.

Field, M.J. [1980] *Equivariant dynamical systems.* Trans. Amer. Math. Soc. **229** 185-205.

Golubitsky, M. and Langford W.F. [1981] *Classification and unfoldings of degenerate Hopf bifurcations.* J. Diff. Eqns. **41**, 375–415.

Golubitsky, M. and Langford, W.F. [1988] *Pattern formation and bistability in flow between counterrotating cylinders.* Physica **32D**, 362-392.

Golubitsky, M., Stewart, I. and Schaeffer, D. G. [1988] *Singularities and Groups in Bifurcation Theory. Vol. 2.* Springer-Verlag, New York.

Iooss G. and Adelmeyer M. [1992] *Topics in Bifurcation Theory and Applications.* World Scientific, Singapore.

Marsden, J.E. [1992] *Lectures on Mechanics.* London Mathematical Society Lecture Note Series **174**, Cambridge University Press, Cambridge.

Newhouse S.E. [1980] *Lectures on Dynamical Systems.* Dynamical Systems, 1-114, C.I.M.E. Lectures, ed. J. Guckenheimer, Birkhauser, Boston.

Poincaré H. [1892] *Les Méthodes Nouvelles de la Mécanique Céleste.* Gauthier-Villars, Paris, 3 vols. Translation [1993] by D.L. Goroff, *New Methods of Celestial Mechanics.* American Institute of Physics, Woodbury, NY.

Scott, S.K. [1991] *Chemical Chaos.* Oxford, New York.

Wiggins, S. [1990] *Introduction to Applied Nonlinear Dynamical Systems and Chaos.* Springer-Verlag, New York.

Fields Institute Communications
Volume **4**, 1995

The Hamiltonian Double-Zero Eigenvalue

H.W. Broer
Department of Mathematics and Computing Science
University of Groningen
P.O. Box 800
9700 AV Groningen
The Netherlands

S-N. Chow
School of Mathematics
Georgia Institute of Technology
Atlanta, GA, USA 30332

Y-I. Kim
P.O. Box 18
Department of Mathematics
University of Ulsan
Ulsan, South Korea 680-749

G. Vegter
Department of Mathematics and Computing Science
University of Groningen
P.O. Box 800
9700 AV Groningen
The Netherlands

Abstract. A universal local bifurcation analysis is presented of an autonomous Hamiltonian system around a certain equilibrium point. This central equilibrium has a double zero eigenvalue, the other eigenvalues being in general position. Main emphasis is given to the 2 degrees of freedom case where these other eigenvalues are purely imaginary. By normal form techniques and Catastrophe Theory unfoldings are obtained having 'integrable' approximations related to the Fold and to the Elliptic and Hyperbolic Umbilic Catastrophes.

1 Setting of the problem, outline

This paper concerns autonomous Hamiltonian systems around an equilibrium point, with a double eigenvalue zero. The main problems already occur in the case of 2 degrees of freedom, where the other eigenvalues form a purely imaginary

1991 *Mathematics Subject Classification.* Primary 58F05; Secondary 58F36, 34C15, 34C20.

pair. Therefore, this research is in the Hamiltonian tradition of e.g. Meyer [1975], Sanders [1977], van der Meer [1985] and Verhulst [1983]. It is our aim to describe versal, viz. generic unfoldings of this equilibrium point. Its codimension depends on whether the linear part is semisimple or not. The non-semisimple case has codimension 1 and the semisimple one, codimension 3; this means that they will only be met generically if at least 1, respectively 3, parameters are present. To fix thoughts we include the following example.

Example 1.1 Consider a system of coupled oscillators

$$\begin{aligned} \ddot{x} + x &= f(x, y, \mu) \\ \ddot{y} + \varepsilon y &= g(x, y, \mu), \end{aligned}$$

where $\mu = (\varepsilon, \cdots) \in \mathbb{R}^p$ is a vector of parameters and where f and g contain nonlinearities. For $|\varepsilon| \ll 1$, such a system goes through many high order resonances. Our unfolding theory here applies. Other examples, also of the semisimple case, are to be expected in the dynamics of the rigid body.

The conjugate pair of imaginary eigenvalues gives rise to a formal rotational symmetry in all unfoldings, i.e. a symmetry in their Taylor series. Here the series is considered in dependence on both the phase space variables and the parameters. This is a standard application of Normal Form Theory, where the terms of the series are changed by canonical coordinate transformations in an inductive process. This rotational symmetry, as in the Kepler problem, yields a (formal) integral and a corresponding reduction to 1 degree of freedom. The reduced system has an equilibrium point with a double eigenvalue 0. For similar approaches see some of the above references, also compare e.g. Arnold [1983], Takens [1974a], Takens [1974b], Broer [1981a], Golubitsky and Stewart [1987] and Broer and Vegter [1992].

This Normal Form will be presented in Section 3. It gives rise to a perturbation problem where the perturbation term is of arbitrarily high order, both in the phase space variables and the parameters.

In Section 2, we begin studying the 1 degree of freedom 'backbone'-problem in its own right. Since in the plane the integral curves of the systems are the level sets of the Hamiltonian functions, here the problem completely reduces to Catastrophe Theory for these Hamiltonians, e.g. compare Bröcker and Lander [1975], Gibson [1979], Martinet [1977], Poston and Stewart [1978] and Thom [1972]. It turns out that the *Fold* and the *Elliptic* and *Hyperbolic Umbilic Catastrophes* contain all the information we want.

Section 4 is the main part of this paper. Here the connection is made between the planar reduction of the symmetric system and the 'backbone'-system of Section 2. It turns out that the integral, obtained by normalization, is a *distinguished* parameter in the sense of Golubitsky and Schaeffer [1985] and Schecter [1992], also compare Damon [1988]. For a similar set-up, see the earlier work of Wassermann [1975]. Now Catastrophe Theory yields new normal forms that are polynomial of degree 3, at least in the phase-space variables. We note that here the normalizing transformations no longer need to be canonical.

Remarks (i) For details we refer to Broer et al. [1993], of which the present paper is a summary. This especially holds for some technical details from Catastrophe Theory with distinguished parameters, that have been collected in an Appendix of Broer et al. [1993].
(ii) Although the original family is *generic,* by Normal Form and Catastrophe Theory, the corresponding unfolding is normalized to an arbitrarily flat perturbation of a form, completely determined by a 1 degree of freedom system, which is polynomial of degree 3. This polynomial character may explain why certain 'integrable' characteristics in the unfoldings are so persistent. This kind of persistence phenomenon has been observed before, for example, see Verhulst [1983].
(iii) The approach of van der Meer [1985] (also compare e.g. Duistermaat [1983]) differs from ours. In fact, after a normal form approximation he reduces the dynamics by the energy-momentum map and the Moser-Weinstein method. This also leads to Catastrophe Theory for planar functions. In this way the dynamical information that is obtained mainly restricts to certain periodic solutions that are important for the whole dynamics.

Our approach just factors out the formal rotational symmetry and seems to get a more direct hold on the global dynamics. It would be interesting to know how both methods compare in this case.

2 The planar backbone

We consider C^∞-functions $H : \mathbb{R}^2 \to \mathbb{R}$ with the origin as a critical point. The corresponding Hamiltonian vector field, to be denoted by X_H, then has the origin as an equilibrium or singular point. We here recall that $dH = \omega(X_H, .)$, where $\omega = dx \wedge dy$ denotes the standard area 2-form on $\mathbb{R}^2$. For this use of notation, e.g. see Abraham and Marsden [1978] and Arnold [1980]. In coordinates, for X_H we get the familiar expression

$$\dot{x} = \frac{\partial H}{\partial y}, \; \dot{y} = -\frac{\partial H}{\partial x}.$$

Our concern is with the situation where the linear part of X_H at the origin has a double eigenvalue zero, and the question is what are generic unfoldings of this singularity both in the semisimple and the non-semisimple case. In the former of these cases the corresponding Hamiltonian H, at the origin, vanishes to third order. Catastrophe Theory (Bröcker and Lander [1975], Gibson [1979], Martinet [1977], Poston and Stewart [1978], Thom [1972]), applied to planar Hamiltonian functions provides a good framework for this problem. This kind of application often is referred to as *dynamical.*

To explain this consider a C^∞-transformation $\Phi : \mathbb{R}^2 \to \mathbb{R}^2$ such that for Hamiltonians H and K one has $H = K \circ \Phi$. Such a map Φ is called a *right equivalence* between H and K. The following lemma compares the Hamiltonian vector field X_K with the transformed vector field $\Phi_*(X_H)$, defined by $\Phi_*(X_H)\,(\Phi(p)) := D\Phi(p)\,(X_H)(p)$.

Lemma 2.1 $\Phi_*(X_H) = \det D\Phi \; X_K$.

The proof is straightforward, e.g. see Broer et al. [1993]. Since the vector fields X_H and $\Phi_*\,(X_H)$ are *conjugate,* the presence of the scalar factor $\det D\Phi$ implies that the vector fields X_H and X_K are *equivalent.* We here recall that conjugacies take integral curves to integral curves in a time-preserving way. However, equivalences do take integral curves to integral curves, but without necessarily preserving this time-parametrisation. Observe that Φ is a conjugacy between X_H and X_K precisely if det $D\Phi \equiv 1$, meaning that Φ is both area and orientation preserving, which in the present setting is the same thing as *symplectic* or *canonical.*

This set-up is widened by also allowing *left equivalences:* C^∞-transformations in the image space $\mathbb{R}$ of the functions. In the case where the systems depend on parameters, we use parameter-dependent (left-right) equivalences on the Hamiltonians, together with C^∞-*reparametrizations*. Such a compound transformation is called a *morphism of unfoldings*.

Catastrophe Theory classifies (germs of) generic unfoldings of functions under the equivalence relation provided by these morphisms. From the above it may be clear that this classification also is relevant for the corresponding Hamiltonian vector fields: it suffices to give a *qualitative* picture of the phase portraits. However, also many quantitative features are kept track of in this way. For similar approaches see Arnold [1983], Broer [1981a], Broer and Vegter [1992], Duistermaat [1983] or van der Meer [1985].

2.1 The linear classification We first briefly discuss the linear classification, speaking in terms of singularities of Hamiltonian vector fields X_H. So we consider its linear part

$$\begin{pmatrix} \dot{x} \\ \dot{y} \end{pmatrix} = L \begin{pmatrix} x \\ y \end{pmatrix},$$

where L is a 2 by 2 matrix. It is easily seen that L has trace zero, i.e., $L \in sl(2,\mathbb{R}) = sp(2,\mathbb{R})$. Here our morphisms induce the following equivalence relation:

$$L \sim M \iff \exists S \in Gl(2,\mathbb{R}), \exists \kappa \in \mathbb{R} \setminus \{0\} : S \circ L \circ S^{-1} = \kappa M.$$

Note that only using right equivalences would imply $\kappa = 1$. Moreover, in the special case where these right equivalences are canonical, one even has det $S = 1$. The corresponding partition into equivalence classes now easily follows: it is a stratification based on the determinant-function. For example, compare Gibson [1979] and Poston and Stewart [1978].

1. The *hyperbolic* case: det $(L) < 0$, corresponding normal forms:

$$L = \begin{pmatrix} 1 & 0 \\ 0 & -1 \end{pmatrix}, H(x,y) = xy.$$

2. The *elliptic* case: det $(L) > 0$, corresponding normal forms:

$$L = \begin{pmatrix} 1 & 1 \\ -1 & 0 \end{pmatrix}, H(x,y) = \frac{1}{2}(x^2 + y^2).$$

3. The *parabolic* case: det $(L) = 0$, $L \neq 0$, corresponding normal forms:

$$L = \begin{pmatrix} 0 & 1 \\ 0 & 0 \end{pmatrix}, H(x,y) = \frac{1}{2}y^2.$$

4. The *zero* case: $L = 0$, corresponding normal forms:

$$L = 0, \ H(x,y) \equiv 0.$$

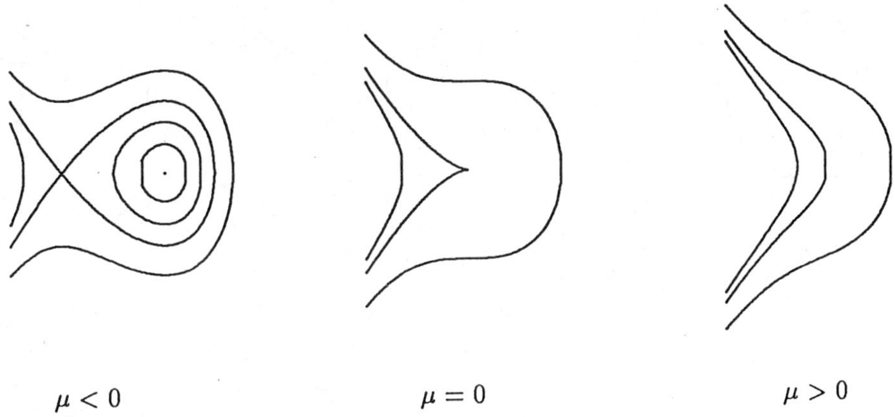

Figure 1 The Hamiltonian Saddle-Node bifurcation

The hyperbolic and elliptic cases 1 and 2 correspond to open (codimension 0) subsets of the 3-dimensional space $sl(2, \mathbb{R})$, and the parabolic case 3 to a codimension 1 cone-surface. The zero case 4 then corresponds to the vertex of this cone, so it has codimension 3. The latter two singularities are the object of our study, case 4 being the semisimple and case 3 the non-semisimple case. In each case this geometric codimension also is the number of unfolding parameters necessary to 'see' the corresponding singularity.

2.2 The fold and the elliptic and hyperbolic umbilic catastrophes What happens to the above linear picture if higher order terms are added? This question is addressed by Catastrophe Theory. For details we again refer to Bröcker and Lander [1975], Gibson [1979], Martinet [1977], Poston and Stewart [1978], and Thom [1972]. The codimension 0 cases 1 and 2—the saddle-point and the center—can be handled by the *Morse Lemma*, which states that the normal forms given above also hold with the higher order terms added. In these cases we have local *structural stability*. (Notice that here we restrict to the class of functions having the value 0 at the origin of $\mathbb{R}^2$.) The situation is less easy in the remaining degenerate cases.

In the *parabolic* case, the *Splitting Lemma* (see the above references) tells us that, after carrying out an appropriate (parameter-preserving) morphism, we only need to consider 1 parameter families of the form $H^\mu(x, y) = \frac{1}{2} y^2 + V^\mu(x)$, where $V^0(x) = O(x^3)$ as $x \to 0$. So, the Splitting Lemma splits the unfolding into a 'Morse part' $\frac{1}{2} y^2$, and a part $V^\mu(x)$ 'containing' the degenerate singularity. A *universal unfolding* is given by

$$H^\mu(x, y) = \frac{1}{2} y^2 + \frac{1}{3} x^3 + \mu x,$$

where the family of 'potential' functions $V^\mu(x) = \frac{1}{3} x^3 + \mu x$ is a *Fold Catastrophe*. Again, this universal unfolding is a normal form, i.e. a preferred member of its

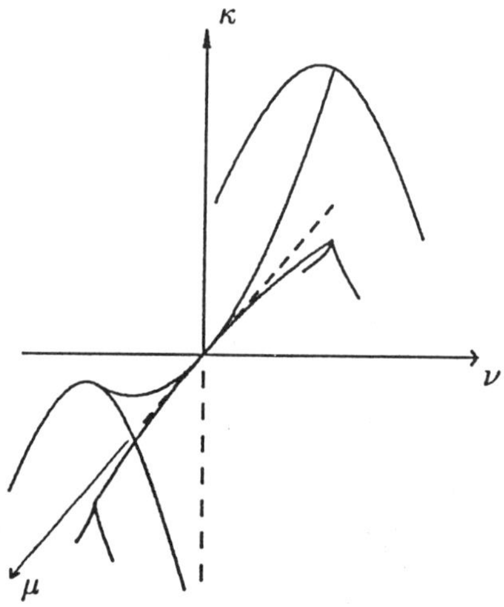

Figure 2 Catastrophe sets of the Hyperbolic Umbilic

equivalence class, now considered in the world of all 1-parameter families. In this world the universal unfolding is structurally stable. Finally, generic unfoldings of the parabolic singularity with more than 1 parameter can be *reduced* to the case with 1 parameter: a suitable morphism leads to the case where the unfolding does not explicitly depend on the other parameters.

In Figure 1 we depict the well-known phase portraits. For $\mu > 0$ no singularities exist, while for $\mu < 0$ a (nondegenerate) saddle-point and center are present, their distance to the origin being of the order $\sqrt{-\mu}$. For $\mu = 0$ the origin occurs as the parabolic singularity. We shall refer to this bifurcation as the *Hamiltonian Saddle-Node bifurcation.*

In the *zero* case the situation is even more complicated. Since the linear part L completely vanishes, we do not have a convenient splitting. Catastrophe Theory provides the following two universal unfoldings:

$$H^{\mu,\nu,\kappa}(x,y) = x^2 y \pm \frac{1}{3} y^3 + \mu(x^2 \mp y^2) + \nu x + \kappa y,$$

the 'upper signs' corresponding to the *Hyperbolic* and the 'lower signs' to the *Elliptic Umbilic Catastrophe.* In another classification, see Arnold [1972], these unfoldings are labeled $D_4^{\pm}$. Again universality means being a preferred member of ones equivalence class of 3-parameter unfoldings, where the number of parameters can be reduced to 3. Also here the families are structurally stable.

We conclude this section with a brief description of the Hyperbolic Umbilic catastrophe, the Elliptic case being somewhat more familiar (e.g. see the cover of Poston and Stewart [1978]). For more infomation see Poston and Stewart [1978], Thom [1972] and Broer et al. [1993].

2.3 Description of the hyperbolic umbilic catastrophe We begin by giving the catastrophe set $\mathcal{C}$ of the Hyperbolic Umbilic Catastrophe in the parameter space. This is the set of parameter-points (μ, ν, κ) where the Hamiltonian $H^{\mu,\nu,\kappa}$ has

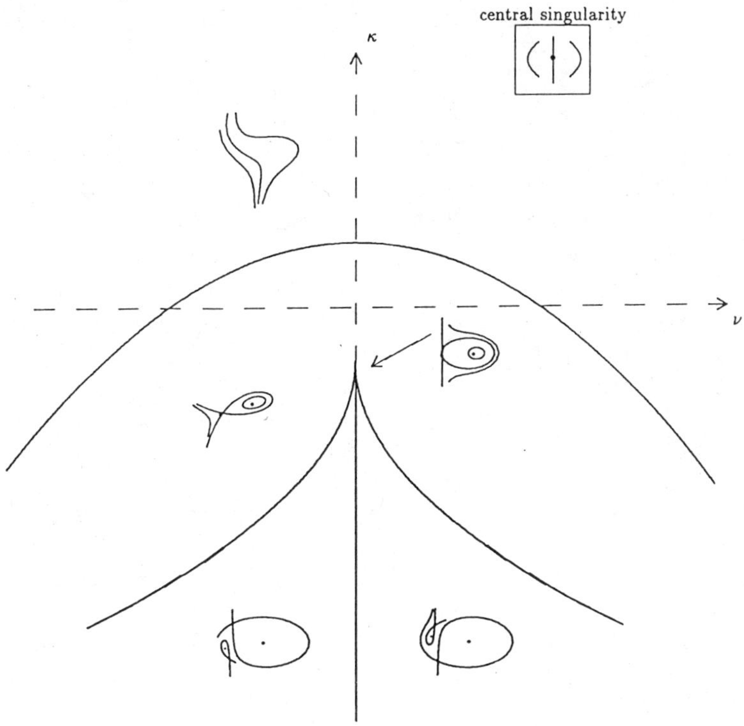

Figure 3 Bifurcation diagram of the Hyperbolic Umbilic, $\mu \neq 0$ fixed

a degenerate critical point. The set $\mathcal{C}$, depicted in Figure 2, is given parametrically by

$$\mu^2 = y^2 - x^2,\ \nu = -2\mu x - 2xy,\ \kappa = 2\mu y - y^2 - x^2,$$

with x and y as parameters. It has a cone-like structure. The axis of the 'cone' is the μ-axis and the base shrinks quadratically in $|\mu|$ as $\mu \to 0$. Moreover, the parameter μ, for $\mu \neq 0$ only affects the characteristic size of the phase portraits, while its sign governs the orientation of the flow. For $\mu = 0$ the central umbilical singularity occurs at the origin.

In Figure 3 the phase portraits are given relative to the 2-dimensional section $\mu = 1$. (Also the central umbilical singularity is depicted, as it occurs for $\mu = 0$.)

1. Upon transversal crossing of the arcs in this section the Hamiltonian Saddle-Node, described above, occurs in a secondary way.

2. There exists a local 2-parameter subfamily, transversal to the cusp. Also here the Splitting Lemma applies. The 'potential' function $V^{\mu,\nu}$ then undergoes a *Cusp Catastrophe*. The corresponding Hamiltonian family, in its own right, has normal form $H^{\mu,\nu}(x,y) = \frac{1}{2}y^2 \pm \frac{1}{4}x^4 + \frac{1}{2}\mu x^2 + \nu x$. For pictures e.g. see Poston and Stewart [1978], Thom [1972].

3. A secondary *heteroclinic bifurcation*, occurs at the 'bisector' of the cusp. By this bisector, the bifurcation set extends the catastrophe set.

Remark 2.2 In a set-up where general 3-parameter families of planar Hamiltonians are studied locally, one would have to extend the list of 'primary' cases. First of all the Cusp Catastrophe—for a normal form see above—would occur as a primary unfolding. Second, a similar remark goes for the Swallowtail Catastrophe with normal form $H^{\mu,\nu,\kappa}(x,y) = \frac{1}{2}y^2 + \frac{1}{5}x^5 + \frac{1}{3}\mu x^3 + \frac{1}{2}\nu x^2 + \kappa x$. For pictures again see the above references.

3 A Hamiltonian normal form: Statement of the perturbation problem

We now return to the 4-dimensional setting. To begin with, we notice that a direct approach applying Catastrophe Theory to the Hamiltonian functions does not give much information on the dynamics, but only on the foliation of their 3 dimensional level sets and its singularities. Therefore, instead, we present a Formal Normal Form permitting the formulation of a perturbation problem related to the 2-dimensional situation of Section 2. The reason is that formally a rotational symmetry can be factored out, which in the approximation leads to an invariant foliation of phase space with 2-dimensional leaves.

The relevant normal form theory started with Poincaré [1928] and Birkhoff [1927], for more recent references see Gustavson [1966], Moser [1968], Moser [1973], Takens [1974a], Takens [1974b], Arnold [1980], Sanders [1977], Broer [1981a], Sanders and Verhulst [1985], van der Meer [1985] and Broer and Vegter [1992].

3.1 The normal form To be more precise, we endow $\mathbb{R}^4$ with coordinates (x_1, y_1, x_2, y_2) and symplectic form $\omega = dx_1 \wedge dy_1 + dx_2 \wedge dy_2$, and consider a C^∞-family of Hamiltonian functions H^μ, $\mu \in \mathbb{R}^p$ being a vector of parameters. As before, for any Hamiltonian H the corresponding Hamiltonian vector field X_H is given by $dH = \omega(X_H, .)$, which in coordinates means

$$\dot{x}_j = \frac{\partial H}{\partial y_j},\ \dot{y}_j = -\frac{\partial H}{\partial x_j},$$

for $j = 1, 2$. We assume that for $\mu = 0$ the origin of $\mathbb{R}^4$ is a singularity. Then we expand as a Taylor series in (x, y, μ)

$$H^\mu(x,y) = H_2(x,y,\mu) + H_3(x,y,\mu) + \dots\,,$$

where H_n is homogeneous of degree n in (x, y, μ). Assuming that the singularity at $(x, y, \mu) = (0, 0, 0)$ has eigenvalues 0 (double) and $\pm i\alpha$, for some real constant $\alpha \neq 0$, it is our aim to normalize or simplify these homogeneous parts, using induction on the degree n. Here 'simple' means 'rotationally symmetric' in a way to be explained below. To this purpose we carry out suitable *symplectic* transformations.

First, by Williamson's Normal Form, compare Galin [1982] and Koçak [1984], abbreviating $I := \frac{1}{2}({x_1}^2 + {y_1}^2)$, we normalize the second order part H_2 to

$$H_2(x,y,\mu) = \alpha I + \Delta {y_2}^2,$$

with $\Delta = \pm 1$ in the non-semisimple and $\Delta = 0$ in the semisimple case. The corresponding (infinitesimally) symplectic matrices, being the linear part of the corresponding Hamiltonian vector fields X_{H^0}, have the form

$$\begin{pmatrix} 0 & \alpha & 0 & 0 \\ -\alpha & 0 & 0 & 0 \\ 0 & 0 & 0 & \Delta \\ 0 & 0 & 0 & 0 \end{pmatrix}.$$

The codimension of these singularities are the same as the corresponding ones in Section 2, i.e. 3 for the semisimple one and 1 for the non-semisimple one. Recalling that our parameter space is $\mathbb{R}^p$, we fix $p = 3$ in the semisimple and $p = 1$ in the non-semisimple case.

The function I, viz. the Hamiltonian vector field X_I, is used to give the rotational symmetry.

Theorem 3.1 *There exists a formal canonical transformation $\hat{\Phi}(x, y, \mu)$, keeping the parameters fixed, and a formal power series $\hat{F}(I, x_2, y_2, \mu)$ such that, formally speaking*

$$(\hat{H} \circ \hat{\Phi})(x, y, \mu) = \hat{F}(I, x_2, y_2, \mu).$$

Remark 3.2 In the non-semisimple (nilpotent) case one may 'improve' the normal form as in e.g. van der Meer [1985]; this means that more terms in the Taylor series can be transformed away. Another way to 'improve' the normal form is to incorporate higher order terms of the Hamiltonian in the normalizing (adjoint) action. However, in the present case, since we can apply Catastrophe Theory in a straightforward manner, only the rotational symmetry is of importance.

Let us see what the formal statement of Theorem 3.1 means on the level of C^∞-functions. Here we use a theorem of É. Borel saying the following, e.g. compare Narasimhan [1968]. Given any formal power series $\hat{\Phi}$ in the variables (x, y, μ), there exists a C^∞-map Φ with $\hat{\Phi}$ as its Taylor series. We apply this theorem to Φ. A careful look at the level of generating functions ensures us that in this particular case it is also possible to choose Φ canonical. So we obtain:

Corollary 3.3 *There exists a C^∞-canonical transformation Φ, which keeps the parameters fixed, and there exist C^∞-functions $F(I, x_2, y_2, \mu)$ and $P(x, y, \mu)$, such that:*

1. *P is infinitely flat at $(x, y, \mu) = 0$;*
2. *$(H \circ \Phi)(x, y, \mu) = F(I, x_2, y_2, \mu) + P(x, y, \mu)$.*

We next consider the vector field associated to the symmetric Hamiltonian $F(I, x_2, y_2, \mu)$. As remarked earlier, one consequence of the rotational symmetry is that I is an integral of the system. This gives us the 2-dimensional foliation discussed at the beginning of this section. We proceed by factoring out the rotational symmetry. To this end we pass to Hamiltonian polar coordinates (I, φ) by

$$x_1 = \sqrt{2I}\cos\varphi, \quad y_1 = \sqrt{2I}\sin\varphi.$$

Since $dI \wedge d\varphi = dx_1 \wedge dy_1$, we directly conclude:

Corollary 3.4 *In the coordinates (I, φ, x_2, y_2) the Hamiltonian vector field X_F has the form*

$$\dot{I} = 0, \qquad \dot{\varphi} = -\frac{\partial F}{\partial I},$$
$$\dot{x}_2 = \frac{\partial F}{\partial y_2}, \qquad \dot{y}_2 = -\frac{\partial F}{\partial x_2}.$$

The latter two equations in Corollary 3.4 form the *reduction to* 1 *degree of freedom*: it is a family of planar Hamiltonian vector fields, parametrized by I and μ.

3.2 The perturbation problem Corollary 3.3 provides the setting for the perturbation problem. This problem can best be phrased in terms of the isoenergetic Poincaré map of the section $\varphi = 0$. These maps are planar and symplectic, i.e. area and orientation preserving.

The 'unperturbed' ones, based on the family F just follows the orbits of planar Hamiltonian vector fields. These orbits are formed by intersecting the 2-dimensional foliation mentioned earlier, with the section $\varphi = 0$, where we restrict to fixed level sets of the Hamiltonian family F. The 'perturbed' maps, based on the family H are generic, flat perturbations of those. For general information on this subject see Broer and Takens [1989] and the references quoted there.

In the present paper we do not go into the perturbation problem at all. For some general remarks and references see Broer et al. [1993], Section 5. The emphasis of this paper is on the integrable family $F = F(I, x_2, y_2, \mu)$ and its Catastrophe-Theory Normal Form, where a good track has to be kept of the special rôle of the distinguished parameter I. This is the subject of the next section.

Remark 3.5 We recall that initially the dynamics of X_F is given restricted to fixed levels of the integral I. We wish to transfer this description to fixed levels of the energy F. In terms of the aforementioned foliation the functions F and I have a quite symmetric rôle, which directly gives the desired transfer. For details see Broer et al. [1993], Section 4.2.

4 Back to the plane

Now we consider the integrable Hamiltonian F for its own sake. Writing

$$F^{I,\mu}(x_2, y_2) := F(x_2, y_2, I, \mu),$$

we obtain a family of functions parametrized by I and μ. Notice that $I \geq 0$, so here the parameter space is a manifold with boundary, being a halfspace of $\mathbb{R}^{1+p}$. From the symmetry it follows that F is smooth at the boundary hyperplane $I = 0$.

From now on we abandon the world of canonical transformations and, as in Section 2, we use the morphisms based on parameter dependent left-right equivalence and reparametrization. For the moment, since they do not influence the phase portraits, we disregard the terms in F that do not depend on (x_2, y_2).

In this planar setting we only allow reparametrizations of the form $(I, \mu) \in \mathbb{R}^{1+p} \mapsto (J, \nu) \in \mathbb{R}^{1+p}$, with $J = J(I, \mu)$ and $\nu = \nu(\mu)$. Both these reparametrizations and the corresponding morphisms will be called *restricted*. The parameter I will be called a *distinguished* parameter. We adopt this terminology from Golubitsky and Schaeffer [1985] and Schecter [1992], who work in the related setting of contact-equivalence.

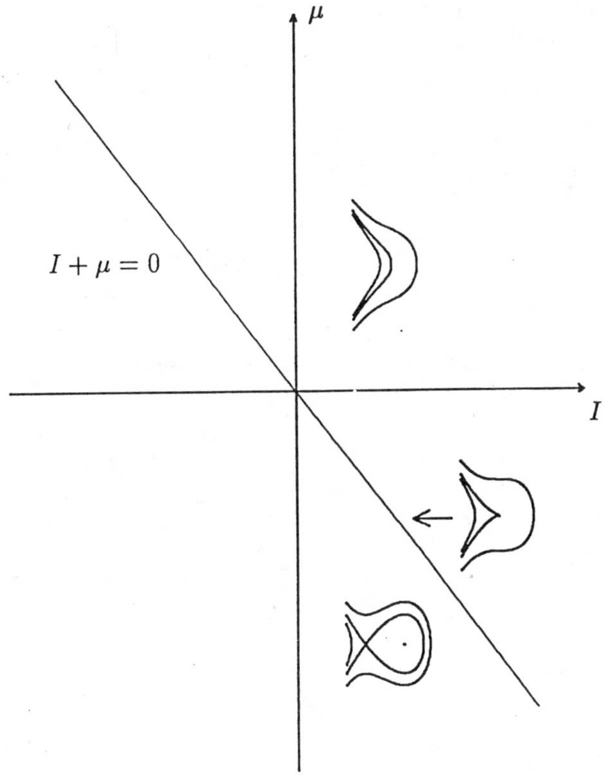

Figure 4 Bifurcation diagram for the reduced integrable Hamiltonian Saddle-Node bifurcation

The distinguished parameter I is non-negative. If this property is preserved by the morphisms, we say that they *respect the boundary*. This means that $J(I,\mu) = I\tilde{J}(I,\mu)$, with $\tilde{J}(0,0) > 0$.

4.1 The reduction to 1 degree of freedom: Catastrophe Theory revisited As our main result, we present normal forms for our planar reductions, viz. the family $F^{I,\mu}$, using restricted morphisms.

Theorem 4.1 *In the non-semisimple case, up to restricted morphisms respecting the boundary, generically the family* $F^{I,\mu}$ *has the form*

$$F^{I,\mu}(x_2, y_2) = \frac{1}{2}y_2^2 + \frac{1}{3}x_2^3 + (\mu \pm I)x_2.$$

This normal form is structurally stable under restricted morphisms: C^3-small changes of the 1-parameter family yield an equivalent family. In fact, we obtained a *universal unfolding* with respect to restricted morphisms. In Figure 4 the bifurcation diagram is depicted. Although it can be extended over the boundary $I = 0$, only the part $I \geq 0$ is of importance to us now. Notice that the Hamiltonian Saddle Node bifurcation occurs upon traversing the line $\mu \pm I = 0$.

In the semisimple case the parameter μ is 3-dimensional, writing $\mu = (\mu_1, \mu_2, \mu_3)$ we have:

Theorem 4.2 *In the semisimple case, up to restricted morphisms respecting the boundary, generically the family* $F^{I,\mu}$ *has one of the forms*

$$\begin{aligned} F^{I,\mu}(x_2, y_2) &= x_2^2 y_2 - \frac{1}{3} y_2^3 + (\mu_1 \pm I)(x_2^2 + y_2^2) + \\ &\quad \delta_1(I,\mu) x_2 + \delta_2(I,\mu) y_2, \\ F^{I,\mu}(x_2, y_2) &= x_2^2 y_2 + \frac{1}{3} y_2^3 + (\mu_1 \pm I)(x_2^2 - y_2^2) + \\ &\quad \delta_1(I,\mu) x_2 + \delta_2(I,\mu) y_2, \end{aligned}$$

for certain coefficient functions $\delta_j : (\mathbb{R} \times \mathbb{R}^3, 0) \to (\mathbb{R}, 0)$, $j = 1, 2$.

Below we roughly indicate how to prove these theorems. As announced before, we apply Catastrophe Theory to the given families $F^{I,\mu}$. A standard application of this theory yields normal forms, modulo general, unrestricted morphisms. The first step of our proof will use this result in order to find preliminary normal forms, modulo restricted morphisms. Secondly, we simplify these forms as far as possible, only using restricted reparametrization.

Remark 4.3 A similar unfolding theory, also in the setting of left-right equivalence, is given by Wassermann [1975]. In the case of Theorem 4.1 his method leads to a simpler normal form

$$F^{I,\mu}(x_2, y_2) = \frac{1}{2} y_2^2 + \frac{1}{3} x_2^3 + I\, x_2.$$

The difference is due to the fact that Wassermann uses a larger class of reparametrizations. Most importantly, he does not need our constraint that the set $\{I \geq 0\}$ has to be preserved. Moreover, he has more left-equivalences at his disposal, since these are allowed to depend on the distinguished parameter I. This is forbidden in our set-up, where I originates from phase space. Similar remarks go when comparing with the results of Golubitsky and Schaeffer [1985] and Schecter [1992].

4.2 Idea of the proofs We here restrict to the non-semisimple case of Theorem 4.1, for more details again referring to Broer et al. [1993].

We apply the Splitting Lemma and compare it to Section 2. This yields a parameter-preserving morphism, giving the form $\overline{F}^{I,\mu}(x_2, y_2) = \frac{1}{2} y_2^2 + V^{I,\mu}(x_2)$, for some family $V^{I,\mu}$ of 'potentials'. From Section 2 we also know that a normal form for these potential functions is the 1-parameter family $N^{\varepsilon}(x_2) := \frac{1}{3} x_2^3 + \varepsilon x_2$, the *Fold*. So, under the usual (generic) transversality conditions, we are given an unrestricted morphism

$$\mathbb{R} \times \mathbb{R}^2 \to \mathbb{R} \times \mathbb{R} : (x_2, I, \mu) \mapsto (H(x_2, I, \mu), \varepsilon(I, \mu)),$$

with $\varepsilon(0,0) = 0$ such that $N^{\varepsilon(I,\mu)}(H(x_2, I, \mu)) = V^{I,\mu}(x_2)$. In other words, if we consider the slightly adapted form

$$\overline{N}^{I,\mu}(x_2) := \frac{1}{3} x_2^3 + \varepsilon(I, \mu) x_2$$

we conclude by inspection that the *restricted* morphism

$$\mathbb{R} \times \mathbb{R}^2 \to \mathbb{R} \times \mathbb{R}^2 : (x_2, I, \mu) \mapsto (H(x_2, I, \mu), I, \mu),$$

satisfies $\overline{N}^{I,\mu}(H(x_2,I,\mu)) = V^{I,\mu}(x_2)$. This procedure leaves us with a preliminary normal form

$$\overline{F}^{I,\mu}(x_2,y_2) = \frac{1}{2}y_2^2 + \frac{1}{3}x_2^3 + \varepsilon(I,\mu)x_2.$$

The remainder of our proof further simplifies to the coefficient-function $\varepsilon(I,\mu) = \mu \pm I$, using restricted reparametrization. This is a straightforward application of the Inverse Function Theorem.

In the semisimple case the preliminary normal forms read

$$F^{I,\mu}(x_2,y_2) = x_2^2 y_2 \pm \frac{1}{3}y_2^3 + \varepsilon(I,\mu)(x_2^2 \mp y_2^2) + \delta_1(I,\mu)x_2 + \delta_2(I,\mu)y_2,$$

for coefficient functions $\varepsilon(I,\mu), \delta_1(I,\mu)$ and $\delta_2(I,\mu)$. Here we only can simplify the first one, again obtaining $\varepsilon(I,\mu) = \mu \pm I$.

4.3 A geometric picture, stuctural stability? We start presenting a geometric picture of the situation at hand. This picture came about in a discussion with Duistermaat. Let us have a closer look at the coefficient-functions $\varepsilon(I,\mu)$ and $(\varepsilon(I,\mu), \delta_1(I,\mu), \delta_2(I,\mu))$ of the above preliminary normal forms. Writing

$$h(I,\mu) := \begin{cases} \varepsilon(I,\mu) & \text{(non-semisimple case)} \\ (\varepsilon(I,\mu), \delta_1(I,\mu), \delta_2(I,\mu)) & \text{(semisimple case)} \end{cases}$$

we obtain a family of half-curves $\{I \geq 0 \mapsto h(I,\mu) \in \mathbb{R}^p\}_\mu$, parametrized by $\mu \in \mathbb{R}^p$. Here $p = 1$ in the non-semisimple and $p = 3$ in the semisimple case. The range $\mathbb{R}^p$ exactly is the parameter-space of the corresponding backbone-system of Section 2. This family of half-curves, by construction is invariant under restricted reparametrization. If $D_\mu h(0,0)$ has maximal rank, we can simplify the starting points $I = 0$ of our half-curves. If also $D_I h(0,0)$ has maximal rank, we can further simplify the I-dependence of exactly one of the components of h: we chose $\varepsilon(I,\mu) = \mu_1 \pm I$. We conclude that, at least in this way, no further simplification of the normal form is possible.

However, notice that we do construct our restricted morphisms in a specific way, first reducing to a polynomial form in (x_2, y_2) and then carrying out suitable reparametrizations. At this point it is not completely clear why further simplification would not be possible in a more general approach. This doubt is taken away in Broer et al. [1993], Section 7 (the Appendix). It then rigourously follows that the umbilical normal forms of Theorem 4.2 are not structurally stable, but that arbitrarily small changes of the coefficient-functions $(\delta_1(I,\mu), \delta_2(I,\mu))$ may yield families that are not equivalent under restricted morphisms.

Remark 4.4 In the normal forms of Golubitsky and Schaeffer [1985], Schecter [1992] and Wassermann [1975], all coefficients are polynomial. If this were also the case here, we would have only finitely many moduli. As said before our means of simplifying are more restricted, and the above argument casts doubt on this possibility.

5 More distinguished parameters

An interesting way to broaden our set-up is the introduction of more distinguished parameters. This is useful in higher degrees of freedom.

5.1 More normal ellipticity In order to motivate this, consider the case, where in the central singularity, next to the double eigenvalue 0, one has pure imaginary eigenvalues $\pm i\alpha_1, \ \pm i\alpha_2, \ \ldots, \pm i\alpha_k$, for $k \geq 2$. So we have $k+1$ degrees of freedom. Here *resonances* between the α_j come into play. To be precise, if for $< m, \alpha > := \sum_{j=1}^{k} m_j \alpha_j$, for a given $N \in \mathbb{N}$ one has

$$1 \leq \sum_{j=1}^{k} |m_j| \leq N \Rightarrow < m, \alpha > \neq 0,$$

a Normal Form result holds completely similar to the conclusion of Corollary 3.3. For general reference also compare, for instance, Birkhoff [1927], Broer [1981a], Takens [1974a], Takens [1974b]. In fact one now finds a canonical transformation Φ, keeping the parameters $\mu \in \mathbb{R}^p$ fixed, such that

$$(H \circ \Phi)^{\mu}(x, y) = F^{\mu}(I_1, I_2, \cdots, I_k, x_{k+1}, y_{k+1}) + P^{\mu}(x, y),$$

where for the perturbation term P one has the finite flatness $P^{\mu}(x, y, \mu) = O(|(x, y, \mu)|^N)$. So, up to an N-flat perturbation, one finds k integrals $I_1, I_2, \cdots, I_k$, providing a k-torus symmetry. Factoring out this symmetry, for $N \geq 4$, gives a reduction as before.

Remarks (i) In the case where strong resonances are present, the analysis becomes more complicated. In that case a straightforward application of Singularity Theory as before, does not apply. As an example consider the 3 degrees of freedom case with eigenvalues 0 and ± 1, each having multiplicity 2. Here the difficulties of the present paper are combined with those of van der Meer [1985].
(ii) In particular all this holds in the case where $N = \infty$, where, by the Borel Theorem, the term P again becomes infinitely flat, compare Section 3.1. In that case the frequencies α_j are independent over the rationals. This result may be of importance for certain problems with infinitely many degrees of freedom where $k = \infty$. It is not yet clear to us, how the asymptotics for $k \to \infty$ can be given a sensible meaning.

We now apply the methods of Section 4 and of Broer et al. [1993], Section 7, to the planar family of functions $F^{I,\mu}(x_{k+1}, y_{k+1}) := F^{\mu}(I_1, I_2, \cdots, I_k, x_{k+1}, y_{k+1})$. We then have k distinguished parameters $I = (I_1, \ I_2, \cdots, I_k)$. The definition of restricted morphism is extended to this context in a straightforward way. In the non-semisimple case the result of Theorem 4.1 still holds. Next let us look at the semisimple case of Theorem 4.2.

Theorem 5.1 *Suppose that the central singularity is semisimple and that* $p \geq 3$. *Then the umbilical families*

$$\begin{aligned} F^{I,\mu}(x_{k+1}, y_{k+1}) \ &= \ x_{k+1}^2 y_{k+1} \pm \frac{1}{3} y_{k+1}^3 + (\mu_1 + I_1)(x_{k+1}^2 \mp y_{k+1}^2) + \\ &\quad (\mu_2 + I_2)\, x_{k+1} + (\mu_3 + I_3)\, y_{k+1} \end{aligned}$$

are universal unfoldings with respect to restricted morphisms, under preservation of the set $\{I = 0\}$.

As before, the unfolding does not explicitly depend on the remaining (both distinguished and external) parameters. Moreover we obtain a structurally stable normal form. Finally, notice that one needs that $k \geq 3$, which rigourously implies that the normal forms of Theorem 4.2 are not versal; this kind of stability only occurs in 4 degrees of freedom!

5.2 Some background In Broer et al. [1993], Section 7, a general, though simple device is provided, for obtaining a universal unfolding with respect to restricted morphisms, given an 'ordinary' universal unfolding, i.e. with respect to unrestricted morphisms. Indeed, let a family of maps $(x, I) \in (\mathbb{R}^n \times \mathbb{R}^k) \mapsto f(x, I) \in \mathbb{R}$ be given, where we consider $f(x, I)$ as an unfolding of $f_0(x) := f(x, 0)$. Now, if $f = f(x, I)$ is a universal unfolding of f_0 with respect to unrestricted morphisms that preserve the set $\{I = 0\}$, then a universal unfolding of $f(x, I)$ in the restricted sense is given by $F : \mathbb{R}^n \times \mathbb{R}^k \times \mathbb{R}^k \to \mathbb{R}$, defined by

$$F(x, I, \mu) = f(x, I) + \sum_{j=1}^{k} \mu_j \frac{\partial f}{\partial I_j}(x, 0),$$

where, as before, I is the distinguished and μ the external parameter. Both the Theorems 4.1 and 5.1 are direct consequences of this.

In our application the distinguished parameters $I_1, I_2, \cdots, I_k$ all are non-negative. Therefore, as in the Theorems 4.1 and 4.2, the restricted morphisms can also be required to 'respect the boundary' $\{I_1 = 0\} \cup \{I_2 = 0\} \cup \cdots \cup \{I_k = 0\}$. However, in that case there is no hope of finding simple universal unfoldings. A simple example for the vector of coefficients (see above) like $h(I_1, I_2, \mu_1, \mu_2) = (I_1 + \mu_1, I_2 + I_1 + \mu_2)$, as may occur for $k = 2$, easily may convince the reader. Therefore, in our set-up, we only require the intersection $\{I = 0\} = \{I_1 = 0\} \cap \{I_2 = 0\} \cap \cdots \cap \{I_k = 0\}$ to be preserved.

Remark 5.2 A drawback of many applications of Catastrophe Theory is the lack of constructiveness regarding the normalizing transformations. This is unlike the situation in the Normal Form Theory of Section 3. The lack of constructiveness is felt the strongest, when dealing with concrete examples as mentioned in Section 1. In such cases, one would, for instance, like to compute the coefficients δ_j, $j = 1, 2$, of Theorem 4.2, which requires keeping track of all normalizing transformations. Similarly, in the above case with k pairs of elliptic eigenvalues, one would like to know where the normalizing morphism takes the boundary set $\{I_1 = 0\} \cup \{I_2 = 0\} \cup \cdots \cup \{I_k = 0\}$.

Therefore, generally speaking, one wishes to have a good (algorithmic) knowledge of these normal form transformations. It turns out that this knowledge exists, even to the level of formula manipulation. In Broer and Vegter (in preparation) we shall come back to this.

6 Further generalizations

We conclude this paper with some general remarks. The case of a double zero eigenvalue with more than one pair of normally elliptic eigenvalues was already discussed in Section 5. Here we also consider the case with a general normal eigenvalue configuration. In both cases, under certain conditions, reduction to the planar case of Section 2 is possible. Finally, we give some remarks on a similar situation concerning symplectic maps.

6.1 Normal hyperbolicity In the case of non-imaginary eigenvalues the usual reduction to the *center manifold* applies, for example, see Hirsch, Pugh, and Shub [1977]. This means that there exists a normally hyperbolic center manifold W^c, tangent to the eigenspace corresponding to the imaginary eigenvalues and invariant under the flow. 'Reduction' then means restriction of the whole bifurcation analysis to this center manifold W^c; all the interesting dynamics takes place here. For example, compare Palis and Takens [1977].

To be precise, let us assume that the phase space is $\mathbb{R}^{2n}$, with symplectic form $\omega = \sum_{k=1}^{n} dx_k \wedge dy_k$. Let X_H be a Hamiltonian vector field on $\mathbb{R}^{2n}$, with the origin as a singularity. We also assume that there are $2m$ purely imaginary eigenvalues, counting multiplicities, so $\dim W^c = 2m$.

Our first aim is to point out the quite familiar fact that W^c inherits a symplectic structure, by restricting ω to it. This restriction is the pull-back $i^*\omega$, where $i : W^c \longrightarrow \mathbb{R}^{2n}$ denotes the Inclusion Map. For this use of notation, again see Abraham and Marsden [1978], Arnold [1980]. We have:

Lemma 6.1 *The restriction $i^*\omega$ defines a symplectic form on W^c, which is preserved by the restriction of X_H to W^c. Moreover, there exist local coordinates $x_1, \ldots, x_m, y_1, \ldots, y_m$ on W^c, such that*

$$i^*\omega = \sum_{k=1}^{m} dx_k \wedge dy_k.$$

The proof of this lemma is elementary, for details see Broer et al. [1993].

Reduction to the center manifold W^c means restriction of H and X_H to W^c. By the local coordinates of Lemma 6.1, we can pass to $\mathbb{R}^{2m}$ with the natural symplectic structure. One problem here is the differentiability of the center manifold. On one hand, for any $k \in \mathbb{N}$ there is a neighbourhood of the origin, where W^c is at least of class C^k. On the other hand examples show that W^c does not have to be C^∞. Nevertheless, with some care, a bifurcation analysis as before can be given, e.g. compare Vegter [1983], also see Broer and Vegter (in preparation).

6.2 Symplectic maps The analogue for symplectic maps is the fixpoint with a double 1 eigenvalue. Now there is a Normal Form Theorem (Broer [1981a], Broer and Takens [1989], Moser [1968], Takens [1974a]) saying that the map formally is the flow of a vector field over time 1. To be precise, assume that $T^\mu : \mathbb{R}^2 \to \mathbb{R}^2$ is a C^∞ family of symplectic diffeomorphisms, again with $\mu \in \mathbb{R}^p$. Also assume that $T^0(0) = 0$ and that D_0T^0 has double eigenvalue 1.

Theorem 6.2 *There exists a C^∞, parameter preserving, symplectic transformation Φ^μ and a C^∞ Hamiltonian vector field X^μ, such that*

$$(\Phi^{-1} \circ T \circ \Phi)^\mu = X_1^\mu + P^\mu,$$

where X_1 denotes the flow over time 1 *of X and where P is infinitely flat at* 0*, both in phase and parameter space.*

As a consequence, the Hamiltonian vector field X^0 has the origin as a singularity with a double zero eigenvalue, so it is subject to the above unfolding theory. In the non-semisimple case this leads to the Saddle Node bifurcation for symplectic maps. For a different approach to this problem, see Meyer [1975]. In the semisimple case the Elliptic and Hyperbolic Umbilic Catastrophes again play a role.

Remarks (i) Theorem 6.2 is a special formulation of a more general result, see Takens [1974a]. Let us sketch how this generalization runs. We consider the fixed point 0 of T^0, where the derivative D_0T^0 has only eigenvalues on the complex unit circle. Note, that this can always be achieved by restricting to a center manifold. Let S denote the semisimple part of this derivative, then a symplectic transformation Φ^μ and a Hamiltonian vector field X^μ exist, satisfying both

1. $S_* X^\mu = X^\mu$, i.e., X^μ is equivariant with respect to the group generated by S;
2. $(\Phi^{-1} \circ T \circ \Phi)^\mu = S \circ X_1^\mu + P^\mu$, where P is flat as before.

(ii) This result especially is of interest in the 'resonant' case, where S has eigenvalues that are roots of unity. For an application with $S = -\text{Id}$, involving period-doubling, see Broer and Vegter [1992]. Also notice that Theorem 6.2 just covers the case where $S = \text{Id}$.
(iii) The validity of this result is not restricted to dimension 2, but holds for arbitrary symplectic fixed points. In fact, as in the above the vector field case, the approach generalizes to any setting where some appropriate structure has to be preserved, compare Broer [1981a] and Takens [1974b]. This preservation is suitably expressed in terms of the Lie algebra of vector fields and the corresponding Lie groups of diffeomorphisms. The structures we have in mind are given by a volume or a symplectic form, or by a symmetry group.

7 Acknowledgements

We thank Richard Cushman, Hans Duistermaat, Marty Golubitsky, Heinz Hanßmann, Martin Krupa, Jeff Xia, André Vanderbauwhede and Ferdinand Verhulst for helpful discussions. Also we are indebted to Martin Krupa for producing the pictures. Moreover, the first author acknowledges hospitality from Michigan State University and the Georgia Institute of Technology during the preparation of this paper.

References

Abraham, R., and Marsden, J.E. [1978] *Foundations of Mechanics,* Benjamin.

Arnold, V.I. [1972] *Normal forms for functions near degenerate critical points, the Weyl groups A_k, D_k, and E_k, and Lagrangian singularities*, Funct. Anal. Appl., **6**.

Arnold, V.I. [1980] *Mathematical Methods of Classical Mechanics*, Springer-Verlag.

Arnold, V.I. [1983] *Geometrical Methods in the Theory of Ordinary Differential Equations*, Springer-Verlag.

Birkhoff, G.D. [1927] *Dynamical Systems*, AMS Publications.

Bröcker, Th., and Lander, L. [1975] *Differentiable Germs and Catastrophes*, Cambridge University Press.

Broer, H.W. [1979] *Bifurcations of Singularities in Volume Preserving Vector Fields*, (thesis), University of Groningen.

Broer, H.W. [1981a] *Formal normal form theorems for vector fields and some consequences for bifurcations in the volume preserving case*, Springer-Verlag LNM, **898**.

Broer, H.W. [1981b] *Quasi-periodic flow near a codimension one singularity of a divergence free vector field in dimension three*, Springer-Verlag LNM, **898**.

Broer, H.W., Chow, S.N., Kim, Y. and Vegter, G. [1993] *A normally elliptic Hamiltonian bifurcation*, **44**, ZAMP.

Broer, H.W. and Takens, F. [1989] *Formally symmetric normal forms and genericity*, Dynamics Reported, **2**.

Broer, H.W. and Vegter, G. [1992] *Bifurcational aspects of parametric resonance*, Dynamics Reported, New Series, **1**.

Broer, H.W. and Vegter, G. *The C^k-preparation theorem with applications in dynamical systems*, (in preparation).

Damon, J. [1988] *Topological equivalence of bifurcation problems*, Nonlinearity, **1**.

Duistermaat, J.J. [1983] *Bifurcations of periodic solutions near equilibrium points of Hamiltonian systems*, Springer-Verlag LNM, **1057**.

Galin, D.M. [1982] *Versal deformations of linear Hamiltonian systems*, Amer. Soc. Trans., **118**.

Gibson, C.G. [1979] *Singular Points of Smooth Mappings*, Pitman.

Golubitsky, M. and Schaeffer, D.G. [1985] *Singularities and groups in bifurcation theory*, Vol. I, Appl. Math. Sc., **51**, Springer-Verlag.

Golubitsky, M. and Stewart, I. [1987] *Generic bifurcation of Hamiltonian systems with symmetry*, Physica D, **24**.

Gustavson, F.G. [1966] *On constructing formal integrals of a Hamiltonian system near an equilibrium point,* Astron. J., **71**.

Hirsch, M.W., Pugh, C.C. and Shub, M. [1977] *Invariant Manifolds*, Springer-Verlag.

Koçak, H. [1984] *Normal forms and versal deformation of linear Hamiltonian systems*, J. Diff. Eq., **51**.

Martinet, J. [1977] *Singularités des Fonctions et Applications Différentiables,* PUC Rio de Janeiro.

Meyer, K. [1975] *Generic bifurcations in Hamiltonian systems*, Springer-Verlag LNM, **468**.

Moser, J.K. [1968] *Lectures on Hamiltonian Systems*, Memoirs AMS.

Moser, J.K. [1973] *Stable and Random Motions in Dynamical Systems*, Princeton University Press.

Narasimhan, R. [1968] *Analysis on Real and Complex Manifolds*, North Holland.

Palis, J. and Takens, F. [1977] *Toplogical equivalence in normally hyperbolic dynamical systems*, Topology, **16**.

Poincaré, H. [1928] *Oeuvres I*, Gauthier-Villars.

Poston, T. and Stewart, I. [1978] *Catastrophe Theory and its Applications*, Pitman.

Sanders, J.A. [1977] *Are higher order resonances really interesting?* Cel. Mech., **16**.

Sanders, J.A. and Verhulst, F. [1985] *Averaging Methods in Nonlinear Dynamical Systems*, Springer-Verlag.

Schecter, S. [1992] *C^p Singularity Theory and heteroclinic bifurcation with a distinguished parameter*, J. Diff. Eq., **99**.

Takens, F. [1974a] *Forced oscillations and bifurcations*, Comm. Math. Inst., University of Utrecht, **3**.

Takens, F. [1974b] *Singularities of vector fields*, Publ. Math. IHES, **43**.

Thom, R. [1972] *Stabilité Structurelle et Morphogénèse*, Benjamin.

van der Meer, J.C. [1985] *The Hamiltonian Hopf bifurcation*, Springer-Verlag LNM, **1160**.

Vegter, G. [1983] *The C^p Preparation Theorem, C^p unfoldings and applications*, Preprint University of Groningen.

Verhulst, F. [1983] *Asymptotic analysis of Hamiltonian systems*, Springer-Verlag LNM, **985**.

Wassermann, G. [1975] *Stability and unfoldings in space and time*, Acta Math., **135**.

Fields Institute Communications
Volume **4**, 1995

Geometric analysis of the effect of symmetry breaking perturbations on an $O(2)$ invariant homoclinic cycle

Pascal Chossat
L'Institute Non-Linéaire de Nice
CNRS-Université de Nice-Sophia
Antipolis, France

Mike Field
Department of Mathematics
University of Houston
Houston, Texas, 77204, USA

Abstract. Using ideas based on blowing-up, we give a geometric analysis of the effect of symmetry breaking perturbations on an $O(2)$-invariant homoclinic cycle previously studied by Armbruster, Holmes and Guckenheimer. We present numerical eveidence to suggest that, at resonant points, breaking symmetry to a reflectional subgroup of $O(2)$ may lead to chaotic dynamics.

1 Introduction

Ordinary differential equations with $O(2)$-symmetry arise naturally in many physical applications and, in recent years, much work has been done on describing the (local) generic bifurcation theory of such systems. (See, for example, Golubitsky, Stewart and Schaeffer [1988].) However, in practice it is often unrealistic to assume perfect symmetry and so there have been a number of investigations into the effect of breaking the full $O(2)$-symmetry to a proper, possibly trivial, subgroup of $O(2)$. In this regard, we mention the recent article by Dangelmayr and Knobloch [1991] on the Hopf bifurcation with broken circular symmetry, where symmetries are broken to a reflectional subgroup $\mathbb{Z}_2$ of $O(2)$; the papers by Crawford and Knobloch [1988] and van Gils and Mallet-Paret [1986], where symmetries are broken to $SO(2)$; and the paper by Golubitsky and Schaeffer [1983], where symmetries are broken to the trivial subgroup. These works are primarily concerned with *local* bifurcation theory.

In another direction, it has been known for several years that differential equations with $O(2)$ symmetry can possess a 'homoclinic cycle' between equilibria on the *same* $O(2)$-orbit. This phenomenon occurs for $O(2)$-equivariant vector

1991 *Mathematics Subject Classification.* 58F14; Secondary 34C37, 58F13.

fields on the $(1,2)$-representation of $O(2)$ on $\mathbb{C}^2$ and is described in the work by Armbruster, Guckenheimer and Holmes [1988]. Specifically, it is shown that there is a normally hyperbolic $O(2)$-orbit α of equilibria contained in the 2-factor of the $(1,2)$-representation of $O(2)$. The unstable manifold $W^u(\alpha)$ of α is 2-dimensional; the stable manifold $W^s(\alpha)$ is 3-dimensional (α is of index 2) and $W^u(\alpha) \subset W^s(\alpha)$. The homoclinic cycle is the set $\Sigma = W^u(\alpha)$ and, in the case of greatest interest, Σ is an attractor. Topologically, Σ is an immersed Klein bottle with self intersection along α.

Very recently, Chossat [1993] showed that under perturbations which break the reflectional symmetry, the attracting homoclinic cycle Σ described in Armbruster, Guckenheimer and Holmes [1988] generically bifurcates to a quasi-periodic flow on a 2-torus. The strategy used by Chossat depends on a reduction to the orbit space of the action followed by an application of Lin's technique based on Melnikov's method.

In this work we present a new technique for the analysis of the effect of symmetry breaking perturbations on $O(2)$-equivariant systems with a homoclinic cycle. Unlike the analysis in Chossat [1993], we do not work on the orbit space. Rather, we use a process of blowing-up to simplify the equations and group action. One advantage of this method is that it leads to a geometrically transparent description of the dynamics near the homoclinic cycle. Moreover, not only does this method provide an alternative analysis of the case where symmetries are broken to $SO(2)$ but it also becomes possible to obtain information about the far more difficult case where symmetries are broken from $O(2)$ to a reflectional subgroup $\mathbb{Z}_2$. In the latter case we can show that the homoclinic cycle typically bifurcates to a hyperbolic periodic orbit.

Our approach to this type of result can be broken into three steps. First of all, we provide, using blowing-up techniques, a geometrically transparent picture of the local dynamics near the bifurcation point. Using this insight, we proceed to analyze a simplified 'affine linear' model and prove existence, local uniqueness and hyperbolicity using the implicit function theorem. Finally, we analyze the general problem using methods based on Shilnikov coordinates, motivated by results of Deng [1989], and the analysis of the affine model. The analytic techniques required here are nontrivial as we have to handle a situation where eigenvalues are small and the homoclinic orbit is degenerate on account of the presence of symmetry. In particular, we cannot use smooth linearization theorems. In this work we present the initial geometric analysis (the analytical theory will appear elsewhere (Field (in preparation))). As we are primarily interested in describing geometric methods, we simplify matters a little by focusing on a closely related problem defined on the unit sphere S^3 of $\mathbb{C}^2$.

Before we give a statement of our main theorem, we need some preliminaries. We shall consider the $(1,2)$-representation of $O(2)$ on $\mathbb{C}^2$ (for a description of this representation, see Section 4). Let $\kappa \in O(2)$ be complex conjugation on $\mathbb{C}^2$ and $\mathbb{Z}_2(\kappa)$ be the two element subgroup of $O(2)$ generated by κ.

Theorem 1.1 *Let X be a smooth $O(2)$-equivariant vector field on $\mathbb{C}^2$ and suppose that α is a nontrivial $O(2)$-orbit of equilibria of X contained in the z_2-plane. Assume that:*

- *α is normally hyperbolic of index 2;*
- *$W^u(\alpha) \subset W^s(\alpha)$. That is, $\Sigma = W^u(\alpha)$ is a homoclinic cycle;*

- *The intersection of* Σ *with* $\mathbb{R}^2 \subset \mathbb{C}^2$ *is topologically a circle;*
- Σ *is an attracting homoclinic cycle.*

We have:

1. *For generic perturbations* Y *of* X *which break symmetry to* $SO(2)$, *there is a unique normally hyperbolic and attracting 2-torus* T *close to* Σ. *The flow of* Y *restricted to* T *is generically quasi-periodic with a slow drift along* $SO(2)$*-orbits;*
2. *For perturbations* Y *of* X *which break symmetry to* $\mathbb{Z}_2(\kappa)$ *and for which certain quantitative estimates hold,* Y *will have a unique hyperbolic and attracting limit cycle* γ *close to* Σ.

Remarks 1.2

(1) We give, at the end of the paper, a precise statement of sufficient conditions on $\mathbb{Z}_2(\kappa)$-equivariant symmetry breaking perturbations Y of X in order that we obtain a periodic orbit γ satisfying the conditions of the theorem. Suffice to say, that for 'most' perturbations Y of X, we satisfy the conclusions of the theorem. We shall also briefly discuss the case where there are resonant eigenvalues and we believe there is bifurcation to chaotic dynamics.

(2) It is also possible to describe the bifurcations of Σ that occur when we break symmetry to other subgroups of $O(2)$. (See also Campbell and Holmes [1992].)

In Section 2, we cover background material on (polar) blowing-up. As we are interested in the possibility of applying our methods to more general groups than $O(2)$, we have included a general description of polar blowing-up and its relation to the classical blowing-up construction of algebraic geometry. Next, in Section 3, we briefly review some basic theory of normally hyperbolic sets needed in the sequel. In Section 4, we describe the effects of blowing-up the homoclinic cycle in the Armbruster-Guckenheimer-Holmes example. In particular, we give a precise description of the dynamics near the blown-up cycle. In Section 5 we investigate the case where we break the reflectional symmetry and show how to recover Chossat's result. Finally, in Section 6, we give a geometric description of the effects of breaking the $SO(2)$-symmetry and a precise statement (without proofs) of the analytic results.

2 Polar blowing-up

Our methods will be based on techniques of *polar blowing-up* similar to those described in Field [1978, 1995] and Field and Swift [1991]. As we believe the methods we describe have the potential for a wide variety of applications in equivariant bifurcation theory, we start by giving a general description of polar blowing-up for G-manifolds.

2.1 Local theory Let $n \geq 2$ and take a positive definite inner product $(\ ,\ )$ on $\mathbb{R}^n$. Let S^{n-1} denote the unit sphere of $\mathbb{R}^n$. Let $\sigma : \mathbb{R}^n \times S^{n-1} \to \mathbb{R}^n \times S^{n-1}$ be the involution defined by

$$\sigma(x, u) = (x, -u).$$

Let $\nu : \mathbb{R}^n \times S^{n-1} \to \mathbb{R}^n$ denote the projection on the first factor. Let $\Sigma_n \subset \mathbb{R}^n \times S^{n-1}$ denote the real analytic submanifold defined by the equations

$$x_i u_j = x_j u_i, \; 1 \leq i < j \leq n, \; x \in \mathbb{R}^n, \; u \in S^{n-1}.$$

We let σ_n denote the restriction of σ to Σ_n. It is an elementary exercise to verify that the map $a : S^{n-1} \times \mathbb{R} \to \Sigma_n$ defined by

$$a((u_1, \dots, u_n), t) = ((tu_1, \dots, tu_n), (u_1, \dots, u_n))$$

is a real analytic diffeomorphism. Furthermore, we have

$$a \cdot \tilde{\sigma} \quad = \quad \sigma_n \cdot a,$$

where $\tilde{\sigma}(u, t) = (-u, -t)$, $(u, t) \in S^{n-1} \times \mathbb{R}$. In the sequel, we set $\tilde{\sigma} = \sigma_n$ and note that, with this convention, a is $\mathbb{Z}_2$-equivariant, where $\mathbb{Z}_2$ is generated by σ_n. We shall also let $\Sigma_n^+ = a(S^{n-1} \times [0, \infty))$. Thus, Σ_n^+ is a manifold with boundary S^{n-1} and $\Sigma_n = \Sigma_n^+ \cup \sigma_n \Sigma_n^+$.

Definition 2.1 *The* polar blowing-up *of* $\mathbb{R}^n$ *at zero consists of the real analytic manifold* Σ_n *together with the analytic projection map* $\pi : \Sigma_n \to \mathbb{R}^n$ *defined by the restriction of* ν *to* Σ_n*. We call* $0 \in \mathbb{R}^n$*, the* center *of the blowing-up.*

Remarks 2.2

(1) The map $\pi : \Sigma_n \to \mathbb{R}^n$ restricts to a 2:1 local analytic diffeomorphism of $\Sigma_n \setminus \pi^{-1}(0)$ onto $\mathbb{R}^n \setminus \{0\}$. Indeed, if we identify Σ_n with $S^{n-1} \times \mathbb{R}$, it is clear that $\pi(u, R) = \pi(-u, -R) = Ru$, $(u, R) \in S^{n-1} \times \mathbb{R}$. In particular, $\pi \sigma_n = \pi$ and $\pi^{-1}(0) = S^{n-1} \times \{0\}$.

(2) Up to real analytic diffeomorphism, the polar blowing-up of $\mathbb{R}^n$ at zero is independent of the choice of inner product on $\mathbb{R}^n$.

(3) The blowing-up of $\mathbb{R}^n$ at zero defined in algebraic geometry may be given as the polar blowing-up of $\mathbb{R}^n$ at zero followed by the quotient of Σ_n by the free $\mathbb{Z}_2$-action with generator σ_n. For example, the polar blow-up of $\mathbb{R}^2$ is the cylinder $S^1 \times \mathbb{R}$, whilst the blow-up of $\mathbb{R}^2$ is the Möbius band.

(4) Polar blowing-up is essentially the result of taking (generalized) spherical polar coordinates. However, we allow the radius to take negative values and so end with a 2:1 map onto $\mathbb{R}^n$ outside zero. If we restrict to positive values of the radius, we obtain the manifold Σ_n^+ with boundary S^{n-1}. In our subsequent applications, we shall (initially) allow the radius to take negative values as this avoids the introduction of boundaries and corners.

It is rather easy to describe the topological effect of blowing-up $\mathbb{R}^n$ at zero: Excise a (small) closed disc, center the origin, from $\mathbb{R}^n$ to obtain a manifold with boundary diffeomorphic to S^{n-1}. We then take two copies of this manifold and glue them together along the boundary with the identity map to obtain the polar blowing-up.

Next we extend Definition 2.1 to allow for more general centers. We start with the local theory. Let $m \geq 0$. We regard $\mathbb{R}^m$ as embedded in $\mathbb{R}^m \times \mathbb{R}^n$ as $\mathbb{R}^m \times \{0\}$.

Definition 2.3 *Set* $M = \mathbb{R}^m \times \mathbb{R}^n$*. The* polar blowing-up *of* M *with* center $\mathbb{R}^m$ *consists of the real analytic manifold* $\tilde{M} = \mathbb{R}^m \times \Sigma_n$ *together with the projection* *map* $\pi : \tilde{M} \to M$ *defined as the product of the identity map on* $\mathbb{R}^m$ *with the projection of* Σ_n *onto* $\mathbb{R}^n$.

We let $\tilde{M}^+ = \mathbb{R}^m \times \Sigma_n^+$.

Remark 2.4 The map $\pi : \tilde{M} \to M$ is a 2:1 local analytic diffeomorphism off the center $\mathbb{R}^m$. If we let $\tilde{\sigma}$ be the involution of $\tilde{M}$ defined as the product of the identity map on $\mathbb{R}^m$ with σ_n on Σ_n then $\pi = \pi\tilde{\sigma}$. If $(x, 0) \in \mathbb{R}^m \times \mathbb{R}^n$, then $\pi^{-1}(x, 0)$ is analytically diffeomorphic to $\{x\} \times (S^{n-1} \times \{0\})$.

Suppose that $W = U \times V$ is an open product neighborhood of $(x, 0) \in \mathbb{R}^m \times \mathbb{R}^n$. It is evident that we may give a local version of polar blowing-up and define the polar blowing-up $\pi : \tilde{W} \to W$ by setting $\tilde{W} = \pi^{-1}(W)$ and restricting the map π of Definition 2.3 to $\tilde{W}$. We may similarly define $\tilde{W}^+$.

Lemma 2.5 *Let* $W_i = U_i \times V_i$, $i = 1, 2$, *be open product neighborhoods of* $(0, 0) \in \mathbb{R}^m \times \mathbb{R}^n$. *Suppose that* $f : W_1 \to W_2$ *is an analytic diffeomorphism and that* $f(U_1 \times \{0\}) = U_2 \times \{0\}$. *Then there exists a unique analytic diffeomorphism* $\tilde{f} : \tilde{W}_1 \to \tilde{W}_2$ *satisfying*

$$\begin{aligned} \tilde{f}(\tilde{W}_1^+) &= \tilde{W}_2^+ \\ \pi\tilde{f} &= f\pi \\ \tilde{\sigma}\tilde{f} &= \tilde{f}\tilde{\sigma}. \end{aligned}$$

Proof Write $f = (f_U, f_V) : U_1 \times V_1 \to U_2 \times V_2$. Identify the polar blowing-up of $\mathbb{R}^m \times \mathbb{R}^n$ along $\mathbb{R}^m$ with $\mathbb{R}^m \times (S^{n-1} \times \mathbb{R})$. We define $\tilde{f}$ by

$$\begin{aligned} &\tilde{f}(x, (u, R)) = \\ &\quad (f_U(x, (u, R)), \operatorname{sgn}(R) f_V(x, (u, R))/|f_V(x, (u, R))|, \operatorname{sgn}(R)|f_V(x, (u, R))|). \end{aligned}$$

It is a straightforward computation, using the fact that f is a diffeomorphism and $f_V(x, 0) \equiv 0$, to show that $\tilde{f}$ is a well-defined analytic diffeomorphism between $\tilde{W}_1$ and $\tilde{W}_2$ which maps $\tilde{W}_1^+$ onto $\tilde{W}_2^+$. Clearly, $\pi\tilde{f} = f\pi$ and this relation characterizes $\tilde{f}$ up to composition with $\tilde{\sigma}$. Since we require $f(\tilde{W}_1^+) = \tilde{W}_2^+$, it follows that $\tilde{f}$ is unique. □

Remark 2.6 Vector fields which are tangent to the center will also lift. This can either be proved directly or by working with the flow and using Lemma 2.5.

Example 2.7 Suppose that $X : \mathbb{R}^2 \to \mathbb{R}^2$ is a smooth vector field and that

$$X(x, y) = (\lambda x, \mu y) + O(|(x, y)|^2).$$

We suppose that $\lambda > 0 > \mu$. We describe the lift $\hat{X}$ of X to $S^1 \times \mathbb{R}$. An easy computation verifies that $\hat{X}$ has four equilibria on $S^1 \times \{0\}$ at $\theta = 0, \pi/2, \pi, 3\pi/2$ (see Figure 1(b)). Each equilibrium is a saddle point. The equilibria at $\theta = 0, \pi$ have radial eigenvalue λ whilst those at $\theta = \pi/2, 3\pi/2$ have radial eigenvalue μ. Since $S^1 \times \{0\}$ is invariant by the flow of $\hat{X}$, the remaining eigenvalues are associated to tangential directions to $S^1 \times \{0\}$. These eigenvalues are $\mu - \lambda < 0$ at $\theta = 0, \pi$ and $\lambda - \mu > 0$ at $\theta = \pi/2, 3\pi/2$. In particular, we observe that the saddle point at $0 \in \mathbb{R}^2$ generates four saddle points on $S^1 \times \{0\}$.

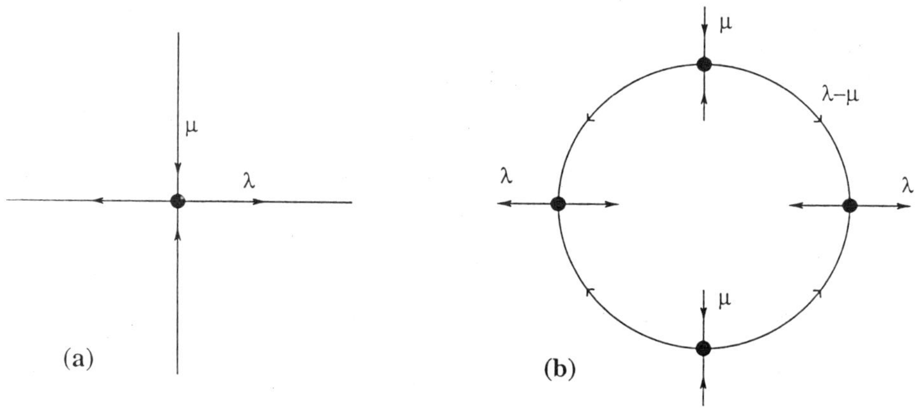

Figure 1 Blowing-up a saddle point

2.2 Global theory Suppose that M is a compact connected real analytic manifold. Let C be a closed real analytic submanifold of M. We shall define the polar blowing-up of M with center C. Although the analytic definition requires some technicalities, geometrically it is very easy to describe the blowing up of M along C: We excise an open tubular neighborhood U of C from M and glue two copies of $M \setminus U$ along ∂U via the identity map.

Analytically, the blowing up of M along C will consist of a real analytic manifold $\tilde{M}$ together with a projection map $\pi : \tilde{M} \to M$.

Formally, we construct the real analytic structure on $\tilde{M}$ by blowing-up the transition functions for a real analytic structure on M. More specifically, let $\mathcal{A}$ be an atlas of charts for M compatible with the submanifold C. That is, if $\phi : D \to \mathbb{R}^m \times \mathbb{R}^n$ is a chart for M containing points of C we assume:

(a) $\phi(D)$ is an open product neighborhood $W = U \times V$ of $(0,0) \in \mathbb{R}^m \times \mathbb{R}^n$;

(b) $\phi(D \cap C) = U \times \{0\}$.

Let $\phi_i : D_i \to W_i = U_i \times V_i$, $i = 1,2$ be charts in $\mathcal{A}$ containing points of C (other cases are trivial). Let $\pi_i : \tilde{W}_i \to W_i$ be the polar blowing-up of W_i with center U_i, $i = 1,2$. Set $W_{12} = W_2 \cap \phi_2(D_1)$. It follows from Lemma 2.5 that the coordinate change map $\phi_{12} = \phi_1 \phi_2^{-1}$ lifts to an analytic diffeomorphism $\tilde{\phi}_{12}$ defined on $\tilde{W}_{12}$. Repeating this construction for all pairs of overlapping charts of M yields the transition functions for a real analytic manifold $\tilde{M}$. The required projection map $\pi : \tilde{M} \to M$ is defined in the obvious way (via the local blowing-ups). Since the local maps $\tilde{\phi}_{12}$ preserve $\tilde{W}_{12}^+$, it follows that we may define $\tilde{M}^+$ as a submanifold of $\tilde{M}$ with boundary $\pi^{-1}(C)$.

Lemma 2.8 *Let M be a compact connected real analytic manifold and C be a closed real analytic submanifold of M. Let $\pi : \tilde{M} \to M$ be the polar blowing-up of M with center C. Then $\pi : \tilde{M} \setminus \pi^{-1}(C) \to M \setminus C$ is a 2:1 local analytic diffeomorphism.*

Example 2.9 Suppose that C is an embedded (unknotted) circle in S^3. Topologically, the polar blowing-up of S^3 with center C is obtained by removing an open neighborhood of C from S^3 and gluing two copies of the resulting manifold together along their common boundary. More precisely, suppose that N is a tubular neighborhood of C in S^3. Obviously, N is diffeomorphic to a solid torus. Since we are assuming C is unknotted, the closure of the complement of N in S^3 is again a solid torus. Gluing two copies of the complementary torus together along their common boundary yields $S^2 \times S^1$ which is, up to diffeomorphism, the polar blowing-up of S^3 along C.

Lemma 2.10 *Let $\pi : \tilde{M} \to M$ be the polar blowing-up of M with center C. Suppose that X is a real analytic vector field on M which is everywhere tangent to C. Then X lifts to a real analytic vector field $\tilde{X}$ on $\tilde{M}$ and*

(1) $T\pi\tilde{X} = X\pi$,

(2) *$\tilde{X}$ is tangent to $\pi^{-1}(C)$,*

(3) *If we choose local analytic coordinate systems at $\tilde{x} \in \tilde{M}$ and $\pi x \in M$ then for all $r \geq 0$, $j^r\tilde{X}(\tilde{x})$ depends only on $j^{r+1}X(x)$.*

We omit the straightforward proofs of Lemmas 2.8, 2.10. (The reader may find details of the computations needed for Lemma 2.10 in either Field [1978] or Field and Richardson [1992], Section 7.)

2.3 Freeing up a G-action For our purposes, one of the most interesting properties of polar blowing-up is that if we repeatedly blow-up along maximal isotropy types we eventually obtain a 'monotypic' G-action.

Suppose that M is a compact connected G-manifold. We shall assume that M has real analytic structure. Note that this is no loss of generality as every compact connected G-manifold can be embedded as a real algebraic submanifold of some G-representation. Denote the set of isotropy types of M by $\mathcal{O}(M, G)$. We define the usual partial order[1] on $\mathcal{O}(M, G)$. Thus, if $\tau, \mu \in \mathcal{O}(M, G)$, we write $\tau < \mu$ if there exists $H \in \tau$, $J \in \mu$ such that $H \subset J$ and $H \neq J$.

As is well known (see Bredon [1972], Chapter IV, Theorem 3.1), there exists a unique minimal isotropy type (the 'principal' isotropy type). In general, there may exist more than one maximal isotropy type. However, for linear actions there is a unique maximal isotropy type corresponding to the fixed point space of the action. We may and shall extend the partial order on $\mathcal{O}(M, G)$ to a total order, which we continue to denote by $<$. Relative to this total order, we shall label the isotropy types as τ_i where $\tau_i > \tau_j$ if and only if $i > j$. Suppose that there are N distinct isotropy types for the action of G on M. For $i \in \{1, \dots, N\}$, we shall let $M_i \subset M$ denote the subset of M consisting of points of isotropy type i. Each M_i is a G-invariant real analytic submanifold of M. However, M_i is a *closed* submanifold if and only if τ_i is a maximal isotropy type.

Set $M = M^N$. We define a sequence $\pi_{N-i+1} : M^{i-1} \to M^i$, $i \in \{2, \dots, N\}$ of polar blowing-ups with the following properties:

(a) $\mathcal{O}(M^i, G) = \{\tau_1, \dots, \tau_i\}$, $1 \leq i \leq N$;

(b) $\pi_{N-i+1} : M^{i-1} \to M^i$ has center $C_i = M^i_i$;

(c) The action of G on M^1 is monotypic (that is, there is only one orbit type—τ_1).

The sequence is characterized by condition (b). All we have to note is that the polar blowing-up construction respects equivariance and that M^{i-1} contains no points of

[1]In fact, the reverse order is usually used in the topological study of G-actions.

isotropy type τ_i. In particular, no new isotropy types are introduced (this would *not* have been the case if we had used (projective) blowing-up).

Remark 2.11 In many cases, the resulting action of G on M^1 will either be *free* or we will be able to factor out by the isotropy of G acting on M^1 to obtain a free action. In any case, M^1 will always be a locally trivial bundle over the *manifold* M^1/G.

Given $i \in \{1, \ldots, N-1\}$, we let $\pi^i : M^{N-i} \to M^N = M$ denote the composition of the maps $\pi_1, \pi_2, \ldots, \pi_i$. We remark that τ_i is a maximal isotropy type of M^i and that M^i_i is a closed real analytic submanifold of M^i.

Lemma 2.12 *Let M be a compact connected real analytic G-manifold. Suppose that X is a real analytic G-equivariant vector field on M. For each $i \in \{1, \ldots, N\}$, X lifts to a real analytic G-equivariant vector field X^i on M^i. Further, if $\tilde{x} \in M^{N-i}$, $x = \pi^i(\tilde{x}) \in M$ and we choose local real analytic coordinate systems at $\tilde{x}$ and x, then for all $r \geq 0$, $j^r X^i(\tilde{x})$ depends only on $j^{r+i}X(x)$.*

Proof It follows by the G-equivariance of X that for $1 \leq j \leq i$, X^j is tangent to M^j_k, $k \geq j$. The result follows from Lemma 2.10. □

Although we have so far only stated our results for compact manifolds, we may easily extend our methods to include G-representations.

Example 2.13 Let (V, G) be an irreducible G-representation. Assume that V is given a G-invariant inner product and denote the unit sphere of V by $S(V)$. Taking $m = 0$, $n = \dim(V)$ in definition 2.3, we see that the polar blowing-up of (V, G) with center $\{0\}$ is given by $\pi : S(V) \times \mathbb{R} \to V$, where $\pi(u, R) = Ru$, $u \in S(V)$, $R \in \mathbb{R}$. Of course, the action of G on $S(V) \times \mathbb{R}$ has no fixed points and so $S(V) \times \mathbb{R}$ has one fewer isotropy type than V. We may apply the preceding results to $S(V)$ and, forming products with the trivial factor $\mathbb{R}$, thereby obtain a sequence $\pi_{N-i+1} : V^{i-1} \to V^i$, satisfying conditions (a,b,c) listed above. The same arguments obviously work for any finite dimensional G-representation.

3 Normal hyperbolicity

In this section, we briefly review some background on normal hyperbolicity for group orbits of equilibria. For further details on the role of normal hyperbolicity in equivariant dynamics, we refer the reader to Field [1980], Field [1989], Krupa [1990]. For the general theory of normal hyperbolicity, see Hirsch, Pugh and Shub [1977].

Let G be a compact Lie group and M be a smooth G-manifold. Suppose that X is a smooth equivariant vector field on M and that $\alpha \subset M$ is a G-orbit which is invariant under the flow Φ^X of X. We refer to α as a *relative equilibrium* of X. In case X is zero on α, α is a group orbit of equilibria of X. For our needs, we shall be able to assume α is a group orbit of *equilibria*. In this case the Hessian, $\mathrm{Hess}_x(X, \alpha)$, of X is defined at each point x of α. We let $\mathrm{index}_x(X, \alpha)$ denote the index of $\mathrm{Hess}_x(X, \alpha)$ at the point x (that is, the number of eigenvalues of $\mathrm{Hess}_x(X, \alpha)$ with strictly negative real part). It follows easily from the equivariance of X that $\mathrm{index}_x(X, \alpha)$ is independent of $x \in \alpha$ and in the sequel we let $\mathrm{ind}_X(\alpha)$ denote the common value of $\mathrm{index}_x(X, \alpha)$ on α.

Let g_α denote the dimension of α. Since $X|\alpha \equiv 0$, it follows that $\mathrm{Hess}_x(X, \alpha)$ has at least g_α eigenvalues *equal* to zero. The G-orbit α is *normally hyperbolic* for the flow of X if $\mathrm{Hess}_x(X, \alpha)$ has precisely g_α eigenvalues with real part zero.

We recall that if α is normally hyperbolic, then there exist smooth equivariantly immersed G-manifolds $W^u(\alpha)$ (the unstable manifold of α) and $W^s(\alpha)$ (the stable manifold of α). Necessarily, $\alpha \subset W^u(\alpha)$, $W^s(\alpha)$ and $W^u(\alpha)$ and $W^s(\alpha)$ meet transversally along α. In addition, $W^u(\alpha)$ has a (smooth) equivariant fibration by strong unstable manifolds $W^{uu}(x, \alpha), x \in \alpha$, where

$$W^{uu}(x, \alpha) = \{y \in W^u(\alpha) \mid \Phi_t(y) \to \Phi_t(x) = x \text{ as } t \to -\infty\}.$$

There is an analogous strong stable fibration $W^{ss}(x, \alpha)$ of $W^s(\alpha)$. It is easy to see that $\mathrm{ind}_X(\alpha)$ is equal to the dimension of a fiber in the strong stable fibration $\{W^{ss}(x, \alpha) \mid x \in \alpha\}$ of $W^s(\alpha)$. This definition of the index also works if α is a relative equilibrium or more generally any normally hyperbolic submanifold (see also Field [1980]).

Finally, we remark that if α is a normally hyperbolic G orbit of equilibria, H is a closed subgroup of G, $r \geq 1$ and U is an open H-invariant neighborhood of α in M, then there exists a neighborhood N of X in the space of C^r H-equivariant vector fields on M α such that:

(N1) For each $Y \in N$ there exists a normally hyperbolic Φ^Y invariant compact C^r submanifold $\alpha_Y \subset U$;

(N2) α_Y depends continuously on Y in the C^r topology;

(N3) α_Y is an H-invariant submanifold of M and α_Y is C^r H-equivariantly diffeomorphic to α;

(N4) $\alpha_X = \alpha$;

(N5) $\mathrm{ind}_Y(\alpha_Y) = \mathrm{ind}_X(\alpha)$.

4 Blowing-up an $O(2)$ orbit of equilibria

We consider the $(1, 2)$ representation of $O(2)$ on $\mathbb{C}^2$. Denote coordinates on $\mathbb{C}^2$ by (z_1, z_2). Regard $SO(2) \subset O(2)$ in the usual way and let $\kappa \in O(2)$ be defined as reflection in the real axis (of $\mathbb{C}$.) In the sequel, we always denote the subgroup of $O(2)$ generated by κ by $\mathbb{Z}_2(\kappa)$. We let $\mathbb{Z}_2 \subset SO(2) \subset O(2)$ denote the subgroup generated by $\exp(\imath\pi)$. The $(1, 2)$-action of $O(2)$ on $\mathbb{C}^2$ is defined by

$$\begin{aligned} \exp(\imath\theta)(z_1, z_2) &= (\exp(\imath\theta)z_1, \exp(2\imath\theta)z_2), \ \exp(\imath\theta) \in SO(2) \\ \kappa(z_1, z_2) &= (\bar{z}_1, \bar{z}_2). \end{aligned}$$

We observe that:

(a) The origin is the unique point with isotropy $O(2)$;

(b) Non-zero points on the z_2 axis have isotropy conjugate to $\mathbb{Z}_2 \times \mathbb{Z}_2(\kappa)$;

(c) Any point of the form $(x_1 e^{\imath\theta}, x_2 e^{2\imath\theta})$, with x_1, x_2 real and $x_1 \neq 0$ has isotropy conjugate to $\mathbb{Z}_2(\kappa)$;

(d) All other points in $\mathbb{C}^2$ have trivial isotropy.

We remark that the $O(2)$-orbits of non-zero points with non-trivial isotropy are circles and that the $O(2)$ orbits of points with trivial isotropy consist of pairs of circles.

Let $X : \mathbb{C}^2 \to \mathbb{C}^2$ be a smooth $O(2)$-equivariant vector field. Since the $O(2)$ representation $\mathbb{C}^2$ contains no trivial factors, it follows that $X(0) = 0$. Arising out of a study of the local bifurcation theory of $O(2)$-equivariant families of vector fields on $\mathbb{C}^2$, Armbruster, Guckenheimer and Holmes [1988], showed that there is a non-empty open subset of vector fields X for which the following properties hold.

(A1) X has a normally hyperbolic group orbit of equilibria consisting of points of isotropy type $\mathbb{Z}_2 \times \mathbb{Z}_2(\kappa)$. (Necessarily, α is contained in the z_2-plane.)

(A2) The index of the equilibria on α is two and, restricted to the z_2-plane, α is an attracting orbit of equilibria.

(A3) The unstable manifold $W^u(\alpha)$ of α is contained in the stable manifold $W^s(\alpha)$. Moreover, if $\theta \in \alpha$, then $W^{uu}(\theta) \subset W^{ss}(-\theta)$.

(A4) $W^u(\alpha)$ is an attracting set for the flow of X (see below).

(A5) (A1–4) persist under smooth $O(2)$-equivariant perturbations of X.

We remark that it follows from (A1–3) that $W^u(\alpha) \setminus \alpha$ consists of points with isotropy conjugate to $\mathbb{Z}_2(\kappa)$.

Let $\Sigma = W^u(\alpha)$. Then $\alpha \subset \Sigma \subset W^s(\alpha)$. We refer to Σ as a *homoclinic connection* of α or a *homoclinic cycle.*

The homoclinic cycle Σ is *attracting* in the sense that we may choose an open $O(2)$-invariant neighborhood W of Σ in $\mathbb{C}^2$ such that for all $z \in W$, $\Phi_t^X(z) \to \Sigma$ as $t \to \infty$. Indeed, we may choose W so that ∂W is smooth and X is everywhere transverse to ∂W and inward pointing. It follows that we may choose a C^1 neighborhood of X in the space of all smooth vector fields on $\mathbb{C}^2$ such that if $Y \in N$ then Y is also transverse to ∂W and inward pointing. Consequently, for small, not necessarily equivariant, perturbations Y of X there will be attracting sets for Y contained in W.

As we pointed out previously, the eigenvalues of $\mathrm{Hess}_z(X, \alpha)$ are independent of $z \in \alpha$. We denote the strictly positive eigenvalue of $\mathrm{Hess}_z(X, \alpha)$ by λ_u. The associated eigenspace of λ_u is the tangent space to $W^{uu}(z, \alpha)$ at z. Since the z_2-plane is invariant by the flow of X, it is easy to verify that the line $\mathbb{R}z$ is an eigenspace of $\mathrm{Hess}_z(X, \alpha)$. We denote the corresponding 'radial' eigenvalue by λ_r. We denote the remaining non-zero eigenvalue of $\mathrm{Hess}_z(X, \alpha)$ by λ_s. It follows from our hypotheses on X that $\lambda_r, \lambda_s < 0$.

The cycle Σ will always be a smooth except along α. If the radial eigenvalue dominates, that is $\lambda_r < \lambda_s$, then Σ can be represented as a C^1-immersed Klein bottle with a normal crossing along α. Indeed, if we take a tubular neighborhood U of α then $U \cap \Sigma$ consists of a pair of Möbius bands intersecting along their center lines $-\alpha$. The band determined by $W^u(\alpha)$ will be smooth, that determined by $W^s(\alpha)$ will be at least C^1. Topologically, Σ is obtained by adjoining a cylinder to $U \cap \Sigma$ along the boundary of the Möbius bands. If $\lambda_r > \lambda_s$ then Σ will be tangent to the z_2-plane along α and will only be homeomorphic to a differentiably immersed Klein bottle.

In Figure 2, we show the intersection of Σ with the fixed point set of κ—the (x_1, x_2)-plane—in case $\lambda_r < \lambda_s$. Note that since the plane is the fixed point set of

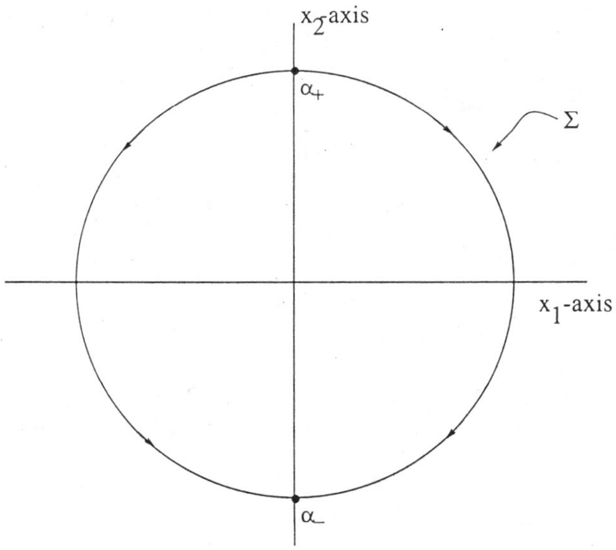

Figure 2 Real points on the homoclinic cycle Σ

κ, it is invariant by the flow of X. We let α_+ (respectively, α_-) denote the point on α with real part greater than zero (respectively, less than zero) and zero imaginary part.

Lemma 4.1 *Let $r \geq 1$ and U be an open neighborhood of α in $\mathbb{C}^2$. There exists an open neighborhood N of X in the space C^r vector fields on $\mathbb{C}^2$ (C^r-topology) such that if $Y \in N$ then Y has a C^r normally hyperbolic submanifold $\alpha_Y \subset U$. Furthermore,*

1. $\mathrm{ind}_Y(\alpha_Y) = \mathrm{ind}_X(\alpha)$,
2. *α_Y depends continuously on Y with respect to the C^r-topology and α_Y is C^r diffeomorphic to a circle,*
3. $\alpha_X = \alpha$,
4. *If Y is $SO(2)$-equivariant, then α_Y is a subset of the z_2-plane.*

Proof The result is an immediate consequence of the normal hyperbolicity of α (see Section 3). □

Our goal is to obtain a good geometric and analytic description of the flow of X near Σ. Subsequently, we use this description to analyze what happens when we break symmetry from $O(2)$ to $SO(2)$ or $\mathbb{Z}_2(\kappa)$.

We let S^3 denote the unit sphere of $\mathbb{C}^2$ and $\mathbb{R}^2$ denote the fixed point set of κ. For the purposes of this exposition, we shall study a slightly simpler problem defined on S^3 rather than $\mathbb{C}^2$. We shall now briefly justify our reduction to S^3. First of all we shall assume that λ_r dominates λ_s and so Σ is at least C^1 immersed. Rescaling we may and shall suppose that α is the unit circle in the z_2-plane. If we blow-up $\mathbb{C}^2$ at the origin, X lifts to an equivariant vector field X_1 on $S^3 \times \mathbb{R}$. The vector field X_1 will have a homoclinic cycle Σ_1, with self-intersection along $\alpha_1 = \alpha \times \{1\} \subset S^3 \times \{1\}$, and Σ_1 will be tangent to S^3 along α_1. Since Σ does not contain the origin of $\mathbb{C}^2$, the flow of X_1 near Σ_1 completely determines the flow of X near Σ. Choose a small closed tubular neighborhood U of α_1. The intersection of Σ with U consists of two circles C_s and C_u, where we shall assume that $X_1|\partial U$ points outward along C_u and inward along C_s. Let V_u be a tubular neighborhood of C_u in ∂U. Provided that V_u is sufficiently small, we can use the flow of X_1 to define a map $P : V_u \to V_s \subset \partial U$, where V_s will be a neighborhood of C_s. Thus, if $x \in V_u$ then $P(x) \in V_s$ will be the point where the forward trajectory of X_1 through x first re-enters U. The dynamics of X_1 near Σ_1 are determined by the map P and the flow of X_1 inside Σ. We may differentiably deform S^3 near α_1 so that $\Sigma_1 \cap U \subset S^3$. Since, as far as the dynamics near Σ are concerned, what matters is the attaching map P it is no loss of generality to assume further that $\Sigma \setminus U \subset S^3$. These arguments indicate that, at least in the case where the radial eigenvalue dominates, we can ignore the radial direction.

In practice, what we shall do is assume that:

(1) S^3 is invariant by the flow of X and normally hyperbolic;

(2) $\Sigma \subset S^3$.

As a consequence of (1,2) and the persistence of normally hyperbolic sets, we may restrict attention to the flow induced by X on S^3. From now on, we shall always assume that X is an $O(2)$-equivariant vector field on S^3 with associated relative equilibrium α and homoclinic cycle Σ. We denote the contracting eigenvalue along α by γ_0 and the expanding eigenvalue by λ_0. We assume always that Σ is an attracting homoclinic cycle. It is not difficult to show that a sufficient condition for Σ to be attracting is given by

$$\lambda_0 + \gamma_0 < 0, \tag{4.1}$$

and we assume that this condition is satisfied.

It follows from Example 2.9 and our previous work that the polar-blowing-up of S^3 along α is an $O(2)$-manifold M diffeomorphic to $S^2 \times S^1$. However, the action of $O(2)$ on M has one less isotropy type than the action of $O(2)$ on S^3 (see Section 2.3). Specifically, M has no isotropy groups containing $\mathbb{Z}_2$ (the subgroup of $SO(2)$ generated by $\exp(\imath\pi)$). In particular, the induced action of $SO(2)$ on M is *free*. Most of our work will be taken up with describing the vector field $\hat{X}$ induced on M from S^3. We let $\pi : M \to S^3$ denote the 'blowing-down' map of M onto S^3.

4.1 The flows of X and $\hat{X}$ near the homoclinic cycle We take an $O(2)$-invariant tubular neighborhood N of $\alpha \subset S^3$. Provided the diameter of N is sufficiently small, $W^u(\alpha) \cap N$ and $W^s(\alpha) \cap N$ define Möbius bands in N intersecting along α. (This is most easily seen by looking at the $O(2)$-orbits of the line segment $W^{uu/ss}(\alpha, \alpha_\pm) \cap N \cap \mathbb{R}^2$.)

The neighborhood N is $O(2)$-equivariantly diffeomorphic to $D \times S^1 \subset \mathbb{C}^2$, where D is the closed unit disc in the z_1-plane and S^1 is the unit circle in the z_2-plane. In the sequel, we frequently identify N with $D \times S^1$. The action of $\mathbb{Z}_2(\kappa)$ on $N = D \times S^1$ is given by

$$(z, \phi) \mapsto (\bar{z}, -\phi).$$

The $SO(2)$ part of the $O(2)$-action is given by

$$\exp(\imath\theta)(z, \phi) = (\exp(\imath\theta)z, \phi + 2\theta).$$

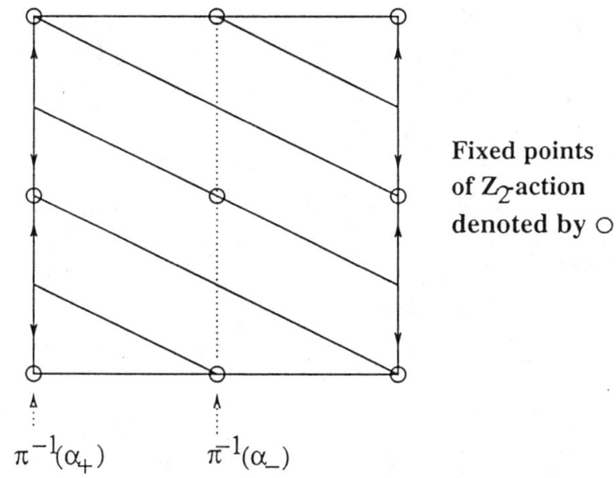

Figure 3 The flow of $\hat{X}$ on $\pi^{-1}(\alpha)$

We want to describe the flow induced by $\hat{X}$ on $\pi^{-1}(\alpha)$. Conceptually, it is probably easiest to do this by first thinking of $\pi^{-1}(\alpha)$ as the boundary of N and then as the boundary of $\overline{S^3 \setminus N} \subset M$.

Lemma 4.2 *The linearized flow of X along α fixes the normal sections $D \times \{\theta\} \subset \mathbb{C}^2$.*

Proof Working in $\mathbb{C}^2$, we write X in component form as (X_1, X_2). It suffices to prove that derivatives in z_1-directions of X_2 along α vanish. Let $P(\mathbb{C}^2)^{O(2)}$ denote the $\mathbb{R}$-algebra of $O(2)$-invariant functions on $\mathbb{C}^2$ and $P_{O(2)}(\mathbb{C}^2, \mathbb{C})$ denote the $P(\mathbb{C}^2)^{O(2)}$-module of $O(2)$-equivariant polynomial maps from $\mathbb{C}^2$ to $\mathbb{C}$, where we take the $(1,2)$-representation of $O(2)$ on $\mathbb{C}^2$ and the 2-representation on $\mathbb{C}$. A simple computation verifies that a set of generators for $P(\mathbb{C}^2)^{O(2)}$ is given by $\{|z_1|^2, |z_2|^2, \mathrm{Re}(z_1^2\bar{z}_2)\}$ and that generators for the $P(\mathbb{C}^2)^{O(2)}$-module $P_{O(2)}(\mathbb{C}^2, \mathbb{C})$ are given by $\{z_1^2, z_2\}$. Hence, by Schwarz' theorem on smooth invariants, we may find smooth functions $h, g : \mathbb{R}^3 \to \mathbb{R}$ such that

$$X_2(z_1, z_2) = h(|z_1|^2, |z_2|^2, \mathrm{Re}(z_1^2\bar{z}_2))z_1^2 + g(|z_1|^2, |z_2|^2, \mathrm{Re}(z_1^2\bar{z}_2))z_2.$$

If $(0, z_2) \in \alpha$ then $X_2(0, z_2) = 0$. Computing, we see that $D_{z_1}X_2(0, z_2) = 0$. □

As a consequence of Lemma 4.2, we see that the induced flow of $\hat{X}$ on $\pi^{-1}(\alpha)$ is determined by by the restriction of the linearized flow to any section $D \times \{\theta\}$ of α.

Noting Example 2.7, we see that the 'radial' eigenvalues for the blown-up section are the same as those for the corresponding directions on S^3 and the tangential eigenvalues are given by $\lambda_0 - \gamma_0, \gamma_0 - \lambda_0$.

We show the flow of $\hat{X}|\pi^{-1}(\alpha)$ in Figure 3, where we have represented the 2-torus $\pi^{-1}(\alpha)$ as a square, with opposite sides identified. Notice that the sets left fixed by the flow—corresponding to the invariant manifolds of α—rotate through π as we go once around the torus.

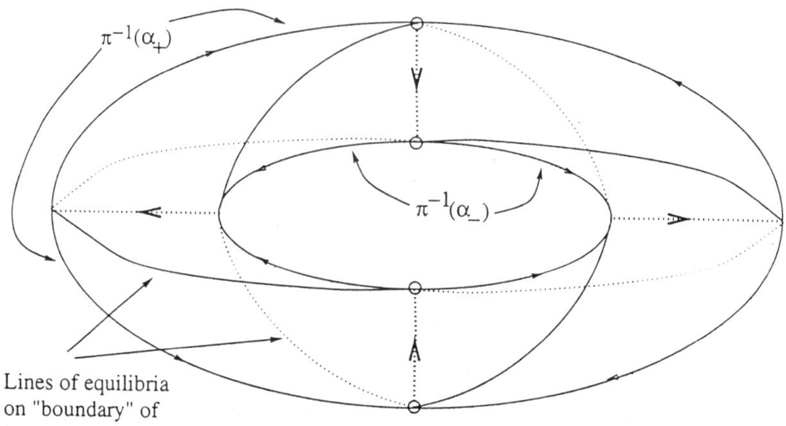

Figure 4 The flow of $\hat{X}$ restricted to M^+

We remark that the $\mathbb{Z}_2(\kappa)$-action on $\pi^{-1}(\alpha)$ has four fixed points and is given by $(\zeta, \phi) \mapsto (-\zeta, -\phi)$.

To represent the flow induced on $\pi^{-1}(\alpha)$—regarded as a boundary of $M \setminus \pi^{-1}(\alpha)$—we merely have to rotate Figure 3 through $\pi/2$. As we have already indicated in Example 2.9, the closure of either of the components $M \setminus \pi^{-1}(\alpha)$ is (topologically) a solid torus. We choose the closure of one such component, say M^+. In Figure 4 and below we describe the main features of the flow on M^+.

We have marked the four fixed points of the $\mathbb{Z}_2$-action on ∂M^+ using circles. Observe that the homoclinic connection joins both pairs of fixed points as shown in the figure. As we move round the torus, the band representing the homoclinic connection undergoes two full twists. Representing the torus M^+ as $D \times S^1$, where $D \subset \mathbb{C}$ is a closed disc, center zero, we see that the $\mathbb{Z}_2(\kappa)$-action on M^+ is given by $(z, \phi) \mapsto (\bar{z}, -\phi)$. The $SO(2)$-action on M^+ is given by

$$\exp(\imath\theta)(z_1, \phi) = (\exp(2\imath\theta)z, \phi + \theta).$$

Define the map $h : D \times S^1 \to D \times S^1$ by

$$h(z, \phi) = (\exp(-2\imath\theta)z, \phi),$$

Obviously, h is a smooth diffeomorphism of the torus. The effect of h is to 'untwist' the $SO(2)$-action. Specifically, if we take the $O(2)$-action on M^+ defined by

$$\begin{aligned} \exp(\imath\theta)(z, \phi) &= (z, \phi + \theta) \\ \kappa(z, \phi) &= (\bar{z}, -\phi) \end{aligned}$$

and for the moment denote M^+, with this new action, by $\hat{M}^+$, then $h : M^+ \to \hat{M}^+$ is an $O(2)$-equivariant diffeomorphism. From now on we use the map h to untwist the $SO(2)$-action and use the new $O(2)$-action on M^+. The homoclinic connection now appears as a separatrix for the flow of $\hat{X}$ on M^+. Cutting the torus through the section $\phi = 0$, we may represent the flow on the torus as in Figure 5.

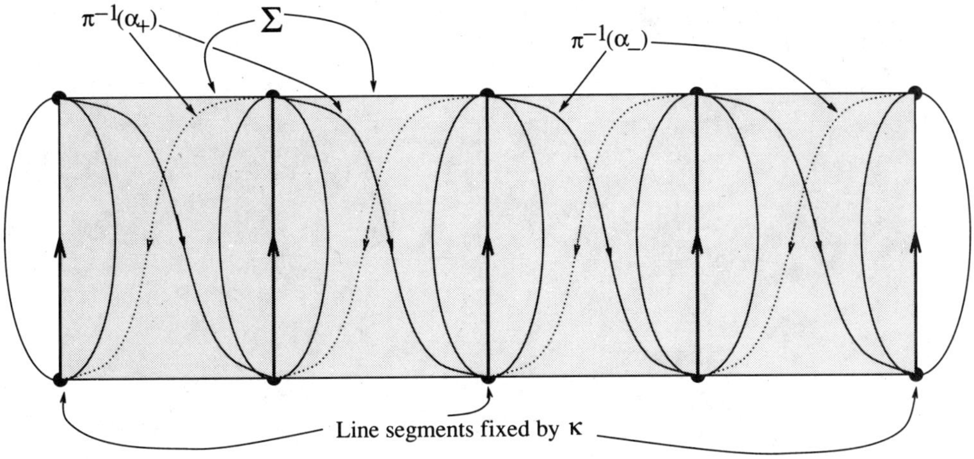

Figure 5 The flow of $\hat{X}$ on M^+ untwisted

4.2 Normalizing families of blowing-ups We conclude this section with a few brief remarks about how we handle the blowing-up of perturbations of X. Let $r \geq 1$ and U be an open $O(2)$-invariant neighborhood of α in S^3. It follows from Lemma 4.1, that we may choose an open neighborhood N of X in the space of C^r vector fields on S^3, such that each $Y \in N$ has a C^r normally hyperbolic submanifold α_Y and that α_Y depends continuously on Y with $\alpha_X = \alpha$. It is an easy exercise (see Bredon [1972], Chapter VI or Hirsch, Pugh and Shub [1977]) to show that there exists a family of isotopies

$$\Xi : U \times S^3 \to S^3$$

satisfying

(I1) For each $Y \in N$, $\Xi_Y : S^3 \to S^3$ is a C^r diffeomorphism which is equal to the identity map on $S^3 \setminus U$,

(I2) $\Xi_Y(\alpha_Y) = \alpha$, for all $Y \in N$,

(I3) If H is any closed subgroup of $O(2)$ and $Y \in N$ is H-equivariant, then Ξ_Y is H-equivariant,

(I4) Ξ_Y depends continuously on Y, C^r-topology.

Suppose $Y \in N$ and define the push-forward $\Xi_Y^\star Y$ of Y by

$$\Xi_Y^\star(Y)(u) = T\Xi_Y Y(\Xi_Y^{-1}(u)),\ u \in TS^3.$$

Clearly, α is normally hyperbolic submanifold for $\Xi_Y^\star Y$ and the dynamics of $\Xi_Y^\star Y$ are C^r topologically equivalent to those of Y. As a consequence of these observations we may and shall always assume in the sequel that perturbations Y of X have $\alpha_Y = \alpha$. In particular, the blowing-up we use for X will work for Y.

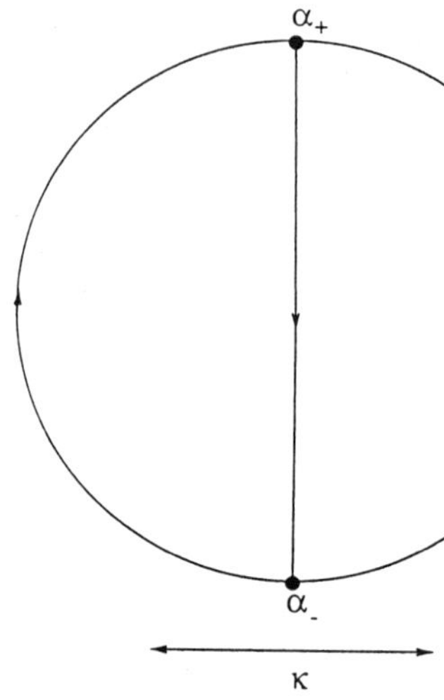

Figure 6 The flow induced by $\hat{X}$ on D

5 Breaking symmetry from $O(2)$ to $SO(2)$

We continue with the notational assumptions of the preceding section. In particular, we consider the vector field $\hat{X}$ defined on the solid torus M^+. The group $SO(2)$ acts freely on M^+. The orbit space $M^+/SO(2)$ is diffeomorphic to the disc D. Moreover, if we identify D with the closed unit disc, center zero, in $\mathbb{C}$, then the $\mathbb{Z}_2(\kappa)$ action on M^+ drops down to the $\mathbb{Z}_2(\kappa)$-action on D defined by $z \mapsto \bar{z}$. The fixed set of this action corresponds precisely to the quotient of Σ by $SO(2)$. In Figure 6, we show the flow induced by $\hat{X}$ on D. We remark that the flow contains two homoclinic cycles. The connection forced by the homoclinic cycle is generic within the class of $\mathbb{Z}_2(\kappa)$-equivariant vector fields on D.

The components of these cycles lying in the boundary of D come from the blowing-up construction and persist under perturbations of $\hat{X}$.

Now we break symmetry from $O(2)$ to $SO(2)$. Generically, we do this by breaking the link forced by the $\mathbb{Z}_2(\kappa)$-symmetry and obtain the flow shown in Figure 7 (or its mirror image). Since the two cycles in D are attracting, it follows easily that for sufficiently small generic $SO(2)$-equivariant perturbations Y of X that break the $\mathbb{Z}_2(\kappa)$-symmetry we obtain a hyperbolic attracting periodic orbit in D close to one of the original pair of attracting cycles. Indeed, if we assume the vector field induced on D is C^2-linearizable at the equilibria, the usual Poincaré map argument yields a hyperbolic attracting orbit depending continuously on perturbations breaking the connection. In general, if we do not assume C^2-linearizability, the same conclusion can be obtained using methods based on Shilnikov coordinates and Deng's exponential expansion theorems (Deng [1989]).

Lifting back up to M^+, we obtain an $SO(2)$-equivariant flow on a 2-torus. Moreover, this torus is normally hyperbolic and attracting for the flow of $\hat{Y}$. We now blow-down to obtain a Φ^Y-invariant 2-torus $T \subset S^3$. Clearly, T is normally hyperbolic and attracting for Φ^Y. Moreover, it follows from Field [1989], Remarks A.11 (3), that the flow induced by Φ^Y on T is generically a quasi-

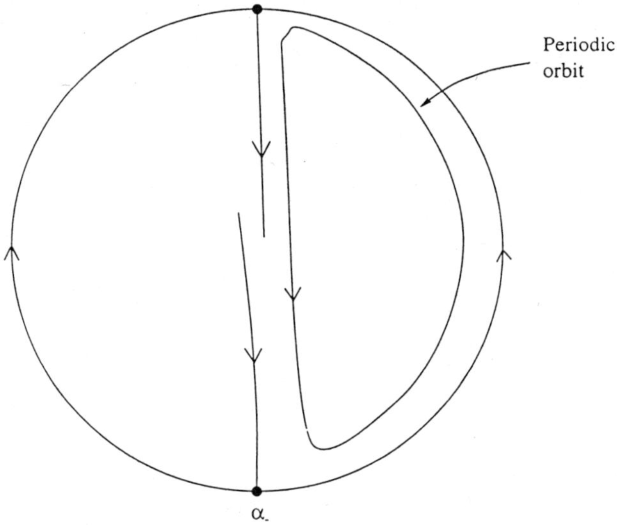

Figure 7 Breaking symmetry to $SO(2)$

periodic flow with a slow drift along $SO(2)$ orbits. Since T is normally hyperbolic, it will persist under general (small) perturbations of Y. Of course, if perturbations are not $SO(2)$-equivariant, we can expect the formation of long-period limit cycles due to phase-locking.

Remark 5.1 Although we assumed that X was defined on S^3, it is easy to carry through our analysis for vector fields defined on $\mathbb{C}^2$. Specifically, we start by blowing-up at the origin of $\mathbb{C}^2$ and then blow-up $S^3 \times \mathbb{R}$ along $\alpha \times \mathbb{R}$. After quotienting out by the free action of $SO(2)$, we obtain a homoclinic cycle on the space $D \times \mathbb{R}$. Using a similar analysis to that given above, it can be shown that breaking symmetry to $SO(2)$ leads to the appearance of an attracting hyperbolic orbit in $D \times \mathbb{R}$. (This result holds irrespective of whether or not the radial eigenvalue λ_r dominates λ_s.)

6 Breaking symmetry from $O(2)$ to $\mathbb{Z}_2(\kappa)$

We continue with the notations of Section 5. In particular, $\hat{X}$ denotes the vector field induced from X on M^+.

We recall that in Figure 5 we showed the flow induced by $\hat{X}$ on M^+ and that the homoclinic connection appears as a separatrix in M^+.

We now break symmetry from $O(2)$ to $\mathbb{Z}_2(\kappa)$. Specifically, we suppose that Y is a $\mathbb{Z}_2(\kappa)$-equivariant perturbation Y of X. It follows by $\mathbb{Z}_2(\kappa)$-equivariance that α_+ and α_- will be equilibria for $Y|\alpha$. We denote the eigenvalues of the linearization of Y at α_- by $\lambda_-^Y, \gamma_-^Y, \nu^Y$ and at α_+ by $\lambda_+^Y, \gamma_+^Y, \delta^Y$. The eigenvalues will depend continuously on Y, C^1-topology, and satisfy

$$\lambda_\pm^X = \lambda_0; \ \gamma_\pm^X = \gamma_0; \ \nu^X = \delta^X = 0.$$

As $\lambda_\pm^Y$ will be close to λ_0 and $\gamma_\pm^Y$ close to γ_0, we generally drop super and subscripts and just set $\lambda_\pm^Y = \lambda$, $\gamma_\pm^Y = \gamma$, $\nu^Y = \nu, \delta^Y = \delta$.

As explained in 4.2, we may and shall suppose that α is always invariant by the flow of Y.

Since α is normally hyperbolic for the flow of X, it follows that for sufficiently small C^1-perturbations Y, α will be normally hyperbolic for Y. In addition, the strong stable and unstable manifolds $W^{ss}(\alpha_\pm)$, $W^{uu}(\alpha_\pm)$ are well defined immersed curves through $\alpha_\pm$.

We shall consider perturbations Y satisfying:

(A) $\nu\gamma - \delta\lambda > 0$;

(B) The invariant manifolds $W^{ss}(\alpha_\pm)$, $W^{uu}(\alpha_\pm)$ meet $\mathbb{Z}_2(\kappa)$-transversally.

Remarks 6.1

(1) It follows from condition (A), that at least one of the equilibria $\alpha_\pm$ for $Y|\alpha$ is hyperbolic and contracting. From the point of view of dynamics, the simplest situation arises when we assume $\alpha_\pm$ are both hyperbolic and the only equilibria for $Y|\alpha$.

(2) Without any group action, we would ask that the invariant manifolds should meet transversally. Essentially, Condition (B) asks that the invariant manifolds meet as transversally as possible, consistent with the $\mathbb{Z}_2(\kappa)$-action. In particular, the intersections of the invariant manifolds will be stable under small $\mathbb{Z}_2(\kappa)$-equivariant perturbations of Y but need *not* be transversal. For more details on this point, we refer the reader to Field [1980]. We shall explain the implication of these conditions for our example shortly.

6.1 Intersections of invariant manifolds Suppose that Y is C^1-close to X and that $\nu < 0$ and $\delta > 0$. Since Y is assumed $\mathbb{Z}_2(\kappa)$-equivariant, we have $W^{uu}(\alpha_+), W^{ss}(\alpha_-) \subset \text{Fix}(\mathbb{Z}_2(\kappa))$. In this case the invariant manifolds $W^{uu}(\alpha_+)$, $W^{ss}(\alpha_-)$ meet $\mathbb{Z}_2(\kappa)$-transversally but non-transversally. Generically speaking, we expect that $W^u(\alpha_+)$ and $W^s(\alpha_-)$ are transverse along $\text{Fix}(\mathbb{Z}_2(\kappa))$. If $W^{uu}(\alpha_-)$ and $W^{ss}(\alpha_+)$ meet $\mathbb{Z}_2(\kappa)$-transversally then they have null intersection. Of course, in this case, $W^u(\alpha_-) = W^{uu}(\alpha_-)$ and $W^s(\alpha_+) = W^{ss}(\alpha_+)$.

If $\nu > 0$ and $\delta < 0$, then $W^{uu}(\alpha_+) = W^u(\alpha_+)$ and $W^{ss}(\alpha_-) = W^s(\alpha_-)$. Both manifolds are contained in $\text{Fix}(\mathbb{Z}_2(\kappa))$ and meet $\mathbb{Z}_2(\kappa)$-transversally but non-transversally. Generically speaking, we expect that the center unstable and center stable manifolds $W^{cu}(\alpha_+)$, $W^{cs}(\alpha_-)$ should meet transversally along $\text{Fix}(\mathbb{Z}_2(\kappa))$. On the other hand if $W^{uu}(\alpha_-), W^{ss}(\alpha_+)$ meet $\mathbb{Z}_2(\kappa)$-transversally then they have null intersection. Generically, we expect $W^u(\alpha_-)$ and $W^s(\alpha_+)$ to meet transversally, possibly with non-empty intersection.

For the remainder of this section we shall assume $\nu < 0 < \delta$ (the analysis and results in case $\delta < 0 < \nu$ are broadly similar).

In Figure 8, we show the generic situation in M^+ near the fixed point set of κ after breaking symmetry from $O(2)$ to $\mathbb{Z}_2(\kappa)$. The $\mathbb{Z}_2(\kappa)$ symmetry forces the connections between the equilibria at A_0, B_0 and A_2, B_2 to persist, though the intersections are only $\mathbb{Z}_2(\kappa)$-transversal and not transversal. In particular, $W^u(A_0) \cap W^s(B_0)$ is 1-dimensional and lies in the fixed point set of $\mathbb{Z}_2(\kappa)$. The connections between A_1, B_1 and A_3, B_3 are broken. It follows from the $\mathbb{Z}_2(\kappa)$ symmetry that the flow in $D \times [\pi, 2\pi]$ is the κ image of the flow in $D \times [0, \pi]$.

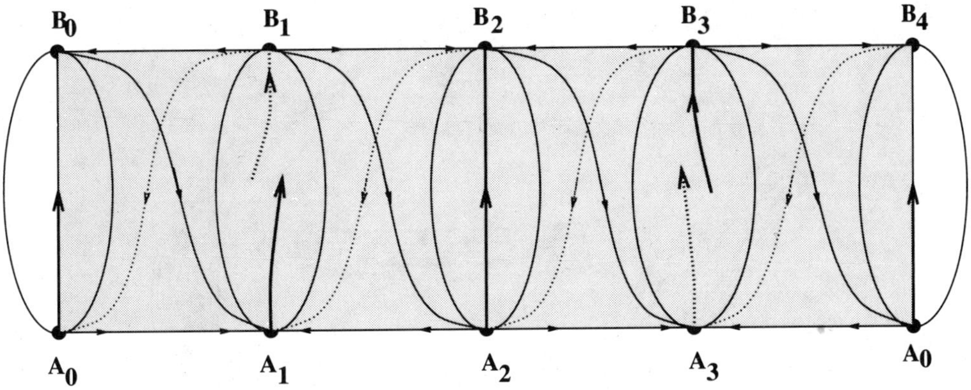

Figure 8 Breaking symmetry to $\mathbb{Z}_2$

In particular, if $W^u(A_1)$ is perturbed upwards then $W^u(A_3)$ will be perturbed downwards. We refer the reader to Figure 8.

6.2 Dynamics Let $\Xi^{\pm}$ denote the two closed regions of M^+ with common boundary the homoclinic connection. We regard Ξ^+ as the 'upper' component and Ξ^- as the 'lower' component. Let $\hat{\Sigma} = \partial\Xi^- \cup \partial\Xi^+$ and observe that $\hat{\Sigma}$ is the blow-up of Σ.

With the notation of Figure 8, we define curves $\hat{\Gamma}_1, \hat{\Gamma}_2 \subset \hat{\Sigma}$ by

$$\hat{\Gamma}_1 = W^u(B_2) \cup W^s(A_2) \cup W^u(A_1) \cup W^{uu}(A_2) \cup W^u(A_3)$$

$$\hat{\Gamma}_2 = (W^u(B_2) \cap \Xi^-) \cup (W^s(A_2) \cap \Xi^+) \cup W^u(A_1) \cup W^{uu}(A_2)$$

Let Γ_1, Γ_2 denote the projections of $\hat{\Gamma}_1, \hat{\Gamma}_2$ in S^3.

Let Y be an $\mathbb{Z}_2(\kappa)$-perturbation of X satisfying (P2,P3). We say that Y is an *upward* perturbation if $W^u(A_1)$ is perturbed into Ξ^+ and a *downward* perturbation if $W^u(A_1)$ is perturbed into Ξ^-. Since Σ, and therefore $\hat{\Sigma}$, are attracting it follows that if Y is sufficiently close to X, trajectories of $\hat{Y}$ with initial conditions close to $\hat{\Sigma}$ stay close to $\hat{\Sigma}$ for all positive time.

We now give a heuristic argument, based on the geometry of the flow on M^+, to show why we expect that $\mathbb{Z}_2(\kappa)$-equivariant symmetry breaking perturbations Y of X should lead to periodic orbits close to either Γ_1 or Γ_2.

Let $y \in \Xi^+$ and suppose that y is close to the midpoint of $W^{uu}(A_2)$. If we follow the forward trajectory of y under the flow $\Phi^{\hat{Y}}$ of $\hat{Y}$, we see that $\Phi_t^{\hat{Y}}(y)$ first passes near the point B_2. The next part of the trajectory will be close to $W^u(B_2)$ and will pass close to A_3. Granted that $W^u(A_3)$ is perturbed down into Ξ^-, it follows that $\Phi_t^{\hat{Y}}(y)$ will enter Ξ^- before getting close to B_3. Since the expansion along α is weak near α_+, the trajectory will then reverse direction near B_3 and approximately follow $W^s(A_2) \cap \Xi^-$. Consequently, the trajectory will approach A_2 and then follow $W^{uu}(A_2)$. At B_2, the trajectory changes direction again and now follows $W^u(B_2)$ approaching A_1. As the trajectory passes by A_1, it will follow $W^u(A_1)$. Provided that the trajectory is close enough to $W^u(A_1)$, it will re-enter Ξ^+. Just as before, the trajectory will now stay in Ξ^+, passing successively close to

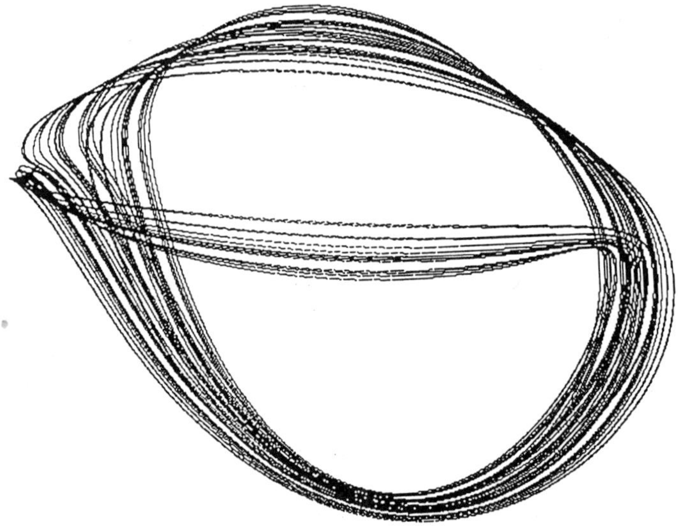

Figure 9 Chaotic dynamics near resonance

B_1, A_2 and the original point y. In this case we can reasonably expect $\hat{Y}$ to have a periodic orbit near $\hat{\Gamma}_1$.

However, this analysis is incomplete and is only valid if the contraction γ is not large compared with λ. Indeed, suppose that $|\gamma| >> \lambda$. Let y be close to the midpoint of $W^s(A_2) \cap \Xi^+$. Then the trajectory of y can be expected to pass very close to $\text{Fix}(\mathbb{Z}_2(\kappa))$. If this is so, then the twist round $W^{uu}(A_2)$ becomes important. If there is a counterclockwise twist around $W^{uu}(A_2)$, the trajectory can enter Ξ^- and so reverse direction at B_2. Indeed, this reversal of direction will be sensitive as to which side of $W^s(B_2)$ the trajectory lies on. If this reversal occurs, we can expect the appearance of a periodic orbit near $\hat{\Gamma}_2$. Of course, this simplified presentation also suggests the possibility of chaotic dynamics arising from resonances of γ and λ.

We now give a precise statement of the main analytical results. We continue to assume that $\nu < 0 \leq \delta$.

Choose a C^r, $r \geq 5$, local coordinate system (x, y, z) at α_+ for X such that the x-axis corresponds to $W^{uu}(\alpha_+)$, the y-axis to $W^{ss}(\alpha_+)$ and the z-axis to α. We remark that we can always do this and refer to Field (in preparation) for details. Choose a non-zero point ζ on the y-axis and let S be a 2-disc with center ζ which is orthogonal to $W^{ss}(\alpha_+)$.

Suppose that Y is a $\mathbb{Z}_2(\kappa)$-equivariant perturbation Y of X. Let $\beta = \beta(Y)$ denote the absolute value of the x-coordinate of the intersection of $W^u(\alpha_-)$ with S. We regard β as our measure of the breaking of the connection between $W^{uu}(\alpha_-)$ and $W^{ss}(\alpha_+)$.

We similarly take a local coordinate system at α_- and let $A(Y) = A$ denote the angle between $W^s(\alpha_-)$ and $W^u(\alpha_+)$.

Suppose that $C_1, C_2, C_3 > 0$. We shall consider the following conditions on $\mathbb{Z}_2(\kappa)$-equivariant perturbations Y of X:

$$\begin{array}{lrcl}
\textbf{(PC1)} & \beta(Y) & \geq & C_1\|Y-X\|_5; \\
\textbf{(PC2)} & (\nu^Y\gamma_+^Y - \delta^Y\lambda_-^Y)/(\lambda_+^Y\lambda_-^Y) & \geq & C_2\beta(Y); \\
\textbf{(PC3)} & A(Y) & \geq & C_3\beta(Y).
\end{array}$$

Theorem 6.2 *Let $C_1, C_2, C_3 > 0$.*

(1) *Suppose that $\lambda_0+\gamma_0 < 0 < 2\lambda_0+\gamma_0$. Then we may find $\varepsilon > 0$ such that if Y satisfies (PC1), (PC2) and $\|X-Y\|_5 < \varepsilon$ then Y has a hyperbolic attracting periodic orbit close to one of Γ_1, $-\Gamma_1$.*

(2) *Suppose that $\lambda_0+\gamma_0, 2\lambda_0+\gamma_0 < 0$. Then we may find $\varepsilon > 0$ such that if Y satisfies (PC1), (PC2), (PC3) and $\|X-Y\|_5 < \varepsilon$ then Y has a hyperbolic attracting periodic orbit close to one of $\pm\Gamma_2, \pm\kappa\Gamma_2$.*

Remark 6.3 As stated, Theorem 6.2 applies to vector fields on S^3. However, the result holds for vector fields defined on $\mathbb{C}^2$ provided that the radial eigenvalue dominates (see Field (in preparation)). The situation when $\lambda_s \leq \lambda_r$ is rather delicate though it is possible that the methods of Field (in preparation) can be extended to this case.

There remains the question of what happens at the resonant point $\gamma_0 = -2\lambda_0$. Our geometric description of the mechanism that leads to different behavior depending on the sign of $2\lambda_0 + \gamma_0$ suggests the possibility of chaotic dynamics near the resonant point. In Figure 9, we present the results of a numerical integration using *dstool* that support this conjecture.

We used the following equations (on $\mathbb{C}^2$) to produce Figure 9:

$$\begin{aligned}
\frac{dz_1}{dt} &= (0.07 - 0.5|z_1|^2 - 0.6114|z_2|^2)z_1 - \bar{z}_1 z_2 + 0.04\bar{z}_1, \\
\frac{dz_2}{dt} &= (0.25 - 0.6|z_1|^2 - 0.5|z_2|^2)z_2 + z_1^2 + 0.01\bar{z}_2,
\end{aligned}$$

and the image shown is the projection of the attractor on the (x_1, y_2)-plane.

The term $(0.04\bar{z}_1, 0.01\bar{z}_2)$ breaks symmetry from $O(2)$ to $\mathbb{Z}_2(\kappa)$. The system has a pair of equilibria $(0, \pm 0.72111)$ on the z_2-axis and it is easy to check that (for the unperturbed $O(2)$-equivariant system) we have the resonance $2\lambda_0+\gamma_0 = 0$. We should also note that, for this example, the radial eigenvalue is not the strong contracting eigenvalue.

References

Armbruster, D., Guckenheimer, J. and Holmes, P. [1988] *Heteroclinic cycles and modulated travelling waves in systems with $O(2)$ symmetry*, Physica D, 257–282.

Bredon, G.E. [1972] *Introduction to Compact Transformation Groups* Pure and Applied Mathematics, **46**, Academic Press, New York and London.

Campbell, S. and Holmes, P. [1992] *Heteroclinic cycles and modulated travelling waves in a system with D_4 symmetry*, preprint.

Chow, S.-N., Deng, B. and Fiedler, B. [1990] *Homoclinic bifurcation at resonant eigenvalues*, J. Dynamics & Differential Equations, **2**(2), 177–244.

Chossat, P. *Forced reflectional symmetry breaking of an $O(2)$-symmetric homoclinic cycle*, Nonlinearity, **6**(5), 723-732.

Crawford, J.D. and Knobloch, E. [1988] *On degenerate Hopf bifurcation with broken $O(2)$-symmetry*, Nonlinearity, **1**(4), 617–512.

Dangelmayr, G. and Knobloch, E. [1991] *Hopf bifurcation with broken circular symmetry*, Nonlinearity, **4**, 399–427.

Deng, B. [1989] *The Sil'nikov problem, exponential expansion, strong λ-lemma, C^1-linearization, and homoclinic bifurcation*, J. Diff. Eq., **79**, 189–231.

Field, M.J. [1978] *Resolving actions of compact Lie groups*, Bull. Austral. Math. Soc., **18**, 243–254.

Field, M.J. [1980] *Equivariant dynamical systems*, Trans. Amer. Math. Soc., **259**, 185–205.

Field, M.J. [1989] *Equivariant bifurcation theory and symmetry breaking*, J. Dynamics and Differential Equations, **1**, 369–421.

Field, M.J. [1989] *Local structure of equivariant dynamics*, in: Singularities, Bifurcations, and Dynamics, Proceedings of Symposium on Singularity Theory and its Applications, Warwick (eds. R. M. Roberts and I. N. Stewart), Lect. Notes in Math. **1463**, Springer-Verlag, Heidelberg (1991), 168-195.

Field, M.J. *Bifurcation and symmetry breaking from a homoclinic cycle with zero eigenvalues*, in preparation.

Field, M.J. [1995] *Symmetry breaking for compact Lie groups,* to appear in Memoirs of the Amer. Math. Soc.

Field, M.J. and Richardson, R.W. [1992] *Symmetry breaking and branching patterns in equivariant bifurcation theory I*, Archive for Rational Mechanics and Analysis, **120**, 147–190.

Field, M.J. and Swift, J.W. [1991] *Static bifurcation to limit cycles and heteroclinic cycles*, Nonlinearity, **4** (4), 1001–1043.

Golubitsky, M. and Schaeffer, D. [1983] *A discussion of symmetry and symmetry breaking*, Proc. Symp. Pure Math., **40**, Part I, 499–515.

Golubitsky, M., Stewart, I. and Schaeffer, D. [1988] *Singularities and Groups in Bifurcation Theory, Vol. II*, Appl. Math. Sci. **69**, Springer-Verlag, New York.

Hirsch, M.W., Pugh, C.C. and Shub, M. [1977] *Invariant Manifolds*, Springer Lect. Notes Math., **583**.

Krupa, M. [1990] *Bifurcations of relative equilibria*, SIAM J. Math. Anal., **21**(6), 1453–1486.

Lauterbach, R. and Roberts, M. [1992] *Heteroclinic cycles in dynamical systems with broken spherical symmetry*, J. Diff. Eq., **100**(1), 22–48.

van Gils, S. and Mallet-Paret, J. [1986] *Hopf bifurcation and symmetry: standing and travelling waves on a circle*, Proc. R. Soc. Edinburgh, **104A**, 279–307.

Fields Institute Communications
Volume **4**, 1995

Bifurcation in a Flow-Induced Vibration Model

Robert M. Corless
Department of Applied Mathematics
University of Western Ontario
London, Canada
N6A 5B7

Abstract. We use the computer algebra language Maple to carry out a detailed stability analysis of the nonzero solutions, obtained by the method of multiple scales, of a flow-induced vibration model. We calculate the stability boundaries in the damping versus coupling parameter plane, present qualitative representations ('cartoons') of the response curves, and locate saddle-node and secondary Hopf bifurcation points (corresponding to bifurcations to tori in the original system). We show that there are nine qualitatively different possible response curves predicted by the model. The approach is classical but new tools, such as Gröbner bases, are used.

1 Introduction

Corless and Parkinson [1988] studied the combined effects of vortex-induced vibration and galloping on a cylinder of square cross-section oscillating purely transversely in two-dimensional flow. They used the method of multiple scales (Nayfeh and Mook [1981]) to analyze a semi-empirical model derived from the Hartlen-Currie model for vortex-induced vibration (Hartlen and Currie [1970]) and from the quasi-steady model of Parkinson and Smith [1964]. They achieved reasonable agreement with the experimental data of Bearman et al.[1987]. However, the primary resonance region solution was less satisfactory, predicting in some cases unbounded solutions (to the order of the computation). This was remedied in Corless and Parkinson [1993] where a more appropriate asymptotic embedding of the model equations was used to make the solution process at once more valid, in that more

1991 *Mathematics Subject Classification.* Primary 34D15; Secondary 76B15.

This work was supported by the Natural Sciences and Engineering Research Council of Canada. The author is indebted to his former supervisor, G. V. Parkinson, for continuing interest in the project. Thanks are also due to W. F. Langford for inviting this work to be presented at the 1992 Fields Institute Workshop on Normal Forms and Chaos, and for encouraging the actual writing-up. An anonymous referee provided many helpful suggestions. The typing of this paper was made much simpler by the use of the Maple Share Library TeX package, for converting Maple expressions to plain TeX, written by Yunliang Yu of Duke University.

nonlinear terms were kept, and simpler, in that only one term in the perturbation solution was needed to achieve useful insights.

This paper takes that solution one step further and examines the changes in the equilibrium solutions that occur as the parameters are varied.

2 The model equations

As in Corless and Parkinson [1988], the model equations are

$$\ddot{Y} + Y = nU^2 \left(A_1 \left(1 - \frac{U_0}{U}\right) \frac{\dot{Y}}{U} - A_3 \left(\frac{\dot{Y}}{U}\right)^3 + A_5 \left(\frac{\dot{Y}}{U}\right)^5 - A_7 \left(\frac{\dot{Y}}{U}\right)^7 + C_v \right) \tag{2.1}$$

$$\ddot{C}_v + \omega_0^2 C_v = a(\omega_0) \left(\dot{C}_v - \frac{4}{3C_{y_0}^2 \omega_0^2} \dot{C}_v^3 \right) + D\dot{Y} + B\ddot{Y} \tag{2.2}$$

where $\omega_0 = U/U_r$ is the Strouhal frequency of the wake oscillator, $U_0 = 2\beta/nA_1$ is the critical reduced wind speed for galloping from the quasi-steady theory, and the A_i are parameters determined by fixed-cylinder experiments.

Remark 2.1 When $A_1 = 0$, note that $U_0 = \infty$, and we thus take the limiting value of the product $A_1 U_0 = 2\beta/n$ in the first term of the right-hand side of the first equation.

Y is the transverse displacement of the cylinder and C_v is the 'displacement' of the (mathematical) wake oscillator. C_{y_0} is the amplitude of the wake oscillator from a fixed cylinder. The parameters $a(\omega_0)$, $D(\omega_0)$, and $B(\omega_0)$ are not well understood and usually fixed in an *ad hoc* fashion. They may be determined by forced-oscillation experiments.

The parameter B is taken as zero for convenience in this paper. Other models, such as that in Berger [1987], achieve good agreement with experiment without this parameter.

Note that A_1, A_3, A_5, and A_7 have no physical meaning when taken separately. It is only the combination $A_1\dot{Y}/U - A_3(\dot{Y}/U)^3 + \cdots$ that summarizes the experimentally measured lift-coefficient data from a fixed cylinder at varying angles of attack. A solution technique that dealt with each term individually (as happened in the resonance region solution of Corless and Parkinson [1988]) is not likely to produce good results. This flaw was repaired in Corless and Parkinson [1993].

A similar model is analyzed in Berger and Plaschko [1993], wherein the method of multiple scales, bifurcation theory, and numerical methods are used to explore the model. The results of this present work are complementary to Berger and Plaschko [1993] in that we study the resonance region in detail, whereas they avoid this region through a suitable choice of scaling of the model parameters. However, their multiple scales solution disagrees with their numerical and bifurcation results in the resonance region, and the results of this present work may show how this can be repaired. For a discussion of other semi-empirical models of flow-induced vibration phenomena, see Parkinson [1989].

We will be restricting our attention here to the resonance region, where ω_0 is close to 1. Mathematically, we will take $\omega_0 = 1 + O(\mu)$ where μ is our small parameter. Hence to first order we may take the parameters a and D as constants.

Remark 2.2 The identically zero equilibrium solution $Y = 0$, $C_v = 0$ is always a solution. However, for Reynolds numbers larger than approximately 100, this solution is physically *unstable*. The Reynolds number ranges that we are interested in are typically many orders of magnitude larger than this critical number, and the present mathematical model predicts this instability as long as $a > 0$. What happens physically, and in the mathematical model, is that even if the cylinder is held fixed the wake oscillator increases to a dynamic equilibrium, with alternating vortices shed from the cylinder forming the Karman vortex street. Thus, bifurcations from the identically zero solution are probably not very interesting, except insofar as they affect the stable solutions with $C_v \neq 0$ (Corless, [1986]). It is thus not immediately clear that the normal forms approach as used in Langford and Zhan [1992] or Namachchivaya and Malhotra [1992] will give useful results. However, if these methods are capable of giving valid predictions of the model solutions with large amplitudes (particularly of solutions with large amplitudes for C_v), then the results will be of interest.

In this paper the method of multiple scales is used, and the solutions with large C_v are obtained directly. This is possible in this case because the proper scales for the solution have already been identified, essentially through a trial and error process (Corless and Parkinson [1993]).

We will take $\mu = a$ and suppose that $n = \mu^2 G$ for some G that is considered to be $O(1)$ as $\mu \to 0$. In Corless and Parkinson [1988] the fact that the A_i did not depend on n was used. This 'embeds' the model equations in a particular family of equations depending on the parameter μ. This family is used to solve the model equations in that an asymptotic series for the solution then can be computed, for example, by the method of multiple scales, which is valid in the limit as $\mu \to 0$. The solution obtained is then used for non-zero μ and can be compared with experimental results.

In the resonance region, the solution for Y obtained by the method of multiple scales was of the form $Y = Y_0 + \mu Y_1 + \mu^2 Y_2 + \cdots$, and it turned out that Y_0 was forced to be zero throughout the resonance region to avoid secularity at the zeroth order solution, as a direct consequence of nonsemisimple 1:1 resonance. This 'rigidity' of the first-order solution caused the nonlinear terms depending on A_3, A_5, and A_7 to disappear from the equation for Y_1. The resulting equations for Y_1 and C_1 could still only be solved numerically, or at best 'semi-numerically' by finding the equilibria numerically. It was then observed that although the solution for Y_1 did agree qualitatively with experimental data, it showed evidence of becoming unbounded, which violated the assumptions of the method of multiple scales. It was suggested that this was probably due to the nonlinear terms being inappropriately neglected (Gartshore [1987]). This was in fact the case, as was shown in Corless and Parkinson [1993].

3 Choice of small parameter for the method of multiple scales

If we put $\alpha_1 = \mu G A_1(U - U_0)$, $\alpha_3 = \mu^3 G A_3/U$, $\alpha_5 = \mu^5 G A_5/U^3$, and $\alpha_7 = \mu^7 G A_7/U^5$, *and suppose that these quantities are all* $O(1)$ *as* $\mu \to 0$, then we can rewrite the model equations as:

$$\ddot{y} + y = \mu(\alpha_1\dot{y} - \alpha_3\dot{y}^3 + \alpha_5\dot{y}^5 - \alpha_7\dot{y}^7 + sc) + O(\mu^2) \tag{3.1}$$

$$\ddot{c} + c = \mu(-\sigma_1 c + \dot{c}(1 - \tfrac{4}{3}\dot{c}^2) + d\dot{y}) + O(\mu^2) \tag{3.2}$$

where $Y = \mu y$, $c = C_v/C_{y_0}$, $s = GU_r^2 C_{y_0}$, $d = D/C_{y_0}$, and $\omega_0^2 = 1 + \mu\sigma_1$, so σ_1 indicates how far we are from exact resonance. Note that this replaces the strong coupling at resonance in equations (2.1) and (2.2) with a weak coupling, and thus these equations have a semisimple 1:1 resonance, unlike the previous equations.

Berger and Plaschko [1993], avoid nonsemisimple 1:1 resonance by an explicit assumption that the coupling parameter d in their (slightly different) model is proportional to $\omega_0^2 - 1$, in our notation. We avoid nonsemisimple 1:1 resonance here by assuming, essentially, that the amplitude of oscillation Y of the cylinder is small.

This scaling assumption is motivated in part by the observation that for a typical data set (Corless and Parkinson [1988]) the A_k grow quite rapidly with k—for example, in one case we had $A_1 = 4.87$, $A_3 = 421$, $A_5 = 17000$, and $A_7 = 194000$. If the solution process dealt with these coefficients at different orders of the perturbation expansion, the growth of these parameters would likely severely restrict the accurate range of a low-order solution anyway. However, note that the *range* of the polynomial in $\dot{Y}$ with these coefficients is itself $O(1)$ over the range of values of $\dot{Y}$ of interest. [We use $O(1)$ in that last sentence in the engineering sense of 'being about the same size as 1', in contrast with the mathematical sense of '$f(\mu) = O(1)$ as $\mu \to 0$ if $\lim_{\mu\to 0}(f(\mu)) = K$ for some constant K' as used previously.] Thus if we can somehow capture all of these coefficients in one term of the perturbation expansion, we would expect better accuracy for larger μ.

We further note that if we define the α_k as above, then for the above typical set of data we would get for $\mu = 0.13$ and $U = U_r = 1.22$ the values $\alpha_3 = 0.758$, $\alpha_5 = 0.348$, and $\alpha_7 = 0.045$. Clearly, embedding our model in a family which assumes that these quantities are (mathematically) $O(1)$ as $\mu \to 0$ is more likely to produce good results for non-zero values of μ than embedding the model in a family that assumes that the A_k are $O(1)$ as $\mu \to 0$. It is true that both will work in the limit of small μ. It is just more likely that the range of μ for which the answers will be acceptable will be larger with the present embedding.

One useful feature of the normal forms approach as an alternative would seem to be its ability to convert this high-order polynomial into an equivalent low-order one, as computations are more difficult with higher-order polynomials. This is not pursued here but is under consideration.

3.1 The slow flow equations Equations (3.1) and (3.2) are easier to solve than the original model equations. In fact, the program TWOVAR (Rand and Armbruster [1987]), written in MACSYMA will handle these equations directly. TWOVAR was used on a Mac II running ALJABR (Gardner and O'Dell [1991]), the Fort Pond Research version of MACSYMA, and took perhaps 15 minutes to derive the following polar-coordinate form of the slow-flow equations:

$$\frac{dr_1}{d\eta} = \tfrac{1}{2}(\alpha_1 r_1 - \tfrac{3}{4}\alpha_3 r_1^3 + \tfrac{5}{8}\alpha_5 r_1^5 - \tfrac{35}{64}\alpha_7 r_1^7 - s r_2 \sin\psi); \tag{3.3}$$

$$\frac{dr_2}{d\eta} = \tfrac{1}{2}\left(r_2 - r_2^3 + r_1 d \cos\psi\right); \tag{3.4}$$

$$\frac{d\psi}{d\eta} = \tfrac{1}{2}(-\sigma_1 - \frac{r_1}{r_2} d \sin\psi - \frac{r_2}{r_1} s \cos\psi); \tag{3.5}$$

where $y = r_1 \cos(t - \theta_1)$, $c = r_2 \cos(t - \theta_2)$, and $\psi = \theta_2 - \theta_1$ is the phase difference, and $\eta = \mu t$ is the slow time. Actually, the program derived the two separate equations

$$\frac{d\theta_1}{d\eta} = \tfrac{1}{2}(\frac{r_2}{r_1} s \cos\psi), \tag{3.6i}$$

$$\frac{d\theta_2}{d\eta} = \tfrac{1}{2}\left(-\sigma_1 - \frac{r_1}{r_2} d \sin\psi\right), \tag{3.6ii}$$

for θ_2 and θ_1, which we have combined for convenience since only the phase difference appeared on the right-hand sides.

Equations (3.3)–(3.5) reduce, in the case $A_i = 0$, to the polar coordinate form of the Hartlen-Currie equations. If instead $r_2 = 0$ and ψ is discarded, then these equations reduce to the polar coordinate form of the quasi-steady model of Parkinson and Smith. When the simplest equilibria are carefully analyzed, as in Chakraborty and Rand [1988] and Corless and Parkinson [1993], we see that phase jumps in time, for fixed wind speeds, are possible, as are phase jumps (at equilibrium) as the wind speed is varied. Figure 1 is reproduced from Corless and Parkinson [1993], showing the various possible equilibria that can occur if one or both the amplitudes are zero. Note the possibility of interesting connections among these equilibria, possibly even a 'double-pulse' heteroclinic connection (Campbell [1992]).

4 Nonzero equilibria

Setting the derivatives to zero gives

$$0 = \alpha_1 r_1 - \tfrac{3}{4}\alpha_3 r_1^3 + \tfrac{5}{8}\alpha_5 r_1^5 - \tfrac{35}{64}\alpha_7 r_1^7 - s r_2 \sin\psi, \tag{4.1}$$

$$0 = r_2 - r_2^3 + r_1 d \cos\psi, \tag{4.2}$$

$$0 = \frac{r_2}{r_1} s \cos\psi + \sigma_1 + \frac{r_1}{r_2} d \sin\psi \ . \tag{4.3}$$

To solve these equations we first use Maple (Char et al. [1992]) to calculate a *Gröbner basis* for the set of algebraic equations. For a discussion of the theory of Gröbner bases, see e.g. (Geddes et al. [1992]). The basic idea of a Gröbner basis is very simple. Using a process akin to Gaussian elimination, we produce from a given set of (multivariate) polynomial equations a 'triangular' set of polynomial equations whose roots are exactly the roots of the original set. By 'triangular' set we mean that the last polynomial is in only *one* variable, and the next to last involves only that one variable and one new one, and so on up to the top of the list. This allows us to count the number of real solutions. We note that this does not always provide a practical numerical procedure for *computing* the roots, because of numerical stability questions which sometimes arise, as well as the inherent difficulty in computing such a basis to begin with.

Equations (4.1)–(4.3) are polynomial in the variables r_1, r_2, $\sin\psi$, and $\cos\psi$. If we are treating $\sin\psi$ and $\cos\psi$ as independent entities, we need one more equation, *viz* $\sin^2\psi + \cos^2\psi = 1$, to determine all quantities.

When we use Maple to calculate a Gröbner basis for the case $A_i = 0$, $i = 1, 3, 5, 7$, which corresponds exactly to the pure Hartlen-Currie case (see Remark 2.1), we get a set of equations which is reasonably compact. However, the presentation can be improved by judicious choice of auxiliary variables. Put

$$\sigma_0 = \frac{sd}{2v} \tag{4.4}$$

and

$$r_c = \sqrt{1-2v} \tag{4.5}$$

where

$$v = \beta/\mu = \beta/a > 0 \tag{4.6}$$

and where we will assume that β and hence v is nonzero. Note that this implies that $v > 0$ and $\sigma_0 > 0$ since $s > 0$ and normally $d > 0$ (the case $d < 0$ may be dealt with by symmetry considerations).

Note that d is taken to be strictly positive, to ensure $\sigma_0 > 0$, although a solution is also available in the less interesting case $d = 0$. Then, putting

$$\sigma_1 = \sigma_0 + \varepsilon\,, \tag{4.7}$$

we can present the last equation in the Gröbner basis as

$$r_2 \cdot p(r_2) = r_2[(r_2^2 - 1)(r_2^2 - r_c^2)^2 + \varepsilon\sigma_0(r_2^2 - r_c^2) + \varepsilon^2(r_2^2 - 1)] = 0\,. \tag{4.8}$$

This equation is cubic in r_2^2 (aside from the trivial solution $r_2 = 0$) and is easily solved given v, σ_0, and ε, although the formula is somewhat cumbersome. In Corless and Parkinson [1992], the equation was solved instead for ε in terms of r_2, to get

$$\varepsilon = \left[\frac{-\sigma_0 \pm \sqrt{\sigma_0^2 - 4(r_2^2-1)^2}}{2(r_2^2-1)}\right](r_2^2 - r_c^2) \tag{4.9}$$

and this provided a more useful way of drawing the response curves. However, it turns out that we can do still better. Put

$$\varepsilon = (r_2^2 - r_c^2)\zeta \tag{4.10}$$

where ζ is to be a parameter. Then equation (4.8) becomes

$$r_2(r_2^2 - r_c^2)^2\left[(r_2^2 - 1) + \sigma_0\zeta + (r_2^2 - 1)\zeta^2\right] \tag{4.11}$$

and this can be solved, if $r_2 \neq r_c$, for r_2 in terms of the parameter ζ, to get

$$r_2^2 = \frac{1 - sd\zeta/(2v) + \zeta^2}{1+\zeta^2} = 1 - \frac{\zeta}{1+\zeta^2}\,\frac{sd}{2v}. \tag{4.12}$$

Equations (4.10) and (4.12) allow us to plot the response of the wake oscillator to various values of ε, i.e. various wind speeds. The remainder of the Gröbner basis gives explicit formulae for the remaining unknowns given both r_2 and ε. This solution is simple enough for the response curves to be graphed on a hand calculator.

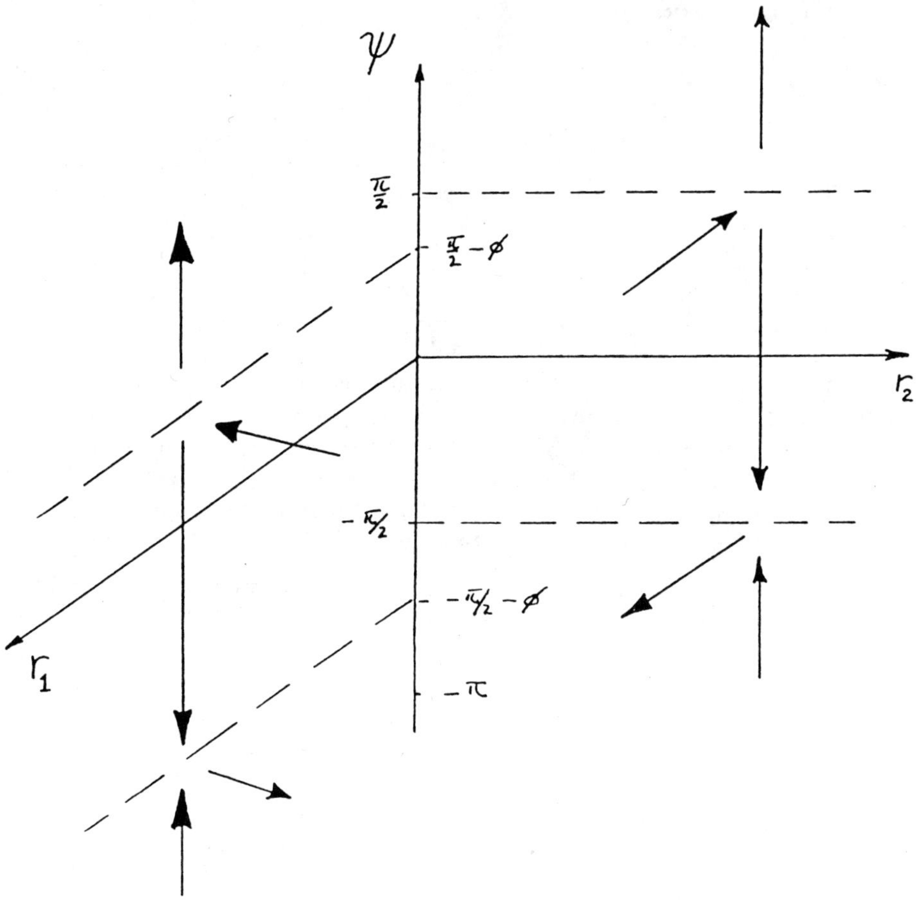

Figure 1 Directions of instability for degenerate equilibria. ϕ is defined by $\tan\phi = -d/b$.

The remainder of the Gröbner basis may be expressed as

$$\cos^2\psi = 1 + (r_2^2 - 1)/(\sigma_0\zeta) \tag{4.13}$$

$$\sin\psi = \cos(\psi)/\zeta \tag{4.14}$$

$$r_1 = -\frac{s}{2v} r_2 \sin\psi\ . \tag{4.15}$$

Equation (4.15) implies that $\sin\psi \le 0$, since all of s, v, r_1 and r_2 are nonnegative. This means that we may take $-\pi \le \psi \le 0$.

Equation (4.14) gives $\tan\psi = 1/\zeta$ and examination of the sign possibilities gives us the parametric solutions

$$\sin\psi = -\frac{1}{\sqrt{1+\zeta^2}} \tag{4.16}$$

$$\cos\psi = -\frac{\zeta}{\sqrt{1+\zeta^2}} \tag{4.17}$$

$$r_1 = \frac{\sigma_0\sqrt{1-\sigma_0\zeta+\zeta^2}}{d(1+\zeta^2)}. \tag{4.18}$$

These equations have real solutions precisely when the right-hand side of equation (4.12) is positive. By inspection of the discriminant (with respect to ζ) of equation (4.12), we see that if $0 \le sd < 4v$ there will be no real values of ζ for which $r_2^2 < 0$, whereas if $sd \ge 4v \ge 0$ then there will be two values of ζ for which $r_2 = 0$, and if ζ is between these values then the solution will not be real. If ζ is such that r_2 is real, we say that ζ is *admissible*.

4.1 Stability analysis We turn now to the procedure for deciding whether or not a given branch of solutions is stable. We compute the Jacobian matrix of the system (3.3)–(3.6):

$$\begin{bmatrix} -2v & s/\sqrt{1+\zeta^2} & sr_2\zeta/\sqrt{1+\zeta^2} \\ -d\zeta/\sqrt{1+\zeta^2} & 1-3r_2^2 & sdr_2/\left(2v\left(1+\zeta^2\right)\right) \\ d/\left(r_2\sqrt{1+\zeta^2}\right) & -sd/\left(2vr_2\left(1+\zeta^2\right)\right) & sd\zeta/\left(2v\left(1+\zeta^2\right)\right) \\ -4v^2\zeta\sqrt{1+\zeta^2}/\left(r_2\, s\right) & +2v\zeta/r_2 & -2v \end{bmatrix}. \tag{4.19}$$

The characteristic polynomial of this matrix is

$$\begin{aligned} &4\,v^3\left(1+\zeta^2\right)^2\lambda^3 \\ &+8\,v^2\left(1+\zeta^2\right)\left(v\zeta^2+2v^2\zeta^2-sd\zeta+v+2v^2\right)\lambda^2 \\ &+v\Big(16v^4+s^2d^2+32v^3\zeta^4+32v^3-32sdv^2\zeta+48v^4\zeta^2+16v^4\zeta^6 \\ &\qquad +3s^2d^2\zeta^2-4vsd\zeta^3+64v^3\zeta^2-4d\zeta sv-32sdv^2\zeta^3+48v^4\zeta^4\Big)\lambda \\ &+8\,v^2\left(-sd\zeta+4v^2\zeta^2+2v^2\zeta^4+2v^2\right)\left(-sd\zeta+2v+2v\zeta^2\right). \end{aligned} \tag{4.20}$$

We use a Maple implementation (Corless [1991]) of the Hurwitz criterion (Levinson and Redheffer [1970]) to decide whether all the roots of this polynomial are in the left-half plane. See the appendix for details.

4.2 Analysis of Hurwitz conditions for the Hartlen-Currie model The Hurwitz conditions for the characteristic polynomial in equation (4.20) are $C_1 > 0$, $C_2 > 0$, and $C_3 > 0$, where

$$C_1 = 2\,v\,(2\,v+1)\,\zeta^2 - 2\,\zeta\,sd + 2\,v\,(2\,v+1), \tag{4.21}$$

$$
\begin{aligned}
C_2 = & -\zeta\left(1+3\zeta^2\right)s^3d^3 + v\left(\zeta^2+1\right)\left(7\zeta^2+34\zeta^2 v+2v+1\right)s^2d^2 \\
& -4\zeta v^2\left(\zeta^2+1\right)^2\left(2\zeta^2 v^2+18v^2+1+16v\right)sd \\
& +32v^4\left(\zeta^2+1\right)^3\left(\zeta^2 v^2+v^2+2v+1\right)
\end{aligned} \tag{4.22}
$$

and

$$
C_3 = \tfrac{1}{2}\Big(2(\zeta^2+1)v - \zeta\, sd\Big)\Big(2\left(\zeta^2+1\right)^2 v^2 - \zeta\, sd\Big)\,. \tag{4.23}
$$

Before we become involved in the analysis of these conditions in an attempt to classify and conveniently summarize all possible types of behaviour of the solutions, some preliminary remarks are in order. The form of the conditions C_1, C_2, and C_3 has already been greatly simplified by the use of the parameter ζ. It is possible that other parameterizations would provide occasionally simpler views, and indeed we will use such as the need arises. Further, s and d only occur as the product sd ($= nU_r^2D/a^2$ in the original, physical, nondimensional variables), and we can effectively suppress the parameter ζ by considering 'qualitative differences' in the response diagrams—that is, in the graphs of the solution in the ε–r_2, ε–r_1, and ε–ψ planes, given by equations (4.10), (4.12), and (4.13)–(4.15). We will draw curves in the sd–v plane which separate it into regions where the response diagrams are qualitatively different; for example, in one region the solution will always be stable, while in another it will have a saddle node bifurcation (Iooss and Joseph [1990]), and in still another it will have two Hopf points, *etc.*

The choice of the sd–v plane for drawing these diagrams was not obvious. The parameter v ($= \beta/a$ in the original variables) corresponds to a ratio of nondimensional damping terms, while sd corresponds to a product of coupling strengths. It is only after the solutions for $C_1 = 0$, $C_2 = 0$, and $C_3 = 0$ are obtained that we can see that this plane gives a convenient summary.

The computations that follow made extremely heavy use of Maple, particularly to factor and simplify the polynomial conditions (it was remarkable that so many very high degree multivariate polynomials factored as nicely as they did) and to compute the discriminants of the conditions C_1, C_2, and C_3 with respect to ζ.

We begin with C_1, though this condition will not turn out to be restrictive.

The condition C_1 is quadratic in ζ, and since the leading and trailing coefficients are both $2v(1+2v) > 0$, the minimum value of C_1 will be positive if the discriminant Δ_{C_1} is negative. By Maple, this discriminant is

$$
\Delta_{C_1} = -4\left(sd+2v+4v^2\right)\left(-sd+2v+4v^2\right) \tag{4.24}
$$

and if $sd < 2v(1+2v)$ then $C_1 > 0$ for all ζ. However, since not all ζ are admissible if $sd \geq 4v$, there is the possibility that we may have $C_1 > 0$ even when $\Delta_{C_1} \geq 0$. It is most convenient to deal with *admissible* ζ by re-parameterizing the condition, using equation (4.12) to replace sd with r_2. When we do this, C_1 becomes

$$
C_1 = 2(1+\zeta^2)(2v+2r_2^2-1)v \tag{4.25}
$$

and we see that since $r_2^2 \geq 0$, $C_1 > 0$ for all real r_2 if and only if $v > 1/2$. Also, $C_1 \leq 0$ for some admissible ζ if and only if $v \leq 1/2$ and $sd \geq 2v(1+2v)$.

We deal with C_3 next, as the next simplest. C_3 is quadratic in sd, but of degree 6 in ζ, and so $\Delta_{C_3} > 0$ is only a necessary condition for real roots (Barbeau [1989]).

However, the zeros of the discriminant do still locate multiple zeros of C_3. The discriminant is

$$\Delta_{C_3} = -\tfrac{1}{16}v^{10}\, s^6 d^6 \left(4\,v + v\, s^2 d^2 - s^2 d^2\right)^2 (4\,v - sd)\,(4\,v + sd)\left(1024\,v^4 - 27\, s^2 d^2\right) . \tag{4.26}$$

It turns out that if $sd \geq 32v^2/3\sqrt{3}$ (from the last factor of the above) then there are two branches of the response curve in the ε–r_2 plane, and it is possible that at least one of them has a saddle-node bifurcation point. The curve $sd = 4v$ divides the sd–v plane into a region where the response curve touches $r_2 = 0$ and a region where it does not, and of course this is the same condition that is derived from equation (4.12). It occurs here because if $r_2 = 0$ then the solution is automatically unstable.

The final curve from this condition is $v = s^2d^2/(s^2d^2 + 4)$, which is a multiple root of Δ_{C_3}. By direct numerical inspection this condition is not restrictive if $v < 1/2$.

When we consider admissible ζ as we did for C_1, we discover that

$$C_3 = 2\,v^2\, r_2^2\left(1 + \zeta^2\right)^2\left((\zeta^2 + 1)v + r_2^2 - 1\right) \tag{4.27}$$

and this is positive when $r_2 \geq 1$ and negative for some admissible ζ if $r_2 < 1$. By equation (4.12), $r_2 < 1$ if and only if $\zeta > 0$. The last factor in equation (4.27) is clearly the critical one (we have already dealt with the condition $r_2 = 0$ in equation (4.12)), and if we extract this factor and substitute in equation (4.12) to eliminate r_2 again, we get

$$2\left(1 + \zeta^2\right)^2 v^2 - sd\,\zeta > 0 \tag{4.28}$$

and this equation has discriminant

$$\Delta'_{C_3} = 4v^4 s^2 d^2 (1024 v^4 - 27 s^2 d^2)\,. \tag{4.29}$$

We come now to the more complicated condition C_2. This condition describes the location of Hopf bifurcation points, and so it is of special interest. From equation (4.22), the discriminant with respect to ζ is

$$\Delta_{C_2} = \sum_{k=0}^{10} c_k(v)(sd)^{2k} \tag{4.30}$$

where the $c_k(v)$ are polynomials in v, just slightly too complicated to be worth printing. Clearly this can only be dealt with numerically.

However, this polynomial may be simpler if we consider only admissible ζ, as before. We use equation (4.12) to eliminate sd from equation (4.22), and get

$$\begin{aligned} C_2 =& 4\,v^2\left(r_2^2 - 1 + 2\,v\right)\zeta^4 \\ &+ \Big(-36\, r_2^2\, v + 34\, v\, r_2^4 + 10\, v - 1 + 6\, r_2^2 - 11\, r_2^4 \\ &\qquad + 6\, r_2^6 + 8\, v^3 + 36\, v^2\, r_2^2 - 20\, v^2\Big)\zeta^2 \\ &+ (r_2 - 1)^2\,(r_2 + 1)^2\left(2\,v + 2\, r_2^2 - 1\right) \end{aligned} \tag{4.31}$$

which is degree 4 in ζ (degree 2 in ζ^2). The discriminant of this equation is slightly simpler, but not much:

$$\begin{aligned}
\Delta'_{C_2} &= \left(-21504 + 28672\, r_2^2\right) v^9 + \left(207872\, {r_2}^4 - 262656\, r_2^2 + 81408\right) v^8 \\
&+ \left(637440\, r_2^6 - 1150592\, r_2^4 + 687104\, r_2^2 - 135296\right) v^7 \\
&+ \left(1080704\, r_2^8 - 2571072\, r_2^6 + 2280000\, r_2^4 - 891456\, r_2^2 + 129472\right) v^6 \\
&+ \Big(1110720\, {r_2}^{10} - 3322304\, r_2^8 + 3946560\, r_2^6 \\
&\qquad - 2325472\, r_2^4 + 679232\, r_2^2 - 78624\Big) v^5 \\
&+ \Big(714464\, r_2^{12} - 2605136\, {r_2}^{10} + 3922720\, r_2^8 - 3122240\, r_2^6 \\
&\qquad + 1385488\, r_2^4 - 324976\, r_2^2 + 31472\Big) v^4 \\
&+ \Big(287520\, r_2^{14} - 1249208\, r_2^{12} + 2302232\, {r_2}^{10} \\
&\qquad - 2334104\, r_2^8 + 1406616\, r_2^6 - 504072\, r_2^4 + 99488\, r_2^2 - 8344\Big) v^3 \\
&+ 4\left(r_2 - 1\right)\left(r_2 + 1\right)\Big(17592\, r_2^{14} - 71736\, r_2^{12} + 124605\, {r_2}^{10} \\
&\qquad - 119562\, r_2^8 + 68471\, r_2^6 - 23411\, r_2^4 + 4426\, r_2^2 - 357\Big) v^2 \\
&+ 2\,\left(603\, r_2^8 - 1410\, r_2^6 + 1242\, r_2^4 - 490\, r_2^2 + 73\right)\left(r_2^2 - 1\right)^2 \left(2\, r_2^2 - 1\right)^3 v \\
&+ \left(24\, r_2^4 - 27\, r_2^2 + 7\right)\left(3\, r_2^4 - 3\, r_2^2 + 1\right)\left(r_2 - 1\right)^3 \left(r_2 + 1\right)^3 \left(2\, r_2^2 - 1\right)^3
\end{aligned} \tag{4.32}$$

Given r_2, we can numerically solve this equation for v. Then, we can use equation (4.31) to numerically find ζ (although this step can introduce extraneous roots, and indeed since the roots of C_2 are multiple this is also numerically awkward), and finally use equation (4.12) to find sd. This will allow us to plot the loci of roots of the discriminant Δ_{C_2} that correspond to admissible ζ.

We see by comparing a graph of the roots of (4.30) with a graph of the roots of (4.32) (neither shown here) that there are some roots in common to both graphs, and there are some roots on each graph that are not on the other. It is the curves in common between the two graphs that divide the sd–v plane into interesting regions.

We draw the bifurcation set in Figure 2, blown up around $sd = 2$ and $v = 1/2$ in Figure 3. In Figure 3 we see evidence of numerical difficulty in accurate computation of multiple roots. These could be resolved by, for example, exact calculation of tight bounds for the roots by using the Maple routine 'rootpoly', but it is not likely that any delicate features possibly uncovered by such careful work would in fact have any corresponding delicate features in the original model. The curves of the preceding analysis divide the plane into more than the nine numbered regions displayed here, but since crossing one of the curves is a necessary but not sufficient condition for qualitative change, extraneous regions were eliminated by sketching a typical response from that region and comparing it with other regions. After this process

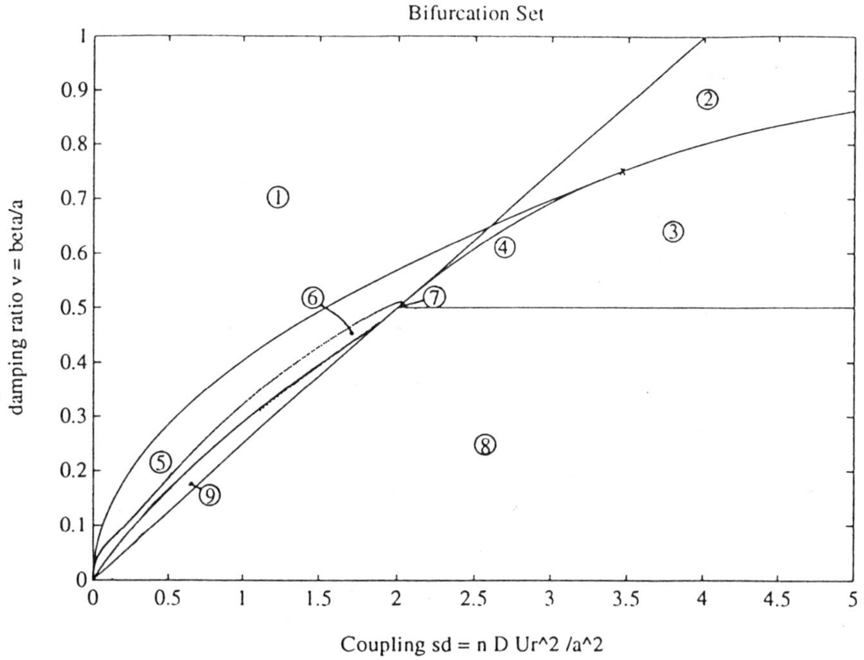

Figure 2 Bifurcation Set. The numbered regions in the figure distinguish the nine qualitatively different types of behaviour sketched in Figure 4.

of elimination, we are left with the nine numbered regions in the figures.

Figure 4 is a set of 'cartoons' which sketch the typical behaviour of the response diagram in each numbered region. These cartoons were hand-drawn to deliberately exaggerate interesting features, but were drawn by working from computer-produced graphs done with Matlab (Moler and Little [1991]), using brute force to evaluate the eigenvalues of the Jacobian along the response diagram to determine stability directly. This provided an independent check on the previous analysis. We use the conventions that solid lines represent stable equilibria and dashed lines represent unstable equilibria and that Hopf bifurcation points are indicated with a circled H near the point. [Of course, Hopf bifurcations in this system correspond to secondary bifurcations to tori in the original system.] We sketch only r_2 versus ε, because the stability behaviour is often clearest in this diagram, but it must be remembered that the response diagrams for r_1 versus ε and the phase difference ψ versus ε are of more direct physical interest. [The Matlab-generated figures included all three diagrams, and are available on request.] It is more usual to draw the amplitude r_1 versus the original frequency ω_0, whereas we are drawing the curves in the amplitude-ε plane. Since $\omega_0 = \sqrt{1+\mu(\sigma_0+\varepsilon)} \sim 1 + \frac{1}{2}\mu(\sigma_0+\varepsilon) + O(\mu^2)$ these curves are approximately linearly related for small μ, and in any case no qualitative information is lost in this transformation.

There may be more types of qualitatively different responses than are sketched here, depending on the meaning of 'qualitatively different' — for example, we made no attempt to discover where 'loops' in the response diagram (such as appear in

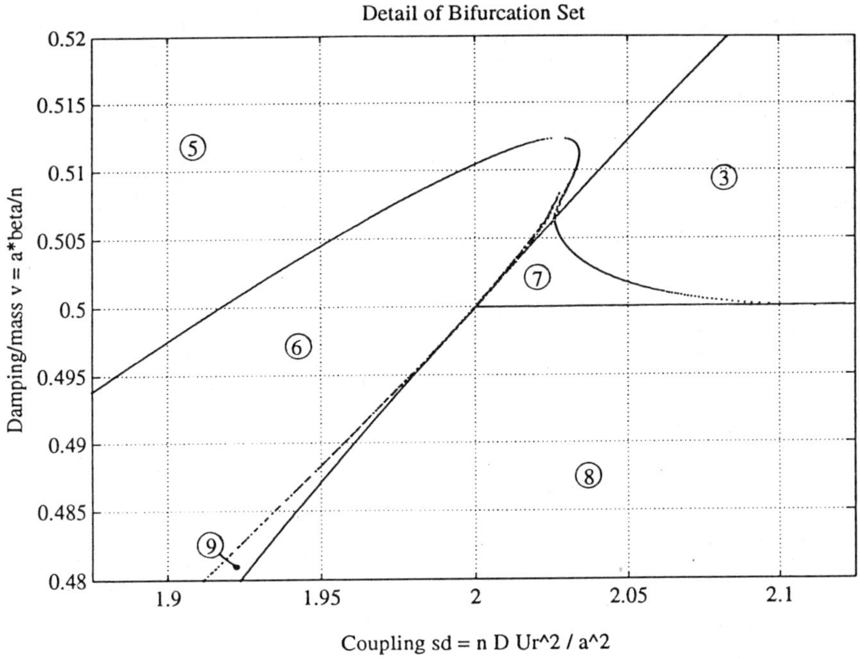

Figure 3 Detail of Bifurcation Set. Near the degenerate point $sd = 2$, $v = 1/2$, there are small regions with interesting behaviour. The poor resolution of the boundary between regions 7 and 6 is due to numerical difficulties.

the sketch for region 9) occur. The sketches here reflect only differences in stability of sections of the response diagrams.

In region 1, the response is always stable. In region 2, again the response is always stable, except when $r_2 = 0$, when the equilibria disappear. There are no stable equilibria in a subset of the small range of wind speeds between the loci $r_2 = 0$. These are pitchfork bifurcations from the trivial solution, corresponding to Hopf bifurcations in the original model. In region 3, the typical response has a saddle-node bifurcation on one branch of the curve, leading to a region with unstable equilibria. In region 4, we have two saddle-node bifurcations on one branch of the response, leading to a response with two stable regions and an unstable region on one branch with another wholly stable branch. In region 5 the response is similar to region 4 except it does not touch $r_2 = 0$, so we only have two saddle-nodes. In region 6 we see our first Hopf bifurcation points, which correspond to bifurcation to tori in the original system. Thus the solution to the original model equations can be expected to have two independent frequencies near these parameter values. There are two Hopf bifurcation points on the typical response diagram, as well as two saddle-node bifurcations, giving rise to two unstable sections of the response (the stable section between the two unstable ones is quite small). Region 7 is a very small region bordering on regions 6, 3, and 8. There are two Hopf points and three saddle-nodes on a typical response diagram from this region, and two of the saddle-node points occur when $r_2 = 0$. The typical response from region 8 has

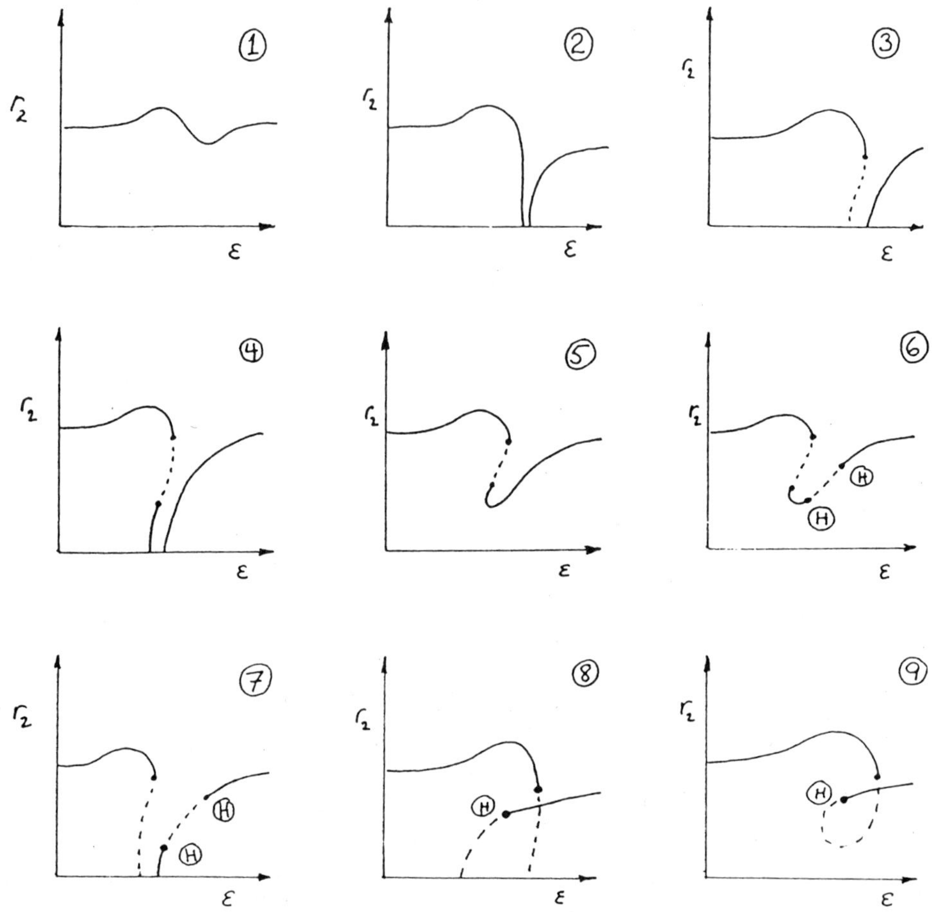

Figure 4 'Cartoons' of the qualitatively different behaviour in the nine different regions. The wake oscillator amplitude is sketched versus ε, which for small μ corresponds to a linear shift of the frequency.

only one Hopf point and three saddle-nodes, two of which occur when $r_2 = 0$. The typical response from region 9 contains one Hopf point and one saddle-node, and the response never touches $r_2 = 0$.

The point $sd = 2$ and $v = 1/2$ is of interest. This is a degenerate point, and some details of the response near $r_2 = 0$ are sketched in Figure 5.

Berger and Plaschko [1993] fail to find, by the method of multiple scales, a set of parameters which has two Hopf bifurcations from the unstable trivial solution of their (very similar) model. Here, if $sd > 4v$, we always have two such points, and the results presented here agree qualitatively with their numerical and (direct Hopf) bifurcation solutions.

Experimentally, only response diagrams that look like those in regions 1, 8, and

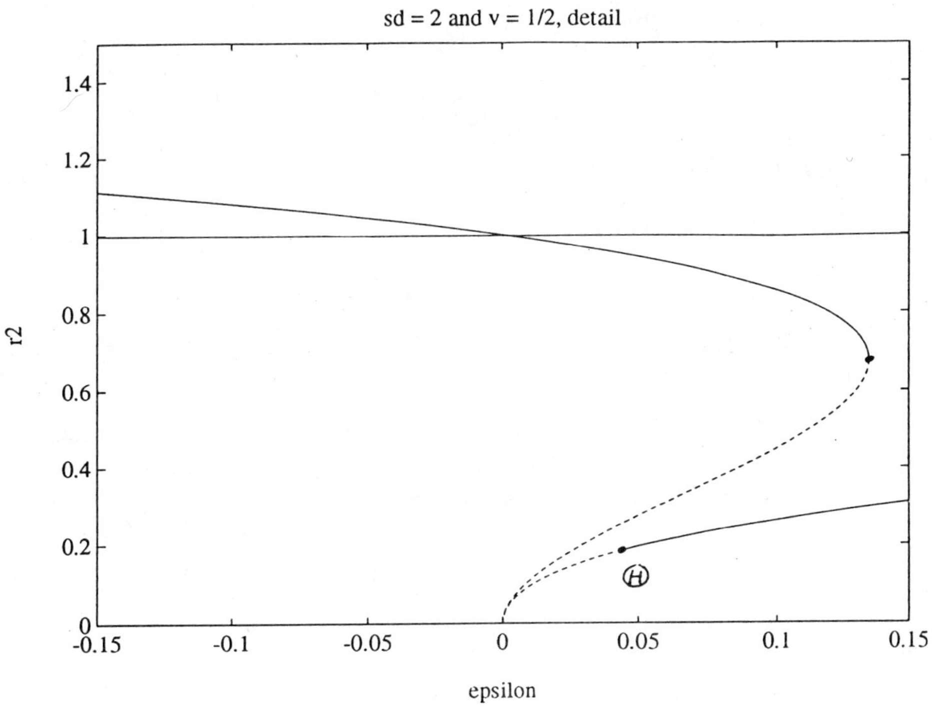

Figure 5 Response for the degenerate case $sd = 2$ and $v = 1/2$. Amplitude of the wake oscillator is graphed versus ε.

perhaps 9 have been seen, to the author's knowledge. It would be very interesting to observe a diagram like that in regions 6 or 7, with the pair of bifurcations to tori giving rise to possibly quite complicated behaviour.

5 Summary

We have used an explicit analytic solution, obtained by use of Gröbner bases, of a multiple-scales approximation to the solution to the Hartlen-Currie model to conveniently summarize the types of response diagrams that can occur. Since the parameters of the flow-induced vibration model are not well understood, this may be helpful in construction of better models. The same techniques as were used here will be attempted for the case $A_3 \neq 0$, which leads to a much more complicated Gröbner basis.

References

Barbeau, E.J. [1989] *Polynomials*, Springer-Verlag, New York.

Bearman, P.W., Gartshore, I.S., Maull, D.J. and Parkinson, G.V. [1987] *Experiments on flow-induced vibration of a square-section cylinder,* J. Fluids and Structures, **1**, 19–34.

Berger, E. [1987] *On a mechanism of vortex-excited oscillations of a cylinder*, Proc. 7th Int. Conf. Wind Eng., Aachen.

Berger, E. and Plaschko, P. [1993] *Hopf Bifurcations and hysteresis in flow-induced vibrations of cylinders, Journal of Fluids and Structures* **7**, 849-866.

Campbell, S. [1992] *private communication.*

Chakraborty, Tapesh and Rand, Richard H. [1988] *The Transition from Phase Locking to Drift in a System of Two Weakly Coupled Van der Pol Oscillators,* Int. J. Non-Linear Mechanics, **23**, no. 5/6, 369–376.

Char, B.W., Geddes, K.O., Gonnet, G.H. Leong, B.L., Monagan, M.B. and Watt, S.M. [1992] *The Maple V Language Reference Manual*, Springer-Verlag, New York.

Corless, R.M. [1986] *Mathematical Modelling of the Combined Effects of Vortex-Induced Vibration and Galloping*, Ph. D. Thesis, University of British Columbia, Vancouver, Canada.

——— [1991] *Hurwitz*, Maple procedure available by anonymous ftp to the Maple share library at daisy.uwaterloo.ca.

Corless, R.M. and Parkinson, G.V. [1988] *A Model of the Combined Effects of Vortex-Induced Vibration and Galloping,* J. Fluids and Structures, **2**, 203–220.

——— [1993] *A Model of the Combined Effects of Vortex-Induced Vibration and Galloping part II,* J. of Fluids and Structures, **7**, 825-848.

Gardner, L.L. and O'Dell, J. [1991] *The ALJABR Language Reference Manual*, Fort Pond Research, Acton, MA.

Gartshore, I.S. [1987] *private communication.*

Geddes, K.O., Czapor, S.R. and Labahn, G. [1992] *Algorithms for Computer Algebra*, Kluwer.

Hartlen, R.T. and Currie, I.G. [1970] *Lift Oscillator model of vortex-induced vibration*, Proceedings of the American Society of Civil Engineers, Journal of Engineering Mechanics, vol. EM5, pp. 557–591.

Iooss, G. and Joseph, D.D. [1990] *Elementary Stability and Bifurcation Theory, 2nd ed.*, Springer-Verlag.

Langford, W.F. and Zhan, K. [1992] *Dynamics of Strong 1:1 Resonance in Vortex-Induced Vibration*, Proceedings of ASME Symposium on Flow-Induced Vibration and Noise, vol. 7, pp. 117–127.

Levinson, N. and Redheffer, R.M. [1970] *Complex Variables*, Holden-Day, San Francisco.

Moler, C. and Little, J. [1991] *The MATLAB User's Guide*, The Mathworks, Inc., Natick, MA.

Nayfeh, A.H. and Mook, D.T. [1981] *Nonlinear Oscillations*, Wiley-Interscience, New York.

Namachchivaya, N.S. and Malhotra, N. [1992] *Parametrically Excited Hopf Bifurcation with Non-semisimple 1:1 Resonance*, Nonlinear Vibrations, Proceedings of the ASME Winter Annual Meeting, vol. DE-50, ASME, Anaheim, pp. 29–46.

Parkinson, G.V. [1989] *Phenomena and Modelling of Flow-Induced Vibrations of Bluff Bodies,* Prog. Aerospace Sci. **26**, 169–224.

Parkinson, G.V. and Smith, J.D. [1964] *The square prism as an aeroelastic nonlinear oscillator*, Quarterly Journal of Mechanics and Applied Mathematics **17**, 225–239.

Rand, R.H. and Armbruster, D. [1987] *Perturbation Methods, Bifurcation Theory, and Computer Algebra*, Applied Mathematical Sciences **65**, Springer-Verlag, New York.

APPENDIX: The Hurwitz Criterion

A polynomial $p = p(\lambda)$ is a *Hurwitz polynomial* if all its roots are in the left half plane. The *paraconjugate* p^* of p is defined as the polynomial whose roots are the roots of p reflected across the imaginary axis. Algebraically, $p^*(z) = \overline{p(-\overline{z})}$.

The auxiliary Maple subroutine 'Hurwitz/data' computes the sequence of partial quotients in the Stieltjes continued fraction of the rational function

$$h(\lambda) = \frac{p(\lambda) - p^*(\lambda)}{p(\lambda) + p^*(\lambda)}$$

to determine if $\Re(h(\lambda)) > 0$ for $\Re(\lambda) > 0$, which will happen if and only if p is Hurwitz. The subroutine returns a list with two elements—the first element of the list is another list, of the partial quotients in the Stieltjes continued fraction, and the second element of the first list is the gcd of p and p^*. If this gcd is nontrivial, then there are pairs of roots symmetric on reflection through the imaginary axis, or else purely imaginary roots, and in either case p is not Hurwitz.

The first element of the returned list of partial quotients is special. If it is of higher degree than 1 in λ, p is not Hurwitz. If it is of the form $a + b\lambda$, where $\Re(a) \neq 0$ or $b < 0$, then p is not Hurwitz. If each subsequent polynomial in the returned sequence is of the form $a + b\lambda$, where $\Re(a) = 0$ and $b > 0$, then p is Hurwitz, and not otherwise. See Levinson and Redheffer [1970] for details.

The Maple routine 'Hurwitz' (Corless [1991]) will return *true* if it can use these rules to decide if p is Hurwitz, *false* if it can decide that p is not Hurwitz, and *FAIL* otherwise, in which case it is up to the user to decide when p is Hurwitz, given the partial quotients. Of course, if the polynomial has symbolic coefficients then there will usually be no way that this routine can tell automatically when the polynomial is Hurwitz, and thus the described mechanism for returning the partial quotients to the user for inspection is required.

Fields Institute Communications
Volume **4**, 1995

Dynamics Near an Irrational Collision of Eigenvalues for Symplectic Mappings

T.J. Bridges
Department of Mathematics
University of Surrey
Guildford, Surrey GU2 5XH
England

R.H. Cushman
Mathematics Institute
Rijksuniversiteit Utrecht
3508 TA Utrecht
The Netherlands

R.S. MacKay
Nuffield Foundation Science
Research Fellow
Mathematics Institute
University of Warwick
Coventry CV4 7AL, England

Abstract. The local dynamics of nonlinear 4D symplectic maps in the neighborhood of a collision of Floquet multipliers of opposite sign at an irrational point—complex instability—is analyzed. The approach is to use nonlinear normal form transformations to simplify the map. The normalization introduces an S^1 symmetry and symmetry reduction leads to a reduced Poisson map. The reduced map is truncated and analyzed two ways: further reduction to an area-preserving map and direct analysis of the Poisson map. A complete local geometrical picture of the local phase space structure near the irrational collision is obtained for the truncated normal form. A similar analysis of the reversible but non-Hamiltonian collision shows that on the orbit space the flow is equivalent to the Hamiltonian case but the reconstructed flow in the 4D phase space differs.

1 Introduction

The local dynamics of nonlinear 4D symplectic maps, in the neighborhood of a collision of Floquet multipliers—of opposite sign—at an irrational point (on the

---

1991 *Mathematics Subject Classification.* Primary 58F05; Secondary 58F08, 58F14.

unit circle), is considered. This bifurcation is the analogue in maps of the 1:-1 resonance or Hamiltonian Hopf bifurcation in Hamiltonian vectorfields when two pure imaginary eigenvalues of opposite sign collide on the imaginary axis (Meyer and Schmidt [1971], van der Meer [1985], Bridges [1991] and references therein). In fact, the simplest physical examples, giving rise to symplectic maps with a collision of Floquet multipliers of opposite sign, are obtained by parametrically forcing, periodically, a Hamiltonian vectorfield with a 1:-1 resonance. Examples of 2-DOF Hamiltonian systems with a 1:-1 resonance are the restricted 3-body problem (Deprit and Henrard [1968]) and the spinning double pendulum (Bridges [1991], Section 4).

Our approach is to work directly with the map, without any particular identification with an underlying Hamiltonian vectorfield. On the other hand, although the dynamical system is discrete in time, many of the features of the local phase space structure, near an irrational collision, are in common with the phase space structure of the Hamiltonian Hopf bifurcation and this will be evident in the pictures in Section 3. But, in the map 1:-1 resonance the phase space is "larger" in the sense that there is no energy manifold; the normal form is not integrable and the dynamics of the unnormalized map can take place in the full four dimensional space.

The idea is to transform the 4D map into nonlinear normal form, then truncate the normal form, reduce and analyze the reduced (non-integrable) system and then reconstruct the full phase space dynamics. It is essential to our analysis that the collision take place at an irrational point on the unit circle ($e^{i\theta}$, $\theta = 2\pi\rho$ with ρ irrational). The irrationality leads, through the normal form transformations, to an S^1-equivariance in the nonlinear normal form.

As far as we are aware, this bifurcation in nonlinear symplectic maps has not been previously analyzed, although numerical studies and observations of the instability associated with an irrational collision abound (see Janssen and Tjon [1983], Pfenniger [1985], Contopoulos [1985, 1986, 1988] and Kook and Meiss [1989] for example). The irrational collision in reversible but non-symplectic maps has been considered by Sevryuk and Lahiri [1991]. Complete analysis of the linear problem, particularly with regard to the role of signature, goes back to Williamson [1937] and Krein [1955]. The collision of multipliers at rational points on the unit circle ($e^{i\theta}$, $\theta = 2\pi\rho$ with $\rho \in \mathbf{Q} \cap (0, \frac{1}{2})$) is more difficult (resonances!) since the normal form symmetry is discrete, requiring analysis in the full 4D phase space. On the other hand a complete theory for existence and stability of periodic points in the unfolding of a rational collision is tractable (Bridges and Furter [1991]).

Our starting point is the iterated symplectic two-parameter family of maps

$$\mathbf{x}' = \Phi(\mathbf{x}; \lambda, \alpha) \qquad \mathbf{x} \in \mathbf{R}^4 \tag{1.1}$$

with $\Phi^*\omega = \omega$ and the symplectic form is taken to be the standard one, $\omega = dx_1 \wedge dx_3 + dx_2 \wedge dx_4$. We also suppose that $\Phi(0; \lambda, \alpha) = 0$ and that the linear map $d\Phi(0; 0, 0)$ has a collision of Floquet multipliers and a non-diagonalizable Jordan normal form. Without loss of generality, by a suitable choice of coordinates,

$$d\Phi(0; 0, 0) = \mathbf{L}_0 = \begin{pmatrix} R_\theta & \epsilon R_\theta \\ 0 & R_\theta \end{pmatrix} \qquad R_\theta = \begin{pmatrix} \cos\theta & \sin\theta \\ -\sin\theta & \cos\theta \end{pmatrix} \qquad \epsilon = \pm 1$$

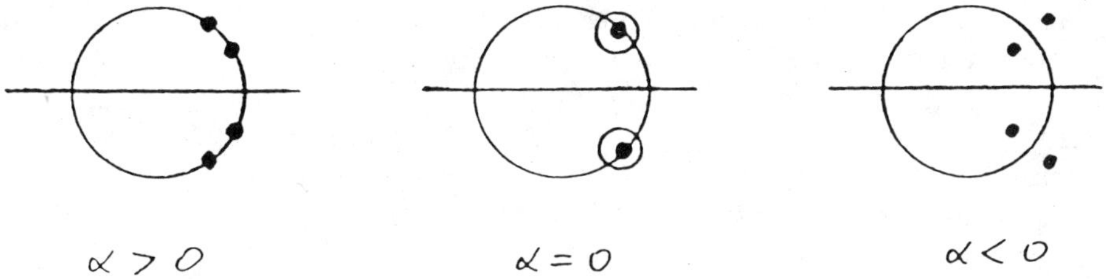

Figure 1 Eigenvalues of $I(\lambda, \alpha)|_{\lambda=0}$ in (1.2) when $\lambda = 0$: a collision of Floquet multipliers of opposite sign.

where $\theta = 2\pi\rho$ with $\rho \in (0, \frac{1}{2})$ and irrational. Note that $\mathbf{L}_0 = \exp(\mathbf{A}_0)$, where $\mathbf{A}_0$ is the normal form for a 1:-1 resonant (linear) Hamiltonian vectorfield,

$$\mathbf{A}_0 = \begin{pmatrix} \theta \mathbf{J}_2 & \epsilon \mathbf{I}_2 \\ \mathbf{0} & \theta \mathbf{J}_2 \end{pmatrix} \qquad \text{where} \quad \mathbf{J}_2 = \begin{pmatrix} 0 & 1 \\ -1 & 0 \end{pmatrix}.$$

The linear map $\mathbf{L}_0$ is in Williamson normal form (Williamson [1937]) for the collision and it has a versal two-parameter unfolding (Bridges and Furter [1991], Section 9),

$$d\Phi(0; \lambda, \alpha) = \mathbf{L}(\lambda, \alpha) = \begin{pmatrix} R_{\theta+\lambda} & \epsilon R_{\theta+\lambda} \\ \alpha R_{\theta+\lambda} & R_{\theta+\lambda} \end{pmatrix}. \tag{1.2}$$

For $\lambda = 0$ the behavior of the eigenvalues of $\mathbf{L}(\lambda, \alpha)$ is shown in Figure 1; for $\lambda \neq 0$ the position of the collision is rotated by angle λ.

Any smooth symplectic map on $\mathbf{R}^4$ with a fixed point at 0, whose Floquet multipliers undergo a collision (of opposite sign) on the unit circle at $\theta = 2\pi\rho$, has the form

$$\mathbf{x}' = \Phi(\mathbf{x}; \lambda, \alpha) = \mathbf{L}(\lambda, \alpha)\mathbf{x} + \Phi_2(\mathbf{x}; \lambda, \alpha) + \Phi_3(\mathbf{x}; \lambda, \alpha) + \cdots \tag{1.3}$$

where Φ_j, $j = 2, 3, \ldots$ are vector-valued polynomials of degree j in $x_1, \ldots, x_4$ with (λ, α) dependent coefficients and (λ, α) are taken to be in a neighborhood of zero.

Further simplification of (1.3) is obtained by transforming (1.3) into nonlinear normal form (Bridges and Cushman [1992]). Let $\mathcal{O}_\Sigma(\phi)$ be the orbit of ϕ with respect to the action of the group of (formal power series) symplectic diffeomorphisms. Then given a map ϕ we can compare $\mathcal{O}_\Sigma(\phi)$ with $\mathcal{O}_\Sigma(\mathbf{A}x)$, the orbit of the linear part of the map. Then, geometrically, the normal form for ϕ, with respect to its linear part, is the intersection of $\mathcal{O}_\Sigma(\phi)$ with the normal space of $\mathcal{O}_\Sigma(\mathbf{A}x)$ at $\phi_o = \mathbf{A}x$.

Algebraically, if $\mathbf{A}$ is semisimple this is just Birkhoff normalization. When $\mathbf{A}$ has a non-trivial unipotent part the above geometric definition leads to unipotent normalization (Bridges and Cushman [1992]). For each $k \in \mathbf{N}$ there exists a symplectic diffeomorphism $\Psi_k(\mathbf{x})$ such that $\Psi_k^{-1} \circ \Phi \circ \Psi_k$ is in normal form to order k. Carrying the theory to arbitrary $k \in \mathbf{N}$, the following nonlinear normal form for (1.3) (see Bridges and Cushman [1992], Section 4) can be achieved formally:

$$\begin{pmatrix} x' \\ y' \end{pmatrix} = \begin{bmatrix} (1-\epsilon h_X)R_{\theta+\phi} & \epsilon R_{\theta+\phi} \\ -h_X\, R_{\theta+\phi} & R_{\theta+\phi} \end{bmatrix} \begin{pmatrix} x \\ y \end{pmatrix} \tag{1.4}$$

where we now take coordinates $(x, y) = (x_1, x_2, y_1, y_2) \in \mathbf{R}^4$ and

$$h = \lambda I + \alpha X + \tfrac{1}{2} b_1 X^2 + b_2 I X + \tfrac{1}{2} b_3 I^2 + \cdots . \tag{1.5}$$

The function h is a smooth function of I and X (assuming the original map Φ is smooth) and

$$\begin{aligned} \phi = \frac{\partial h}{\partial I} &= \lambda + b_2 X + b_3 I + \cdots \\ h_X = \frac{\partial h}{\partial X} &= \alpha + b_1 X + b_2 I + \cdots \end{aligned}$$

where

$$\begin{aligned} I &= \langle y, \mathbf{J}_2 x \rangle = y_1 x_2 - x_1 y_2 \\ X &= \langle x, x \rangle = x_1^2 + x_2^2 . \end{aligned} \tag{1.6}$$

The normal form transformations result in an S^1-equivariant symplectic map with S^1 action

$$t \cdot (x, y) = (R_t x, R_t y) \qquad \forall t \in S^1 . \tag{1.7}$$

Moreover the function $I(x, y)$ defined in (1.6) is an associated invariant for the map (1.4): $I(x', y') = I(x, y)$.

Our goal in this paper is to give an analysis of the normal form (1.4) truncated at the third order, in the neighborhood of the origin in phase space and parameter space, as well as the effects of remainder terms (breaking of normal form symmetry). The first step is to "divide out the symmetry" by reducing the system (1.4) to the S^1 orbit space (Arms, Cushman and Gotay [1991], Bridges and Cushman [1992], Section 5). Introduce coordinates $\sigma = (\sigma_1, \sigma_2, \sigma_3)$ on the S^1 orbit space, defined by

$$\begin{aligned} \sigma_1 &= x_1^2 + x_2^2 \\ \sigma_2 &= y_1^2 + y_2^2 \\ \sigma_3 &= x_1 y_1 + x_2 y_2 \end{aligned} \tag{1.8}$$

(the Hilbert basis for the smooth S^1 invariant functions on $\mathbf{R}^4$). Note that the coordinates σ are related to the invariant I by

$$\sigma_1 \sigma_2 - \sigma_3{}^2 - I^2 = 0 . \tag{1.9}$$

Using (1.4) the map in the new coordinates is

$$\begin{aligned}
\sigma_1{}' &= \langle x', x'\rangle = \langle R_{\theta+\phi}((1-\epsilon h_X)x+\epsilon y), R_{\theta+\phi}((1-\epsilon h_X)x+\epsilon y)\rangle \\
&= a^2\sigma_1 + \sigma_2 + 2\epsilon a\sigma_3 \\
\sigma_2{}' &= \langle y', y'\rangle = \langle R_{\theta+\phi}(-h_X x + y), R_{\theta+\phi}(-h_X x + y)\rangle \\
&= h_X^2\sigma_1 + \sigma_2 - 2h_X\sigma_3 \\
\sigma_3{}' &= \langle x', y'\rangle = \langle R_{\theta+\phi}(ax+\epsilon y), R_{\theta+\phi}(-h_X x + y)\rangle \\
&= -ah_X\sigma_1 + \epsilon\sigma_2 + (1-2\epsilon h_X)\sigma_3
\end{aligned}$$

where $a = 1 - \epsilon h_X$, which can be written as

$$\sigma' = T(\sigma; I, \alpha) = \mathbf{M}(\sigma_1, I, \alpha)\sigma \tag{1.10}$$

with

$$\mathbf{M}(\sigma_1; I, \alpha) = \begin{bmatrix} a^2 & 1 & 2\epsilon a \\ h_X^2 & 1 & -2h_X \\ -ah_X & \epsilon & 1-2\epsilon h_X \end{bmatrix}. \tag{1.11}$$

The map (1.10), defined on the subset of $\mathbf{R}^3$ with σ_1 and σ_2 non-negative, is the map (1.4) reduced to the orbit space.

The map (1.10) is a Poisson map. A map T is a Poisson map with respect to a given Poisson structure, with bracket $\{\, , \}_m$, if

$$\{F, G\}_m \circ T = \{F\circ T, G\circ T\}_m$$

for all smooth functions F and G. To see that the map in (1.10) is a Poisson map we introduce the following Poisson bracket

$$\begin{aligned}
\{F, G\}_m &= \frac{d}{dt} F(\sigma + t\nabla\psi\times\nabla G)\bigg|_{t=0} \\
&= -\nabla\psi\cdot\nabla F\times\nabla G \\
&= \langle\nabla F, \mathbf{J}(\sigma)\nabla G\rangle \qquad (\text{with } \mathbf{J}(\sigma) = \nabla\psi\times)
\end{aligned} \tag{1.12}$$

for smooth functions F and G. We choose the function $\psi(\sigma)$ to be

$$\psi(\sigma) = \tfrac{1}{2}\langle\sigma, \mathbf{B}\sigma\rangle = \sigma_1\sigma_2 - \sigma_3{}^2, \qquad \mathbf{B} = \begin{pmatrix} 0 & 1 & 0 \\ 1 & 0 & 0 \\ 0 & 0 & -2 \end{pmatrix}. \tag{1.13}$$

Note that by (1.9) we can recognize $\psi(\sigma)$ as being I^2. To verify that a map T is a Poisson map we need only verify that

$$dT\,\mathbf{J}(\sigma)dT^T = \mathbf{J}(\sigma)\circ T.$$

But from (1.11), $T(\sigma; I, \alpha) = \mathbf{M}(\sigma_1; I, \alpha)\sigma$, so

$$dT = \begin{pmatrix} 2a\dot{a}\sigma_1 + a^2 + 2\epsilon\dot{a}\sigma_3 & 1 & 2\epsilon a \\ -2(1-a)\dot{a}\sigma_1 + (1-a)^2 + 2\dot{a}\sigma_3 & 1 & -2\epsilon(1-a) \\ \epsilon a(a-1)\sigma_1 + \epsilon(2a-1)\dot{a}\sigma_1 + 2\dot{a}\sigma_3 & \epsilon & 2a-1 \end{pmatrix},$$

where

$$\dot{a} = \frac{\partial a}{\partial\sigma_1} = -\frac{\partial^2 h}{\partial\sigma_1{}^2}.$$

A straightforward but lengthy calculation shows that

$$dT\,\mathbf{J}(\sigma)\,dT^T = \begin{pmatrix} 0 & 2\sigma_3{}' & \sigma_1{}' \\ -2\sigma_3{}' & 0 & -\sigma_2{}' \\ -\sigma_1{}' & \sigma_2{}' & 0 \end{pmatrix} = \mathbf{J}\circ T$$

verifying that T is a Poisson map with respect to the Poisson bracket (1.12).

The Poisson structure for the reduced system has ψ as a Casimir function, since $\mathbf{J}(\sigma)\nabla\psi = \nabla\psi \times \nabla\psi = 0$. Moreover, $\psi(\sigma)$ is preserved by the reduced system. In other words $\psi(\sigma') = \psi(\sigma)$. This follows from the definition in (1.13) since

$$\begin{aligned} \psi(\sigma') &= \tfrac{1}{2}\langle \mathbf{M}\sigma, \mathbf{B}\mathbf{M}\sigma\rangle \\ &= \tfrac{1}{2}\langle \sigma, \mathbf{M}^T\mathbf{B}\mathbf{M}\sigma\rangle \\ &= \tfrac{1}{2}\langle \sigma, \mathbf{B}\sigma\rangle \\ &= \psi(\sigma). \end{aligned}$$

The second to last equality follows from the identity $\mathbf{M}^T\mathbf{B}\mathbf{M} = \mathbf{B}$ which can be verified by direct calculation using the definitions of $\mathbf{B}$ and $\mathbf{M}$. Hence the sets $\psi(\sigma) = I^2$ are invariant. For $I \neq 0$ the surface $\psi(\sigma) = I^2$ is a non-compact two-manifold (a hyperboloid of one-sheet). When $I = 0$ there is a singular point on the Casimir level set $\psi = 0$ at the point $\sigma = 0$.

In Section 2 the Poisson map (1.10) is reduced to an area-preserving map. This does not include the singular case ($I = 0$) and moreover it flattens the reduced phase space eliminating useful geometric information. It is however the most straightforward way to analyze the non-singular case.

In Section 3 a geometric treatment of the reduced Poisson system is given which includes the singular case $I = 0$. We find that up to arbitrarily high order remainder terms, the unfolding of the nonlinear complex instability is reminiscent of the local phase-space geometry of the time-1 map of the Hamiltonian Hopf bifurcation.

Finally in Section 4 some results on the reversible but non-Hamiltonian case are presented. The question here is how much of the analysis of the symplectic case carries over to the reversible but non-Hamiltonian collision. With no attempt at generality we argue that the nonlinear normal form for a map with a reversible collision is approximated by the exponential of the normal form of a reversible vectorfield with a $1:1$ resonance. Analysis of the normal form shows that—when restricted to the orbit space—the normalized map reduces precisely to a Poisson map of the form (1.10); the only substantial difference in the discrete-time flow of the normal form—for the reversible but non-symplectic collision—is in the flow along the group orbit.

2 The reduced area-preserving map

In this section the dynamics of the Poisson map (1.10) are analyzed for (I, α) small but $I \neq 0$ and $|\sigma|$ near zero by further reduction to an area-preserving map. With $\|\sigma\|$ small we truncate h in (1.5) at quadratic order. Then after a suitable rescaling the derivative of h with respect to σ_1 (denoted h_1) is

$$h_1 = \alpha + b\sigma_1 \qquad \text{with} \qquad b = \pm 1\,. \tag{2.1}$$

Taking $\epsilon = +1$, without loss of generality (by a suitable reflection) the Poisson map (1.10) becomes

$$\begin{aligned} \sigma_1{}' &= a^2\sigma_1 + \sigma_2 + 2a\sigma_3 \\ \sigma_2{}' &= h_1^2\sigma_1 + \sigma_2 - 2h_1\sigma_3 \\ \sigma_3{}' &= -ah_1\sigma_1 + \sigma_2 + (1-2h_1)\sigma_3 \qquad \text{where } a = 1 - h_1 \end{aligned} \tag{2.2}$$

with the invariant $I^2 = \sigma_1\sigma_2 - \sigma_3{}^2$, $\sigma_1 \geq 0$ and $\sigma_2 \geq 0$.

Each non-singular leaf of the foliation given by level sets of the Casimir ψ is a smooth symplectic 2-manifold to which the Poisson map (2.2) restricts to an area-preserving map. This is the most straightforward way to analyze the dynamics of (2.2) (although it is not the most informative geometrically). Such an area-preserving map is constructed by charting the level set $\sigma_1\sigma_2 - \sigma_3{}^2 = I^2$. For example when $I \neq 0$, introduce new coordinates u and v defined by

$$\begin{aligned} \sigma_1 &= Ie^v \\ \sigma_2 &= I(1+u^2)e^{-v} \\ \sigma_3 &= Iu\,. \end{aligned}$$

Then $\sigma_1\sigma_2 = I^2(1+u^2) = \sigma_3^2 + I^2$ and the Poisson map (2.2) restricts to

$$\begin{aligned} u' &= a(a-1)e^v + (1+u^2)e^{-v} + (2a-1)u \\ v' &= \log[a^2e^v + (1+u^2)e^{-v} + 2au] \end{aligned} \tag{2.3}$$

which is area-preserving in the standard way: $du' \wedge dv' = du \wedge dv$ and where

$$a = 1 - h_1 = 1 - \alpha - \beta e^v \qquad \text{with} \qquad \beta = bI\,.$$

The map (2.3) is a two-parameter (α, β) area-preserving map and the theory for such maps is well-developed.

The backbone of the phase space structure is given by the fixed points and for (α, β) small the map (2.3) has a family of fixed points given by

$$u = \tfrac{1}{2}h_1e^v\,, \quad h_1 - \tfrac{1}{4}h_1^2 = e^{-2v} \quad \text{with} \quad h_1 = \alpha + \beta e^{-v}\,. \tag{2.4}$$

To leading order the second equation is

$$\beta V^3 + \alpha V - 1 + \cdots = 0 \qquad (\text{taking } V = e^v) \tag{2.5}$$

which has a discriminant variety

$$\left(\frac{\beta}{2}\right)^2 = \left(\frac{\alpha}{3}\right)^3.$$

The requirement $V = e^v > 0$ forces the discriminant variety to have only one branch ($\alpha > 0$, $\beta < 0$) as shown in Figure 2(a). Figure 2(a) also shows the complex V plane for roots of (2.5) in each component of the (α, β) plane. Only real non-negative roots of (2.5) are admissible as fixed points. For $\beta < 0$ only the region inside the cusp has fixed points. Figure 2(b) shows the local phase portraits for $\beta > 0$ and $\beta < 0$ (and inside cusp) in the neighbourhood of the fixed points.

The stability type for the fixed points is verified as follows. Writing the map (2.3) as $(u', v') = f(u, v)$ the Jacobian of the map is

$$df = \begin{pmatrix} r_u - 1 & r_v - (a + a_v)e^v \\ r_u/r & r_v/r \end{pmatrix}$$

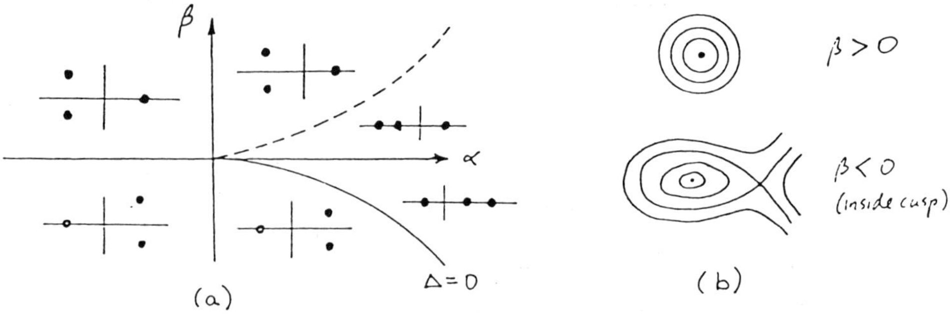

Figure 2 (a). Fixed points of the reduced area preserving map (2.3) in the complex plane as functions of α and β. (b). Phase portraints surrounding the positive real fixed points of (2.3).

with

$$r(u,v) = e^{v'} = a^2 e^v + (1+u^2)e^{-v} + 2au\,.$$

df has unit determinant and

$$\begin{aligned}\mathrm{Tr}(df) &= r_u - 1 + \frac{r_v}{r} \\ &= a^2 + 2a - 1 - (1+a)\beta e^v \\ &= 2 - 4\alpha - 6\beta e^v + \cdots\end{aligned}$$

when evaluated at a fixed point (2.4). Therefore the residue evaluated at a fixed point is

$$\mathcal{R} = \frac{1}{4}(2 - \mathrm{Tr}(dT)) = \alpha + \frac{3}{2}\beta e^v + \cdots\,. \tag{2.6}$$

For (α, β) sufficiently small it is the sign of $\mathcal{R}$ that determines linear stability: $\mathcal{R} > 0$ (and small) corresponds to an elliptic fixed point and $\mathcal{R} < 0$ to a hyperbolic fixed point. Substitution of (2.4) and (2.5) into (2.6) verifies the fixed point type shown in Figure 2. Fixed points of the reduced area preserving map are "relative fixed points" in the 4D normalized map; they lift to invariant circles in the normalized map with flow along the group orbit.

3 Geometry of the normalized map

The reduction to a 2-parameter family of area-preserving maps as in Section 2 has three disadvantages: it doesn't include the singular case $I = 0$, it gives the misleading impression that highly non-integrable dynamics will be found (since the reduced area-preserving map is not integrable) and it flattens the orbit space

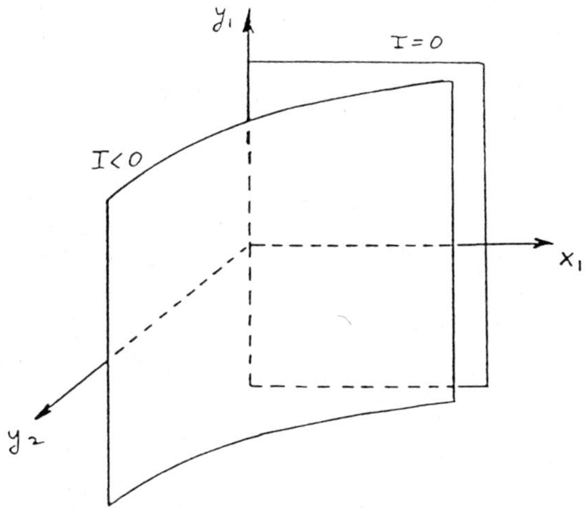

Figure 3 Invariant manifold $I = y_1x_2 - y_2x_1$ in the slice H_+.

eliminating useful geometric information about the structure of the phase space. In this section we take as the starting point the Poisson map (2.2) on the orbit space.

In addition to the σ coordinates, the coordinates (x_1, y_1, y_2) of the unreduced map (1.4) will also be used. They correspond to the places where the S^1 orbits cut the half space $H_+ = \{(x, y) : x_1 \geq 0 \text{ and } x_2 = 0\}$. It is easy to show that H_+ is a slice for the S^1 action (1.7). In these coordinates the invariant manifold $I = y_1x_2 - y_2x_1$ sits in the phase space as shown in Figure 3. When $I = 0$ it is possible for $x_1 = 0$ in which case we take without loss of generality $y_2 = 0$ and $y_1 \geq 0$.

The picture in $\mathbf{R}^4$ for the unreduced system is obtained by simultaneously rotating in the x and y planes at a rate dependent on I and σ_1. For example, taking $x = R_\gamma \hat{x}$ and $y = R_\gamma \hat{y}$ and substituting into (1.4) requires that

$$\begin{aligned} R_{\gamma'}\hat{x}' &= R_{\theta+\phi+\gamma}(a\hat{x} + \epsilon\hat{y}) \\ R_{\gamma'}\hat{y}' &= R_{\theta+\phi+\gamma}(\epsilon(a-1)\hat{x} + \hat{y})\,. \end{aligned}$$

The flow along the S^1 group orbit can then be factored out by taking

$$\gamma' = \gamma + \theta + \phi(I, \sigma_1) \tag{3.1}$$

where $\phi(I, \sigma_1)$ is defined in equations (1.4)–(1.6). The picture for the original map (breaking normal form symmetry) is obtained by adding a small perturbation resulting in weak chaos associated with break-up of the separatrices and tori.

By setting $x_2 = 0$ and using the definition of σ_1, σ_2 and σ_3 the Poisson map (2.2) has the following representation in terms of the coordinates x_1, y_1 and y_2 in the slice H_+

$$\begin{aligned}
\sigma_1'\Big|_{x_2=0} &= {x_1'}^2 = (ax_1+y_1)^2+y_2^2 \\
\sigma_2'\Big|_{x_2=0} &= {y_1'}^2+{y_2'}^2 = (h_1x_1-y_1)^2+y_2^2 \\
\sigma_3'\Big|_{x_2=0} &= x_1'y_1' = (ax_1+y_1)^2-x_1(ax_1+y_1)+y_2^2
\end{aligned} \tag{3.2}$$

and the set $I=-x_1y_2$ is invariant.

The singular case $I=0$ is considered first. Setting $I=0$ means, by convention, restricting to the plane $y_2=0$. In this case (3.2) reduces to

$$\begin{aligned}
x_1' &= |ax_1+y_1| \\
y_1' &= (-h_1x_1+y_1)\,\mathrm{sgn}(ax_1+y_1)
\end{aligned} \tag{3.3}$$

(the absolute value and sgn function can be removed by continuing to $x_1<0$ and dividing out by the reflection $(x_1,y_1)\mapsto(-x_1,-y_1)$ afterwards).

For (x_1,y_1) in a neighborhood of zero the map (3.3) has two fixed points: $(0,0)$ and $(\sqrt{-\alpha/b},0)$. The trivial fixed point is the trivial fixed point of the original 4D map. To determine the stability type of the fixed points write (3.3) as $(x',y')=f(x,y)$ then the derivative of f, evaluated at fixed points, is

$$df=\begin{pmatrix} 1-h_1-2bx_1^2 & 1 \\ -h_1-2bx_1^2 & 1 \end{pmatrix}$$

with residue

$$\mathcal{R}=(2-\mathrm{Tr}(df))/4=(\alpha+3bx_1^2)/4$$

and for linear stability it is required that $\mathcal{R}\in(0,1)$. For the two families of fixed points we find

$$\mathcal{R}\Big|_{(0,0)}=\frac{1}{4}\alpha \qquad \text{and} \qquad \mathcal{R}\Big|_{(\sqrt{-\alpha/b},0)}=-\frac{1}{2}\alpha\,.$$

The fact that the trivial fixed point loses stability for $\alpha\le 0$ recovers the basic picture in Figure 1. The non-trivial fixed point goes from unstable to stable for $\alpha<0$. Using the stability type the phase portraits surrounding the fixed points are sketched in Figures 4a,b. Note that breaking the normal form symmetry will introduce weak localized chaos into the picture; but, since $\mathcal{R}$ is near zero for $|\alpha|$ small the chaos will be small.

When $I\neq 0$ we can eliminate y_2 using

$$y_2=-\frac{I}{x_1}\,.$$

Then (3.2) can be recast into

$$\begin{aligned}
({x_1}')^2 &= (ax_1+y_1)^2+I^2/x_1^2 \\
{x_1}'{y_1}' &= (ax_1+y_1)(-h_1x_1+y_1)+I^2/x_1^2\,.
\end{aligned}$$

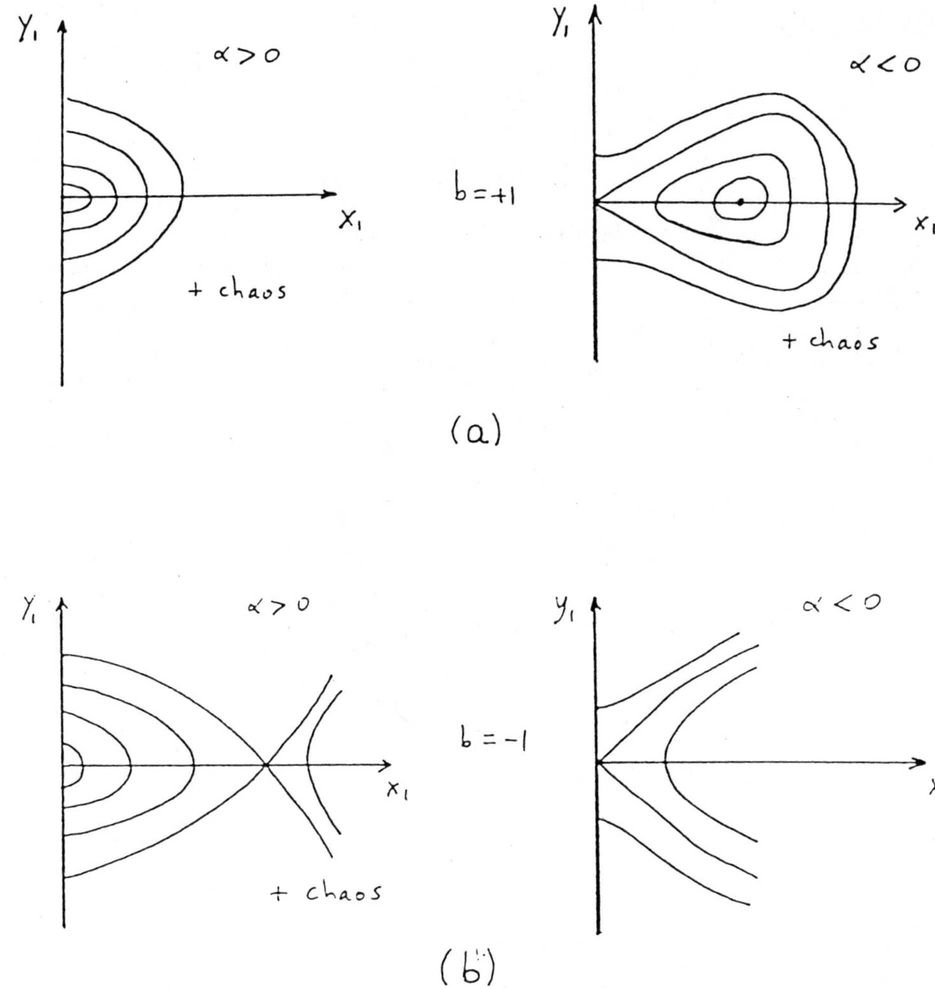

Figure 4. Phase portraits for the Poisson map (3.2) in the x_1–y_1 plane when $I = 0$ as a function of α. (a) when $b = +1$. (b) when $b = -1$.

So in the slice H_+ we find

$$\begin{aligned} x_1{}' &= \sqrt{(ax_1+y_1)^2+I^2/x_1^2} \\ y_1{}' &= \frac{(ax_1+y_1)(y_1-h_1x_1)+I^2/x_1^2}{\sqrt{(ax_1+y_1)^2+I^2/x_1^2}}\,. \end{aligned} \tag{3.4}$$

This is another variant of the reduced area-preserving map of Section 2. The advantage of (3.4) is that it is clear how to relate orbits of the reduced mapping (3.4) to the phase space geometry of the unreduced system. Again we determine the fixed points of the reduced area-preserving map and their stability type and use this information to construct the local phase space geometry.

Fixed points of the map (3.4) satisfy

$$\begin{aligned} x_1^2 &= (ax_1+y_1)^2+I^2/x_1^2 \qquad x_1 \geq 0 \\ x_1y_1 &= x_1^2-(ax_1+y_1)x_1\,. \end{aligned} \tag{3.5}$$

Therefore, given x_1 it easily follows that

$$\begin{aligned} y_1 &= \frac{1}{2}h_1x_1 = \frac{1}{2}\frac{I^2}{x_1^3}+\cdots \\ y_2 &= -\frac{I}{x_1}\,. \end{aligned} \tag{3.6}$$

To determine an expression for x_1 we substitute (3.6) into the first equation of (3.5) to obtain

$$x_1^4(h_1-\tfrac{1}{4}h_1^2) = I^2$$

or, for $|h_1| << 1$,

$$bx_1^6+\alpha x_1^4-I^2+\cdots = 0\,. \tag{3.7}$$

(Note that (2.5) is recovered by taking $V = x_1^2/I$ and $b = \beta/I$.) The discriminant variety of (3.7) is given by $\Delta = (\frac{I}{2})^2 - (\frac{\alpha}{3})^3$ and is shown in the (I, α) plane in Figure 5.

The pictures are symmetric about the α-axis and differ for $b = \pm 1$. There are three real roots of (3.7), inside the cusp, but only non-negative roots (of the cubic in x_1^2) are admissible. When $b = +1$ there is one positive root and when $b = -1$ there are two positive roots. Outside the cusp in (I, α) space there are no real fixed points of (3.4) when $b = -1$. Another way of seeing this is to plot the function (3.7) in the (I^2, x_1^2) plane as shown in Figure 6. Using the fixed points as a backbone we can construct the geometric picture in the slice H_+. In Figures 7a,b the case $b = +1$ is shown. This is a supercritical bifurcation in the sense that a non-trivial fixed point bifurcates for $\alpha \leq 0$ (which is the region in which the trivial fixed point of the original 4D map is complex unstable).

When $\alpha \geq 0$ the nontrivial fixed point takes the asymptotic form

$$\left\{\begin{aligned} x_1 &= |I|^{1/2}\alpha^{-1/4}+\cdots \\ y_1 &= \tfrac{1}{2}|I|^{1/2}\alpha^{3/4}+\cdots \\ y_2 &= -\mathrm{sgn}(I)\,|I|^{1/2}\alpha^{1/4}+\cdots \end{aligned}\right. \qquad \text{for } |I| << |\alpha|^{2/3}$$

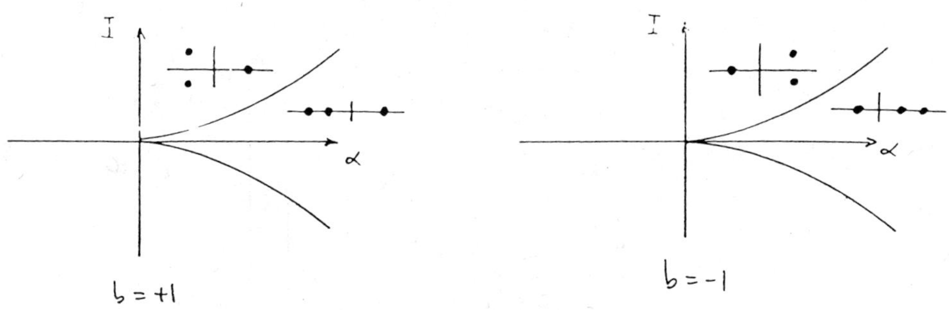

Figure 5 Fixed points of the reduced area preserving map (3.4) in the complex plane as a function of (I, α).

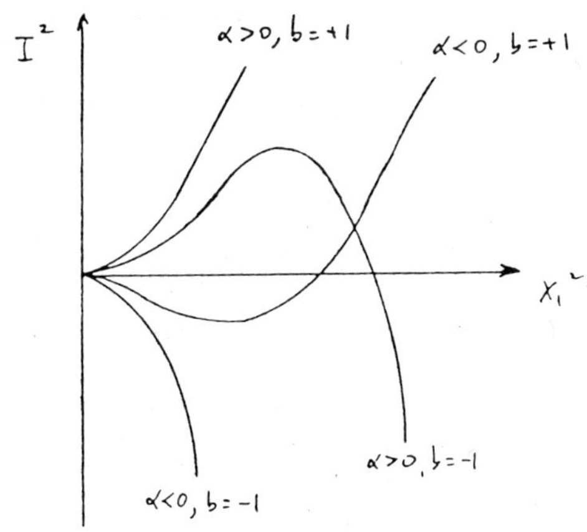

Figure 6 Plot of the function in equation (3.7).

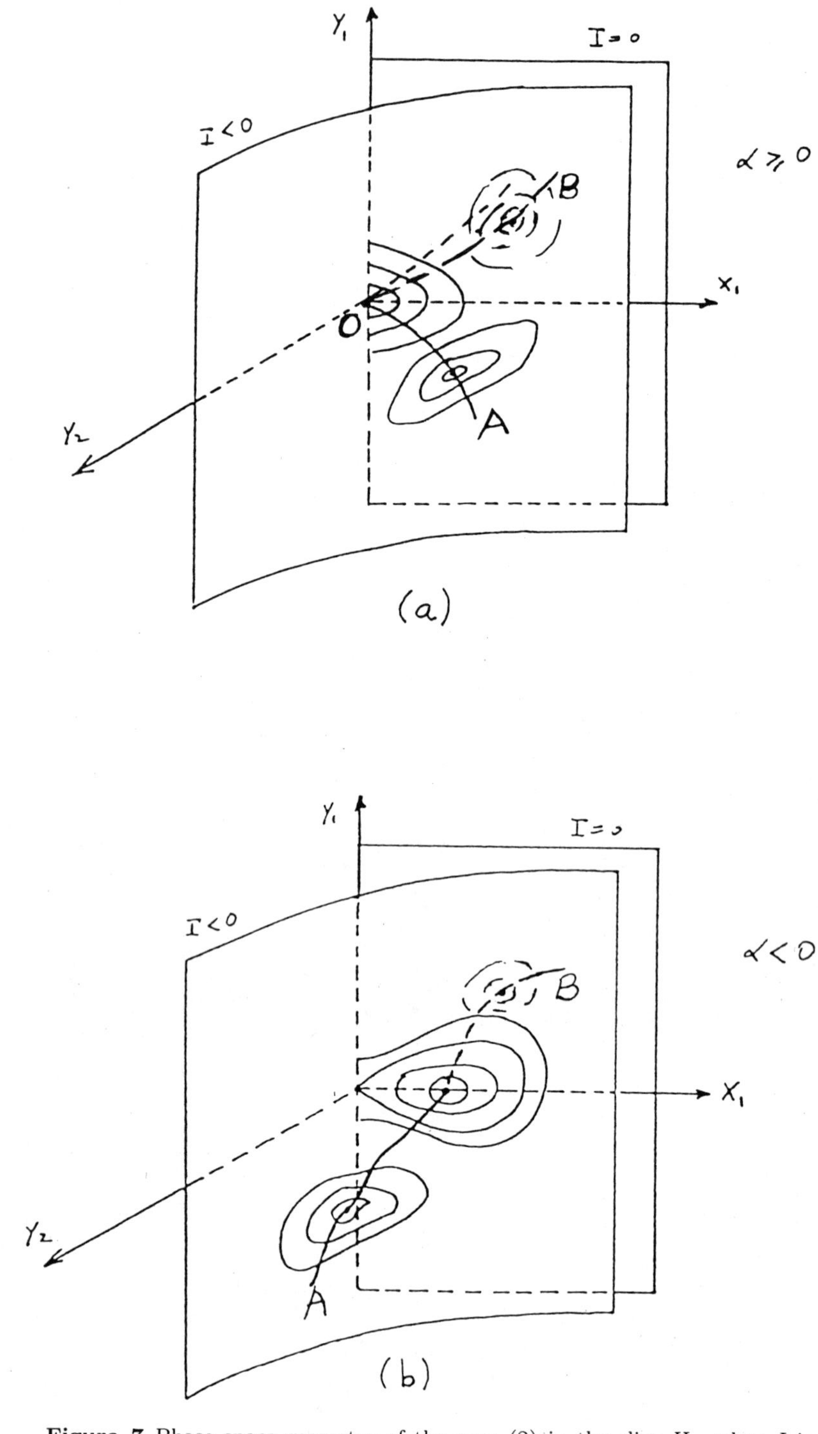

Figure 7 Phase space geometry of the map (2) in the slice H_+ when I is perturbed away from zero for the case when $b = +1$. (a) $\alpha \geq 0$. (b) $\alpha < 0$.

or

$$\begin{cases} x_1 = |I|^{1/3} + \cdots \\ y_1 = \frac{1}{2}|I| + \cdots \\ y_2 = -\mathrm{sgn}(I)\,|I|^{2/3} + \cdots \end{cases} \qquad \text{for } |I| >> |\alpha|^{2/3}$$

and when $\alpha < 0$ the nontrivial fixed point takes the asymptotic form

$$\begin{cases} x_1 = (-\alpha)^{1/2} + \cdots \\ y_1 = -\frac{1}{2}|I|^2(-\alpha)^{-3/2} + \cdots \\ y_2 = -I(-\alpha)^{-1/2} + \cdots \end{cases} \qquad \text{for } |I| << |\alpha|^{2/3}.$$

Note that I may be positive or negative with $(x, y) \in H_+$.

When $\alpha > 0$ (before the trivial fixed point loses stability) the reduced map in the slice H_+ includes the formally stable (trivial) fixed point as well as two arcs OA and OB of elliptic fixed points, see Figure 7a. In phase space each of these arcs becomes an invariant cone of the original map. These cones join at O. The reason for this is that through every point of OA except O there is an S^1 orbit of the action (1.7) which degenerates to the point O at O. Hence when $b = +1$ and as α goes from positive to negative, the invariant cones of the original map become an invariant cylinder which corresponds to the arc AB of elliptic fixed points of the reduced map in Figure 7b. As the cone and elliptic cylinder are made up of invariant circles only bits of them are likely to survive perturbations that break the normal form symmetry.

The phase space geometry for the subcritical case, when $b = -1$, is shown in Figures 8a,b.

When $\alpha > 0$ (Figure 8a) there is a saddle-center at $I^2 \approx \frac{4}{27}\alpha^3$ with asymptotic coordinates

$$\begin{aligned} x_1 &= \sqrt{\frac{2}{3}}\alpha^{1/2} + \cdots \\ y_1 &= 3\sqrt{\frac{3}{2}}\alpha^{3/2} + \cdots \\ y_2 &= -\frac{\sqrt{2}}{3}\alpha\,\mathrm{sgn}(I) + \cdots . \end{aligned}$$

Note that the arc AB in H_+ is made up of hyperbolic fixed points of the reduced map (3.2). In the original phase space this corresponds to an invariant cylinder $\mathcal{C}$ of hyperbolic periodic orbits of the original map. $\mathcal{C}$ is normally hyperbolic and hence persists. The arcs OA and OB in H_+ correspond to invariant cones of the original map. As α goes from positive to negative values, the invariant cylinder annihilates the elliptic cones leaving only the stable and unstable manifolds of the trivial fixed point.

Much of this phase space geometry is reminiscent of the phase space for the two classes (subcritical and supercritical) of the Hamiltonian Hopf bifurcation (see van der Meer [1985], pp. 72–76). That is, the phase space structures are similar to small perturbations of the time-1 map of the unfoldings of the vectorfield cases. However, in the present case there is no conserved quantity in the un-normalized system (such as the Hamiltonian function in the vectorfield case). Therefore, when the normal form symmetry is broken, orbits can in principle explore four dimensions rather than being confined to three-dimensional energy surfaces.

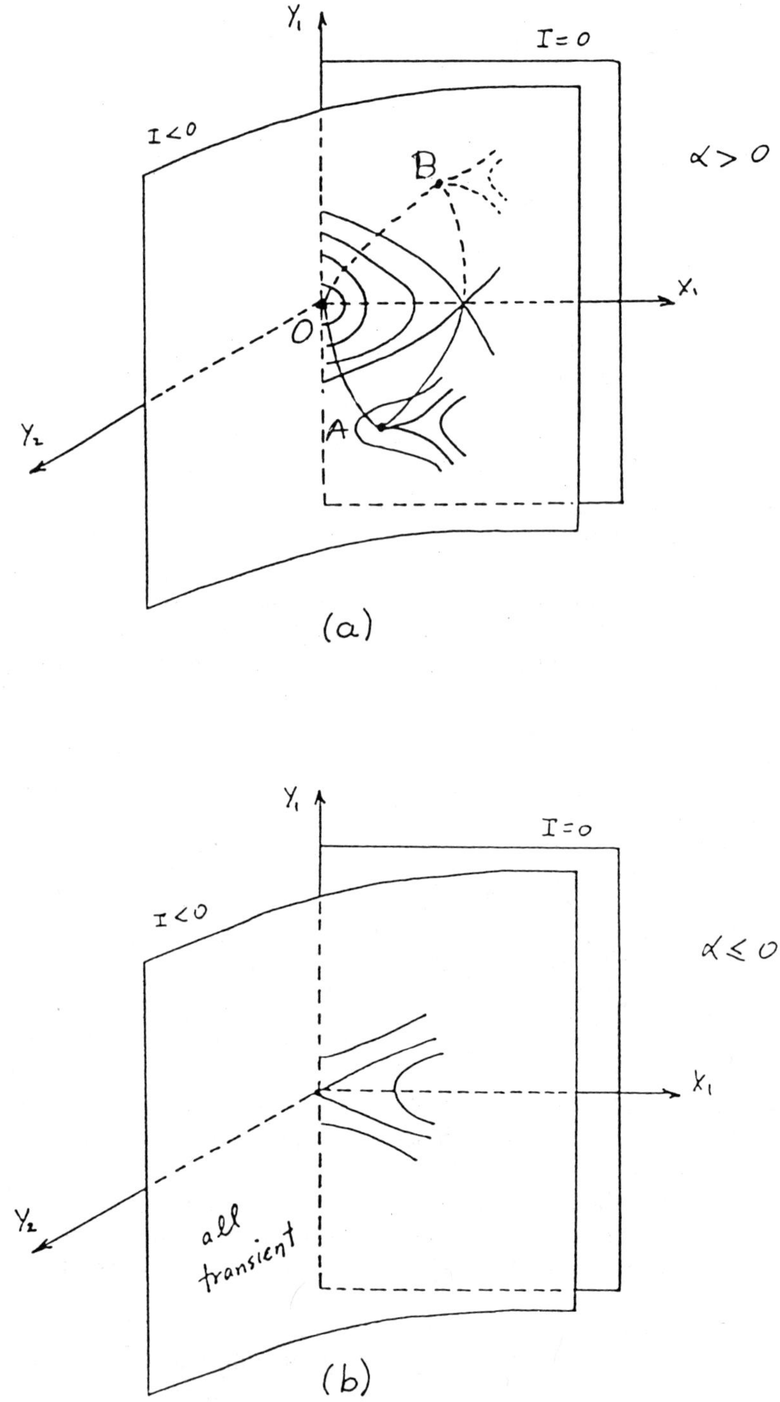

Figure 8 Phase space geometry of the map (2) in the slice H_+ when I is perturbed away from zero for the case when $b = -1$. (a) $\alpha \geq 0$ (b) $\alpha < 0$.

4 A reversible but non-Hamiltonian collision

In this section the starting point is the iterated two-parameter family of maps

$$\mathbf{x}' = \Phi(\mathbf{x};\lambda,\alpha) \qquad \mathbf{x}\in\mathbf{R}^4 \tag{4.1}$$

with $\Phi(0;\lambda,\alpha) = 0$ as in (1.1) but now the only hypotheses on Φ are that it is a reversible diffeomorphism with respect to a standard reversor (diagonal matrix with two $(+1)$ and two (-1) eigenvalues) and $d\Phi(0;\lambda,\alpha)$ is a linear map in general position for a collision of Floquet multipliers at an irrational point.

Without any attempt at a general theory for normal forms for reversible maps we argue that the nonlinear normal form for (4.1) can be approximated by the time-1 map of the normal form for the reversible $1:1$ resonant vectorfield. The reversible $1:1$ is not to be confused with the Hamiltonian $1:1$ (which is diagonalizable). The reversible $1:1$ has a non-diagonalizable linear part and is similar to the Hamiltonian $1:-1$ (in the Hamiltonian case $1:-1$ indicates pure imaginary eigenvalues of opposite signature).

The nonlinear normal form for a vectorfield with a reversible $1:1$ resonance has been given by Iooss and Perouème [1990],

$$\begin{aligned}\frac{dx}{dt} &= [\omega_o + P(I,X)]\mathbf{J}x + y\\ \frac{dy}{dt} &= [\omega_o + P(I,X)]\mathbf{J}y + Q(I,X)x\end{aligned} \tag{4.2}$$

where $\omega_o\in\mathbf{R}$, P,Q are smooth functions satisfying $P(0,0) = Q(0,0) = 0$, I,X carry the same definitions as in Section 1 and $(x,y)\in\mathbf{R}^4$.

The time-1 map of the linear part of the vectorfield (4.2) (obtained by setting $P=Q=0$) is easily seen to be

$$\mathbf{L}_0 = \begin{pmatrix} R_\theta & R_\theta\\ 0 & R_\theta\end{pmatrix} \tag{4.3}$$

where $\theta=\omega_o$ and R_θ is the rotation matrix defined in Section 1. Note that $\epsilon=\pm1$ does not appear in the nilpotent part of (4.3) since a reversible transformation can be used to eliminate the sign. The formal nonlinear normal form for Φ in (4.1) is then taken to be

$$\phi(x,y) = \mathbf{L}_0\exp(\mathbf{X}(x)) \tag{4.4}$$

where $\mathbf{X}(x)$ is just the nonlinear part of the vectorfield in (4.2); that is

$$\mathbf{X}(x) = P(I,X)\begin{pmatrix}\mathbf{J}x\\ \mathbf{J}y\end{pmatrix} + Q(I,X)\begin{pmatrix}0\\ x\end{pmatrix}. \tag{4.5}$$

Exponentiating the vectorfield $\mathbf{X}(x)$ in (4.4)–(4.5) we find the following formal expression for the nonlinear normal form for (4.1)

$$\begin{pmatrix}x'\\ y'\end{pmatrix} = \begin{pmatrix}(1-q(I,X))R_{\theta+\phi(I,X)} & R_{\theta+\phi(I,X)}\\ -q(I,X)R_{\theta+\phi(I,X)} & R_{\theta+\phi(I,X)}\end{pmatrix}\begin{pmatrix}x\\ y\end{pmatrix}. \tag{4.6}$$

It is interesting to compare the reversible normal form (4.6) with the symplectic normal form in (1.4). The expressions are almost identical; the difference lies in the fact that, in the reversible but non-symplectic case, the functions $\phi(I,X)$ and $q(I,X)$ are not related. Nevertheless the map (4.6) still has an associated invariant

$I(x', y') = I(x, y)$; just as in the symplectic case. This is quite surprising but is also true of the reversible 1:1 vectorfield (Bridges [1992]).

In spite of the differences between the maps (1.4) and (4.6) they reduce to identical forms on the orbit space. This is easily seen by introducing orbit space coordinates $\sigma \in \mathbf{R}^3$. Following the analysis in Section 1 the map (4.6) reduces to a map $\sigma' = T(\sigma)$ on the orbit space. In fact, since the function ϕ drops out on the orbit space, the reduced map T is exactly the Poisson map (1.10) with h_X replaced by q. The non-symplectic part of (4.6) disappears on the orbit space and all the analysis in Sections 2 and 3 carries over to the reversible non-symplectic case. However the reconstructed four-dimensional phase space for (4.6) will differ from the symplectic case because the flow along the S^1 group orbit will in general result in non-symplectic dynamics.

This result is in general agreement with the conjecture of Sevyruk and Lahiri [1991], page 107 and also goes much further in illustrating the relation with the symplectic case.

References

Arms, J., Cushman, R. and Gotay, M. [1991] *A universal reduction procedure for Hamiltonian group actions*, in: The Geometry of Hamiltonian Systems, (T. Ratiu, ed.), MRSI Workshop Proceedings, Springer-Verlag, 33–51.

Bridges, T.J. [1991] *Stability of periodic solutions near a collision of eigenvalues of opposite signature*, Math. Proc. Camb. Phil. Soc. **109**, 375–403.

——— [1992] *Poisson structure of the reversible 1:1 resonance*, preprint, submitted to proceedings of the Diepenbeek Meeting on Dynamical Systems.

Bridges, T.J. and Cushman, R.H. [1993] *Unipotent Normalization for Symplectic Mappings*, Physica D, **65**, 211–241.

Bridges, T.J. and Furter, J.E. [1993] *Symmetries, singularities and periodic points of symplectic maps*, University of Warwick, to appear in Lect. Notes in Math., Springer-Verlag.

Contopoulos, G. [1985] *Bifurcation and stability in 3-dimensional systems*, in: Stability of the solar system and its minor natural and artificial bodies, (V. G. Szebehdy, ed.), Reidel, 97–113.

——— [1986] *Qualitative changes in 3-dimensional dynamical systems*, Astron. Astrophy. **161**, 244–256.

Contopoulos, G. and Giorgilli, A. [1988] *Bifurcations and complex instability in a 4-dimensional symplectic mapping*, preprint.

Deprit, A. and Henrard, J. [1968] *A manifold of periodic orbits*, Adv. Astron. and Astrophysics **6**, 1–124.

Iooss, G. and Pérouème, M.C. [1990] *Perturbed homoclinic solutions in reversible 1:1 resonant vector fields*, University of Nice, (to appear in J. Diff. Eqns., preprint (282).)

Janssen, T. and Tjon, J.A. [1983] *Bifurcations of lattice structures*, J. Phys. A: Math. Gen. **16**, 673–696.

Kook, H. and Meiss, J.D. [1989] *Periodic orbits for reversible-symplectic mappings*, Physica D **35**, 65–86.

Krein, M.G. [1955] *On criteria for stability and boundedness of solutions of periodic canonical systems*, Prikl. Mat. Mekh. **19**, No. 6, 641–680.

Meyer, K. and Schmidt, D. [1971] *Periodic orbits near L_4 for mass ratios near the critical mass ratio of Routh*, Celestial Mechanics **4**, 99–109.

van der Meer, J. [1985] *The Hamiltonian Hopf Bifurcation*, Lect. Notes in Math. **1160**, Springer-Verlag.

Pfenniger, D. [1985] *Numerical study of complex instability 1. Mappings*, Astron. and Astrophys. **150**, 97–111.

Sevryuk, M.B. and Lahiri, A. [1991] *Bifurcations of families of invariant curves in four-dimensional mappings*, Physics Letters **A154**, 104–110.

Williamson, J. [1937] *On the normal forms of linear canonical transformations in dynamics*, Amer. J. Math. **59**, 599–617.

Fields Institute Communications
Volume **4**, 1995

The Constrained Liapunov-Schmidt Procedure and Periodic Orbits

Martin Golubitsky*
Department of Mathematics
University of Houston
Houston, TX 77204-3476, USA

Jerrold E. Marsden**
Department of Mathematics
University of California, Berkeley
Berkeley, CA 94720, USA

Ian Stewart[†]
Mathematics Institute
University of Warwick
Coventry CV4 7AL, UK

Michael Dellnitz[‡]
Department of Mathematics
University of Hamburg
D-20146 Hamburg, Germany

Abstract. This paper develops the Liapunov-Schmidt procedure for systems with additional constraints such as having a first integral, being Hamiltonian, or being a gradient system. Similar developments for systems with symmetry, including reversibility, are well known, and the method of this paper augments and is consistent with that approach. One of the results states that the bifurcation equation for Hamiltonian systems is actually a Hamiltonian vector field. In general, we use "implicit constraints" to encode the information constraining the system. The method is applied to the Liapunov center theorem for reversible systems and systems with an integral, as well as to the Hamiltonian Hopf bifurcation and resonance bifurcations for Hamiltonian and reversible systems.

1991 *Mathematics Subject Classification.* Primary 58F05; Secondary 58F14, 70K30.
*Research partially supported by NSF Grant DMS-9101836, the Texas Advanced Research Program (003652037) and The Fields Institute.
**Research partially supported by NSF Grant DMS-9302992 and The Fields Institute.
[†]Research partially supported by The Fields Institute, the Science and Engineering Research Council of the UK, and a European Community Laboratory Twinning grant (European Bifurcation Theory Group).
[‡]Research partially supported by The Fields Institute, and a European Community Laboratory Twinning grant (European Bifurcation Theory Group).

Introduction

The Liapunov-Schmidt procedure is widely used to reduce bifurcation problems that satisfy appropriate hypotheses to the solution of equations defined on finite-dimensional spaces. This procedure has been adapted to systems with symmetry (see Golubitsky and Schaeffer [1985] Chapter 7 Section 2). As observed in Hale [1969], the Liapunov-Schmidt procedure is especially effective for finding periodic orbits arising through Hopf bifurcation and may be adapted to Hopf bifurcation with symmetry (see Golubitsky, Stewart and Schaeffer [1988] and references therein).

In this paper we observe that the Liapunov-Schmidt procedure preserves additional structure, of a kind that we call an "implicit constraint". We show that this observation can be used to derive several known results (listed below) in a uniform and simple manner. The constraint is implicit in the sense that it becomes an implicit condition on the Liapunov-Schmidt reduced bifurcation equation. However, it may be an explicit "system constraint" to begin with, such as the conditions of having a first integral, being Hamiltonian, or being a gradient.

Implicit constraints can greatly simplify the solution of the Liapunov-Schmidt reduced equations because of the restrictions they impose on their form. The Hamiltonian case is especially well behaved because, under appropriate hypotheses, the reduced bifurcation equation defines a Hamiltonian vector field. In this sense the Liapunov-Schmidt procedure performs a Hamiltonian reduction analogous to the well known "orbit space" reduction method for Hamiltonians with a symmetry group; see for example Abraham and Marsden [1978].

The paper divides into three main parts:

- In Part I, we set the stage by adding the extra ingredient of a constraint to the Liapunov-Schmidt procedure for systems with symmetry. We illustrate how to use this extra ingredient by giving a new proof of the Liapunov center theorem for systems with a first integral and extending it to an equivariant context.
- In Part II, we apply the constrained Liapunov-Schmidt procedure to give a simple and direct determination of the periodic orbit structure of the Hamiltonian-Hopf bifurcation.
- In Part III, we apply this method to study the bifurcation structure of periodic orbits near a $k:\ell$ resonance for both reversible and Hamiltonian systems.

We will give a more detailed introduction to each of the applications and discuss some of the relevant literature throughout the text.

Acknowledgments We would like to express our sincere thanks to Tom Bridges, Hans Duistermaat, Vivien Kirk, Victor LeBlanc, James Montaldi, Jan-Cees van der Meer, Tudor Ratiu, Jürgen Scheurle, Alan Weinstein, and, in particular, André Vanderbauwhede, for useful comments and discussions.

Part I The Constrained Liapunov-Schmidt Procedure

1 Introduction

The basic idea of the method is as follows. Particular types of solutions in a differential equation, such as a fixed point, relative equilibrium, or a periodic orbit, can be found by determining the zeros of an appropriate map and applying the

Liapunov-Schmidt procedure. For example, the problem of finding periodic orbits can be formulated as the search for zeros of a map $F(u,\tau) = 0$ near a known zero, where u is an element of a loop space (a candidate periodic orbit) and τ is the perturbed period, an idea that goes back to Hale [1969]. Additionally, F may depend on bifurcation parameters λ, but we suppress these for the moment. The Liapunov-Schmidt procedure leads to a *reduced bifurcation equation* $g(k,\tau) = 0$ defined for k in the kernel of the linearization of the original equation, and the *reduced map* g inherits the symmetries of F. If we are looking for periodic orbits, then the map F has a natural symmetry group $\mathbf{S}^1$ representing phase shifts along the periodic orbits, as well as any other symmetries that may be present in the original problem.

When the system under consideration has extra structure, we can often encode this structure in an identity of the form $\Phi(u,\tau,F(u,\tau)) \equiv 0$, for a suitable mapping Φ. We call such an identity an *implicit constraint* (implicit because of the way F occurs in Φ). We show below that an implicit constraint on F imposes a related implicit constraint on the reduced map g, so that $\varphi(k,\tau,g(k,\tau)) = 0$ for a function φ derived from Φ.

Our first illustration of the method is an application to the Liapunov center theorem. Here the extra structure that leads to an implicit constraint is given by a first integral. The kernel is two-dimensional and is identified with the complex numbers $\mathbb{C}$; the circle group $\mathbf{S}^1$ may be viewed as the unit modulus complex numbers with its usual action on $\mathbb{C}$; and $\mathbf{S}^1$-equivariance implies that $g : \mathbb{C}\times\mathbb{R} \to \mathbb{C}$ has the form

$$g(z,\tau) = p(|z|^2,\tau)z + q(|z|^2,\tau)iz \tag{1.1}$$

for real-valued functions p and q. The first integral condition implies that p and q are related in such a way that p vanishes if q vanishes. Therefore, to find solutions, it is enough to solve the equation $q = 0$. As in Hopf bifurcation, it can be shown that q_τ is nonzero at the origin, so we can solve for zeros by the implicit function theorem. We remark that this method using the Liapunov-Schmidt procedure together with symmetries and an implicit constraint avoids the procedure of blowing up the singularity.

If we make the stronger assumption that the system is Hamiltonian or reversible, then it can be proved directly that the function p in (1.1) is identically zero. In the reversible case, this is seen directly from the fact that the map g inherits the reversible symmetry structure. In the Hamiltonian case, the implicit constraint states that the map g is a Hamiltonian vector field (depending parametrically on τ), using the usual symplectic structure on $\mathbb{C}$ with the real and imaginary parts of z as conjugate variables. We use this Hamiltonian structure in place of having a first integral. In fact, it is easy to see that the map g in (1.1) is Hamiltonian if and only if $p = 0$. Vanderbauwhede and van der Meer [1994] have also proved, independently and by a different method, that the reduced bifurcation equation inherits a Hamiltonian structure.

2 Implicit constraints

This section develops the constrained Liapunov-Schmidt procedure. In practice, a key step is to translate conditions such as having a first integral, being variational, or being Hamiltonian, into constraints and to pass these constraints to the bifurcation equation. We illustrate this procedure in various applications, but for the moment we concentrate on the abstract formulation.

Let $\mathcal{X}$ and $\mathcal{Y}$ be Banach spaces and let $F:\mathcal{X} \to \mathcal{Y}$ be a C^∞ map whose zeros we seek. Assume that $F(0) = 0$. Let dF denote the (Fréchet) derivative of F and assume that $dF(0)$ is Fredholm. Let Γ be a compact Lie group acting linearly on both $\mathcal{X}$ and $\mathcal{Y}$ and suppose that F is Γ-equivariant. We say that there is an *implicit constraint* if there is another Banach space $\mathcal{W}$ on which Γ also acts, and a C^∞ equivariant map

$$\Phi:\mathcal{X} \times \mathcal{Y} \to \mathcal{W}$$

such that

$$\Phi(X, F(X)) = 0 \tag{2.1}$$

for all $X \in \mathcal{X}$ (or in a neighborhood of zero, since the constructions here are local).

For a system of ODEs, the possession of a first integral can be phrased as an implicit constraint, as we now explain. Consider a system of ODEs on $\mathbb{R}^n$ of the form

$$\dot{X} = f(X),$$

and assume that it has a first integral; that is, there exists a function $H:\mathbb{R}^n \to \mathbb{R}$ that is constant along trajectories:

$$H(X(s)) = H(X(0))$$

for all solutions $X(s)$. Differentiation with respect to s shows that this condition is equivalent to

$$dH(X) \cdot f(X) = 0$$

for all X. Suppose that we are looking for equilibria of this system; that is, looking for zeros of f. Then we can choose $\mathcal{X} = \mathcal{Y} = \mathbb{R}^n$ and define

$$\Phi(X, Y) = dH(X) \cdot Y$$

and so the condition that H be an integral is equivalent to

$$\Phi(X, f(X)) = 0,$$

which is an implicit constraint. Looking for equilibria, while conceptually simple, is an important problem in Hamiltonian bifurcation theory, and this remark lets us investigate the basic bifurcations (Hamiltonian pitchfork, Hamiltonian saddle node, *etc.*). The problem of finding periodic orbits is a bit more subtle, but the basic idea is similar, as we shall see in the next section.

We now investigate the condition imposed on the bifurcation equation by an implicit constraint. Since $dF(0)$ is assumed to be Fredholm, $\mathcal{K} := \ker dF(0)$ has a Γ-invariant closed complement $\mathcal{M}$ so that $\mathcal{X} = \mathcal{K} \oplus \mathcal{M}$. Similarly $\mathcal{R} :=$ range $dF(0)$ is closed and has a Γ-invariant closed complement $\mathcal{N}$, so that $\mathcal{Y} = \mathcal{R} \oplus \mathcal{N}$. Let $\mathbf{P} : \mathcal{Y} \to \mathcal{R}$ denote the projection with kernel $\mathcal{N}$ and (locally) define a C^∞ map $w:\mathcal{K} \to \mathcal{M}$ with $w(0) = 0$ by solving

$$\mathbf{P}F(k + w) = 0 \tag{2.2}$$

for w as a function of k using the implicit function theorem. Uniqueness of the solution $w(k)$ and invariance of the splitting under Γ imply that w is Γ-equivariant. Moreover,

$$dw(0) = 0. \tag{2.3}$$

Higher derivatives of w at zero are determined, as usual, by implicit differentiation. Clearly, solutions of the equation $F(X) = 0$ are given by $X = k + w(k)$, where k is a solution of the bifurcation equation

$$g(k) := (I - \mathbf{P})F(k + w(k)) = 0. \tag{2.4}$$

To see how the extra structure (2.1) affects the bifurcation equation, define $\varphi :$ $\mathcal{K} \times \mathcal{N} \to \mathcal{W}$ by

$$\varphi(k, n) = \Phi(k + w(k), n). \tag{2.5}$$

Observe that φ is Γ-equivariant because w and Φ are. The derivatives of φ at $(0, 0)$ can be calculated in terms of the derivatives of Φ and w. From (2.2) and (2.4), we get

$$\begin{aligned} \varphi(k, g(k)) &= \Phi(k + w(k), (I - \mathbf{P})F(k + w(k))) \\ &= \Phi(k + w(k), F(k + w(k))), \\ &= 0 \end{aligned}$$

by (2.1). Thus the implicit constraint

$$\varphi(k, g(k)) = 0 \tag{2.6}$$

encodes the implications for the bifurcation equation (2.4) of the extra structure in the original problem. This imposes conditions on the bifurcation equation similar to those imposed by equivariance. We will see how to deal with such conditions and how to sometimes make them explicit (for example in the Hamiltonian case) in the following sections. We begin by showing that this technique gives an easy and natural proof of the Liapunov center theorem.

3 The Liapunov center theorem

The Liapunov center theorem is one of the classical periodic orbit theorems for Hamiltonian systems. We give a proof of it that illustrates the method of the preceding section. We prove it for systems with a first integral, which includes the Hamiltonian case. In the Hamiltonian case there is additional structure in the bifurcation equation, but it leads to the same conclusion. As we have indicated, this extra structure states that the map g is a Hamiltonian vector field, a methodology we shall deal with shortly. Expositions of the Liapunov center theorem are found in, for example, Kelley [1967] and Abraham and Marsden [1978, pages 496–499]; the latter giving a proof based on blowing up the singularity due to Duistermaat. We also note the approach of Alexander and Yorke [1978] (see also Schmidt [1976a,b]) which links the Liapunov center theorem with the Hopf bifurcation. See also the proof in Vanderbauwhede [1982b]. In addition, Liapunov center theorems for reversible systems can be found in Vanderbauwhede [1982a], Sevryuk [1986] and Golubitsky, Krupa and Lim [1991].

Let f be a C^∞ vector field on a C^∞ manifold M. Assume:

(H1) X_0 is a fixed point of the system, that is, $f(X_0) = 0$.

(H2) The linearization $A_0 = df(X_0)$ has simple eigenvalues $\pm\omega_0 i$, and $k\omega_0 i$ is not an eigenvalue for $k = 0, 2, 3, \ldots$. Let V denote the eigenspace associated to the eigenvalues $\pm\omega_0 i$.

(H3) The vector field f has a C^∞ *first integral* $H : M \to \mathbb{R}$ (that is, $dH(X) \cdot f(X) = 0$) with the following two properties:

(a) $dH(X_0)|V = 0$;

(b) $d^2H(X_0)|V$ is nondegenerate.

Theorem 3.1 *Under conditions* (H1)–(H3) *there is a one-parameter family of periodic orbits in a neighborhood of* X_0 *with periods close to* $2\pi/\omega_0$. *They locally fill out an invariant* C^∞ *manifold of dimension two that is tangent to* V *at* X_0.

Proof Rescaling time, we can assume that $\omega_0 = 1$. Passing to a coordinate chart in M we can assume that $M = \mathbb{R}^n$, since the theorem, while intrinsic for manifolds, is local. Let $\mathcal{C}^1_{2\pi}$ be the space of C^1 maps $u : \mathbf{S}^1 \to \mathbb{R}^n$ and let $\mathcal{C}^0_{2\pi}$ be the space of C^0 maps $v : \mathbf{S}^1 \to \mathbb{R}^n$. Define

$$F : \mathcal{C}^1_{2\pi} \times \mathbb{R} \to \mathcal{C}^0_{2\pi}$$

by

$$F(u, \tau) = (1 + \tau)\frac{du}{ds} - f(u). \tag{3.1}$$

The map F is C^∞ by the 'Ω-lemma': see, for example, Abraham, Marsden and Ratiu [1988], Section 2.4.

The group $\mathbf{S}^1$ acts on $\mathcal{Y} = \mathcal{C}^0_{2\pi}$ by $(\theta \cdot v)(s) = v(s + \theta)$ and similarly on $\mathcal{C}^1_{2\pi}$. Further, $\mathbf{S}^1$ acts on $\mathcal{X} = \mathcal{C}^1_{2\pi} \times \mathbb{R}$ with trivial action on $\mathbb{R}$. With these actions, F is $\mathbf{S}^1$-equivariant.

The derivative of F at $u \equiv X_0$ is L, where

$$Lv \equiv dF(X_0, 0) \cdot v = v' - A_0 v. \tag{3.2}$$

By (H2),

$$\begin{aligned} \ker L &= \operatorname{span}\{\operatorname{Re}(e^{is}v_0), \operatorname{Im}(e^{is}v_0)\} \\ &= \{\operatorname{Re}(ze^{is}v_0) \mid z \in \mathbb{C}\} \end{aligned}$$

where $A_0 v_0 = i v_0$. Thus we can identify the kernel of $dF(X_0, 0)$ with $\mathbb{C}$ via the map $z \mapsto \operatorname{Re}(ze^{is}v_0)$.

Choose orthonormal coordinates on $\mathbb{R}^n$ so that the first two coordinates, when complexified, span the eigenspace corresponding to eigenvalues $\pm i$ and that on this space we have

$$A_0 = \begin{bmatrix} 0 & 1 \\ -1 & 0 \end{bmatrix}.$$

The components of the (complex) vector v_0 are $(1, i, 0, \ldots, 0)$. Differentiating the conservation condition $dH(X) \cdot f(X) = 0$ twice with respect to X and evaluating at $X = X_0$ shows that the matrices representing A_0 and $d^2H(X_0)$ commute. A 2×2 matrix calculation shows that on the kernel, the matrix representing $d^2H(X_0)$ is a nonzero multiple of the identity, say $d^2H(X_0) = \mu I$.

Now we use the Fredholm alternative relative to the L^2 inner product obtained by integrating the standard inner product on $\mathbb{R}^n$. Let $\mathcal{M} = \operatorname{range} L^*$ be the orthogonal complement of $\mathcal{K} = \ker L$, and let $\mathcal{N} = \ker L^*$ be the orthogonal complement to $\mathcal{R} = \operatorname{range} L$. Notice that $L^*v = -v' - A_0^*v$. It is important to be careful with this splitting, because L and L^* are differential operators and a derivative is lost. However these operators are in fact elliptic, so this causes no difficulties: see Golubitsky and Schaeffer [1985], p. 332 for details.

The Liapunov-Schmidt procedure with constraints, as described in the preceding section, may be applied. This procedure gives a bifurcation equation

$$g: \mathbb{C} \times \mathbb{R} \to \mathbb{C}.$$

Being $\mathbf{S}^1$-equivariant, g has the form

$$g(z,\tau) = p(|z|^2,\tau)z + q(|z|^2,\tau)iz; \tag{3.3}$$

see Golubitsky and Schaeffer [1985], p. 344. Note that $p(0,0) = 0$ and $q(0,0) = 0$ since the linear terms vanish in any Liapunov-Schmidt reduced map. In the reversible case $p = 0$ identically, so it suffices to solve the equation $q = 0$. We now show that a similar approach works for systems with a first integral; more precisely, we can use the first integral to show below that $q = 0$ *implies* $p = 0$.

In the case of Hamiltonian systems, our general results below will show that the map g is a Hamiltonian vector field on $\mathbb{C}$. Since any Hamiltonian vector field on $\mathbb{C}$ has the form $f(z) = -2i\partial H/\partial \bar{z}$ for some *real valued* function $H(z,\bar{z})$, it follows that g is purely imaginary, and hence $p = 0$ identically (see Bridges [1990] for another argument along these lines). In this respect, the Hamiltonian and reversible cases are similar, as is well known. The resemblance is less close in other situations; see Roberts and Quispel [1992].

The calculation in Golubitsky and Schaeffer [1985], pp. 344–349, shows that

$$q_\tau(0,0) = -1.$$

Thus, we can solve $q = 0$ for $\tau = \tau(|z|^2)$. We claim that this automatically defines the required manifold of periodic orbits. To prove this, we must show that $\tau = \tau(|z|^2)$ implies $p = 0$ for z small. To do this we use the implicit constraint afforded by the first integral. Define

$$\Phi: \mathcal{X} \times \mathcal{Y} \to \mathbb{R}$$

by

$$\Phi((u,\tau),v) = \int_0^{2\pi} dH(u(s)) \cdot v(s)ds. \tag{3.4}$$

Note that

$$\begin{aligned} \Phi((u,\tau),F(u,\tau)) &= \int_0^{2\pi} dH(u(s)) \cdot \left[(1+\tau)\frac{du}{ds} - f(u(s))\right] ds \\ &= (1+\tau)\int_0^{2\pi} \frac{d}{ds}H(u(s))ds - \int_0^{2\pi} dH(u(s)) \cdot f(u(s))ds. \end{aligned} \tag{3.5}$$

The first term in (3.5) vanishes since $H(u(2\pi)) - H(u(0)) = 0$ by the periodicity of u, while the integrand in the second term vanishes pointwise because H is a first integral of f as in (H3). Thus (2.1) holds, so the bifurcation equation satisfies (2.6). In other words,

$$\varphi((z,\tau), g(z,\tau)) = 0,$$

so that

$$\Phi(z + w(z,\tau), \tau, g(z,\tau)) = 0,$$

whence

$$\int_0^{2\pi} dH(u_z(s)) \cdot g(z,\tau) ds = 0.$$

Here $u_z(s)$ is the element of $\mathcal{X} = \mathcal{C}^1_{2\pi}$ corresponding to the implicitly defined function $z + w(z,\tau)$ and so has the form $\mathrm{Re}(ze^{is}v_0) + O(|z|^2)$, and $g(z) =: \zeta = p(|z|^2,\tau)z + q(|z|^2,\tau)iz$. We regard ζ as an element of $\mathcal{Y}$ by the identification

$$\zeta \mapsto \mathrm{Re}(\zeta e^{is} v_0^*) \tag{3.6}$$

where v_0^* is the eigenvector for A_0^*, that is, $A_0^* v_0^* = -iv_0^*$. The first two components of v_0 and v_0^* are the same.

Choose $\tau = \tau(|z|^2)$ so that $q = 0$ as above. Then

$$0 = p(|z|^2, \tau(|z|^2)) \int_0^{2\pi} dH(\mathrm{Re}(ze^{is}v_0 + O(|z|^2))) \cdot \mathrm{Re}(ze^{is}v_0^*) ds. \tag{3.7}$$

We show that the second factor (that is, the integral term) is nonzero for small $|z|$. To do so let $z = r$ be real and let

$$\Psi(r) = \int_0^{2\pi} dH(\mathrm{Re}(re^{is}v_0 + O(r^2))) \cdot \mathrm{Re}(e^{is}v_0^*) ds.$$

Note that $\Psi(0) = 0$ and that

$$\Psi'(0) = \int_0^{2\pi} d^2H(0) \cdot (\mathrm{Re}(e^{is}v_0), \mathrm{Re}(e^{is}v_0^*)) ds.$$

By the choice of inner product,

$$\Psi'(0) = \mu \int_0^{2\pi} \langle(\mathrm{Re}(e^{is}v_0), \mathrm{Re}(e^{is}v_0^*))\rangle ds = 2\pi\mu$$

and so $\Psi(r) \neq 0$ for $r > 0$ and small. Identity (3.7) now implies that $p(|z|^2, \tau(|z|^2)) = 0$ for z small.

The smoothness of the manifold of periodic orbits follows from the smoothness of $\tau(|z|^2)$, which is guaranteed by the implicit function theorem. □

4 An equivariant Liapunov center theorem

The proof of the Liapunov center theorem given in the previous section automatically produces an analogous equivariant theorem, whose hypotheses we now explain.

Let $\Gamma \subset \mathbf{O}(n)$ be a group acting in the usual way on $\mathbb{R}^n$. Assume that the vector field $f: \mathbb{R}^n \to \mathbb{R}^n$ is Γ-equivariant and that

(E1) $f(X_0) = 0$ *where* X_0 *is fixed by* Γ.

As before, let $A_0 = df(X_0)$ and observe that Γ-equivariance of f implies that A_0 commutes with Γ. It follows that eigenspaces of A_0 are Γ-invariant. In particular, eigenvalues of high multiplicity may be expected. Our second assumption is the following.

(E2a) *The eigenvalues of A_0 include $\pm\omega_0 i$, and $k\omega_0 i$ is not an eigenvalue for $k = 0, 2, 3 \ldots$.*

Let V denote the generalized eigenspace associated to the eigenvalues $\pm\omega_0 i$. We replace the previous assumption that the eigenvalues $\pm\omega_0 i$ are simple by requiring V to be Γ-simple. This notion is the Γ-equivariant version of simple eigenvalues (see Golubitsky, Stewart and Schaeffer [1988]) and is defined as follows:

Definition 4.1 *The action of the group Γ on V is said to be* Γ-simple *if either*
(a) $V = W \oplus W$ *and Γ acts absolutely irreducibly on W, or*

(b) Γ *acts irreducibly, but not absolutely irreducibly, on V.*

Our next assumption is the following:

(E2b) Γ *acts Γ-simply on V.*

There is an action of $\mathbf{S}^1$ on V induced by the one-parameter group $\exp(sA_0^T)$. Since A_0 commutes with Γ, so does this action of $\mathbf{S}^1$. Thus, we have a well defined action of $\Gamma \times \mathbf{S}^1$ on V. We now determine families of periodic solutions by looking for their symmetries. This involves two more conditions, as follows:

(E3a) *Let $\Sigma \subset \Gamma \times \mathbf{S}^1$ be an isotropy subgroup of the action of $\Gamma \times \mathbf{S}^1$ on V. Assume that* $\dim \mathrm{Fix}_V(\Sigma) = 2$.

(E3b) *f has a C^∞ first integral $H : \mathbb{R}^{2m} \to \mathbb{R}$ such that*

(i) $dH(X_0)|\, \mathrm{Fix}_V(\Sigma) = 0$;

(ii) $d^2H(X_0)|\, \mathrm{Fix}_V(\Sigma)$ *is nondegenerate.*

To state the main theorem, we define what is meant by the symmetry of a periodic solution. An element $\sigma = (\gamma, \theta) \in \Gamma \times \mathbf{S}^1$ is called a *symmetry* of a 2π-periodic function $X(t)$ if $\gamma X(t) = X(t - \theta)$. That is, the periodic orbit is preserved as a set by γ; uniqueness of solutions to ordinary differential equations guarantees the existence of a unique θ.

We may now prove an equivariant version of the Liapunov center theorem:

Theorem 4.2 *Under conditions* (E1)–(E3), *there is a one-parameter family of periodic orbits in a neighborhood of X_0 with Σ symmetry and periods close to $2\pi/\omega_0$. They locally fill out an invariant C^∞ manifold of dimension two that is tangent to* $\mathrm{Fix}_V(\Sigma)$ *at X_0.*

Proof Use the same Liapunov-Schmidt reduction technique as in the preceding section, and observe that the reduced map g is $\Gamma \times \mathbf{S}^1$-equivariant, where the $\mathbf{S}^1$-action induced by phase shift is identical to the $\mathbf{S}^1$-action induced by A_0. Equivariance implies that g: $\mathrm{Fix}_V(\Sigma) \times \mathbb{R} \to \mathrm{Fix}_V(\Sigma)$ has the form (3.3). Using (E3a) we can identify $\mathrm{Fix}_V(\Sigma)$ with $\mathbb{C}$. Now proceed as in the proof of Theorem 3.1. □

Remark 4.3 It is also possible to prove this equivariant version of the Liapunov center theorem by using the trick of adding dissipation to the Hopf theorem; see Golubitsky, Stewart and Schaeffer [1988], Exercise 4.4, p. 283.

5 Remarks and generalizations

(a) The entire machinery of finding two-dimensional fixed-point subspaces in Γ-equivariant Hopf bifurcation can now be applied to finding periodic solutions in systems with first integrals. For example, in $\mathbf{O}(2)$-equivariant systems with $\dim V = 4$, Theorem 4.2 produces both rotating and standing waves.

(b) As noted in Golubitsky and Stewart [1993], there are surprising group-theoretic conditions restricting those Γ that can produce two-dimensional fixed-point subspaces. Let $\pi_\Gamma : \Gamma \times \mathbf{S}^1 \to \Gamma$ be the projection and let $H = \pi_\Gamma(\Sigma)$. Note that H is isomorphic to Σ. Let $K = (\Gamma \times \{0\}) \cap \Sigma$. Indeed, we can think of Σ as a *twisted* subgroup of $\Gamma \times \mathbf{S}^1$; that is, uniqueness of solutions guarantees the existence of a group homomorphism $\theta : H \to \mathbf{S}^1$ such that $(\gamma, \theta(\gamma)) \in \Sigma$ for all $\gamma \in H$. Note that $K = \ker \theta$, from which it follows that K is normal in H and H/K is isomorphic to a Lie subgroup of $\mathbf{S}^1$ and hence is either cyclic or $\mathbf{S}^1$. In any case H/K is abelian. The following theorem is proved in Golubitsky and Stewart [1993]: *If there is any representation V in which Σ has a two-dimensional fixed-point subspace, then H/K is a maximal abelian subgroup of $N_\Gamma(K)/K$ where $N_\Gamma(K)$ is the normalizer of K in Γ.*

(c) In the case of a Hamiltonian system with symmetry group G, the method of the preceding section may be applied to the new Hamiltonian system obtained by symplectically reducing by the continuous part of a symplectic group action (namely, by the connected component of the identity of G) at a regular value of the momentum map. (See, for example, Marsden [1992] for the general set-up. The case of a singular value of the momentum map is very interesting and the above ideas would need to be modified to cover it. It is probable that a symplectic slice theorem will be useful for this situation.) After reduction by the continuous group, one is in general left with a discrete symmetry group, which can be used for the group Γ in the above result. The application of the Liapunov center theorem to the reduced problem will produce relative equilibria, so that one will get relative periodic orbits (often tori) in the original space.

(d) Montaldi, Roberts and Stewart [1988] deduce Theorem 4.2 from an equivariant version of the Weinstein-Moser theorem — see their Remark 1.2(b). However, they do not give the proof of local smoothness of the manifold in any detail.

(e) First integrals coupled with Liapunov-Schmidt reduction can be used to recover the periodic solutions found in the zero eigenvalue bifurcations studied in Golubitsky and Stewart [1987]. The advantage of the approach of the present paper is that these families of periodic solutions will exist even when the Hamiltonian has many degrees of freedom. A similar conclusion can be obtained using the splitting lemma.

(f) The existence of integrals in the Liapunov-Schmidt reduction works in finite differentiability since the proof depends only on the standard implicit function theorem. Proofs of results like the Liapunov center theorem in finite differentiability can be based on a finite differentiability version of (3.3), which we have not derived. (A related approach may be found in Vanderbauwhede [1990].) It is clear that if the original vector field is of class C^{2k+1} then the invariant functions p and q will be only of class C^k. (See, for example, Abraham, Marsden and Ratiu [1988] Section 2.4 for this type of argument.) This information is sufficient to prove the existence of the manifold of periodic solutions in the Liapunov center theorem, but some care is needed to determine the exact degree of differentiability of that manifold.

(g) As mentioned, possession of a first integral is just one kind of extra structure that a system may have. Of course, symmetry provides extra structure, but this can be exploited directly: the symmetry of the system transfers to the bifurcation equation. In Remark (i) we show how to formulate the condition of being Hamiltonian as an implicit constraint. In the next section we show how, in some cases, this implicit constraint can be recast into the condition that g is Hamiltonian and use this observation to lay the foundation for an approach to finding periodic solutions of Hamiltonian systems.

(h) The condition of being a gradient system can also be formulated as an implicit constraint. Suppose that we seek equilibria of a gradient system on $\mathbb{R}^n$ (or more generally on a Riemannian manifold), say

$$\dot{x} = \nabla h(x).$$

We use the standard inner product on $\mathbb{R}^n$ denoted $\langle\,,\rangle$, and use it to identify $\mathbb{R}^n$ with its dual space. We seek zeros of the map $F(x) = \nabla h(x)$, assuming, as usual that $F(0) = 0$. In this case, the linearized operator $L = dF(0)$ is self-adjoint, so the splitting in the domain and range space are identical, namely $\mathbb{R}^n = \ker L \oplus \operatorname{range} L$. Let $\mathcal{X} = \mathbb{R}^n$, $\mathcal{Y} = \mathbb{R}^n$ and $\mathcal{W} = \mathbb{R}$. Define the map

$$\Phi : \mathcal{X} \times \mathcal{Y} \to \mathcal{W}$$

by

$$\Phi(x, y) \cdot u = \langle y, u\rangle - dh(x) \cdot u.$$

The implicit constraint

$$\Phi(x, F(x)) = 0$$

holds. The condition imposed on the bifurcation equation is

$$0 = \varphi(k, g(k)) \cdot u = \langle g(k), u\rangle - dh(k + w(k)) \cdot u.$$

However, by definition of the implicit function w, we have

$$\begin{aligned} dh(k + w(k)) \cdot (dw(k) \cdot u) &= \langle \nabla h(k + w(k)), (dw(k) \cdot u)\rangle \\ &= \langle \nabla h(k + w(k)), \mathbf{P}(dw(k) \cdot u)\rangle \\ &= \langle \mathbf{P}\nabla h(k + w(k)), (dw(k) \cdot u)\rangle \\ &= 0. \end{aligned}$$

Therefore, the implicit constraint can be written as

$$0 = \varphi(k, g(k)) \cdot u = \langle g(k), u\rangle - dh(k + w(k)) \cdot (u + dw(k) \cdot u).$$

By the chain rule, we can rewrite this as

$$0 = \langle g(k), u\rangle - d\tilde{h}(k) \cdot u$$

where $\tilde{h}(k) = h(k + w(k))$, so that g is a gradient field, which recovers a result of Rabinowitz [1977].

(i) We can modify the above example to find equilibria of a Hamiltonian system, by replacing the inner product by the symplectic structure. By definition a Hamiltonian vector field X_H on a symplectic vector space (V, Ω) satisfies

$$\Omega(X_H(z), u) = dH(z) \cdot u.$$

Let $F(z) = X_H(z)$, and choose $\mathcal{X} = V$, $\mathcal{Y} = V$ and $\mathcal{W} = V^*$, the dual space. Define the map

$$\Phi : \mathcal{X} \times \mathcal{Y} \to \mathcal{W}$$

by

$$\Phi(z, y) \cdot u = \Omega(y, u) - dH(z) \cdot u.$$

Again the bifurcation equation inherits an implicit constraint reflecting the Hamiltonian condition. In fact, as we shall see in the next section, under appropriate hypotheses that play the role of the self-adjointness in the preceding example, the map g will be a Hamiltonian vector field. This way of phrasing the implicit constraint will play an important role in what follows.

6 The Hamiltonian constraint

We begin this section by amplifying the remark at the end of the last section and then we consider the problem of the existence of periodic orbits for Hamiltonian systems (possibly with symmetry) near equilibria.

6.1 Conditions for the bifurcation map to be Hamiltonian Let V be a symplectic vector space (possibly infinite-dimensional) with symplectic form Ω and let $F : V \times \Lambda \to V$ be a Hamiltonian vector field with Hamiltonian $H : V \times \Lambda \to \mathbb{R}$. Here, Λ denotes a space of parameters that we will, as usual, suppress for the moment.

Assume, as in the general Liapunov-Schmidt procedure, that $F(0) = 0$. Let $L : V \to V$ be the linearization of F at the origin. Recall that the generalized kernel of L is a symplectic subspace of V. (See Golubitsky and Stewart [1987], for example.) As is well known, and is easy to check, the operator L is Hamiltonian with Hamiltonian the quadratic form associated with $d^2H(0)$, the second derivative of H at the origin. It follows that L^* is also Hamiltonian with Hamiltonian the quadratic form associated to $Jd^2H(0)J^*$. Another way to express the fact that L is Hamiltonian is to say that L is Ω-skew: $\Omega(Lu, v) = -\Omega(u, Lv)$ for all $u, v \in V$.

Assume that

$$\Omega(u, v) = \langle\langle Ju, v \rangle\rangle$$

where $J : V \to V$ is a complex structure and where $\langle\langle\, , \rangle\rangle$ is a *real* inner product on V. This is the standard relation between complex and symplectic structures (see, for example, Abraham and Marsden [1978].) In these terms, the skew symmetry of L reads

$$\langle\langle JLu, v \rangle\rangle = - \langle\langle Ju, Lv \rangle\rangle .$$

A check of the definitions shows that $J(\ker L) = \ker L^*$. That is, the complex structure maps the kernel of L to that of L^*. To show this, let $u \in \ker L$. To show that $Ju \in \ker L^*$ we write, for any $v \in V$,

$$\langle\langle L^* Ju, v \rangle\rangle = \langle\langle Ju, Lv \rangle\rangle = - \langle\langle JLu, v \rangle\rangle = 0.$$

Thus, $L^* Ju = 0$, as we claimed.

Consistent with the notation in the general Liapunov-Schmidt procedure, let $\mathcal{K} = \ker L$, $\mathcal{N} = \ker L^*$, $\mathcal{R} =$ range L and $\mathcal{M} =$ range L^*. A main hypotheses that will be appropriate for the Liapunov center theorem and resonance bifurcations is as follows:

(S) The complex structure J leaves $\ker L$ invariant: $\ker L = \ker L^*$.

The formula relating Ω and J shows that if a subspace $W \subset V$ is invariant under J, then it is automatically symplectic. Conversely, if it is symplectic, then writing $V = W \oplus W^\Omega$ where W^Ω is the Ω orthogonal complement of W, and choosing an associated inner product and complex structure on each summand, one sees that it is possible to choose the inner product and complex structure on V so that W is invariant under J. Because of this remark, we see that the above assumption is automatic if the kernel of L equals the generalized kernel (and, if necessary, the inner product and the complex structure are adjusted). In this sense, the assumption is akin to a semisimplicity assumption.

Proposition 6.1 *Under hypothesis (S), the spaces $\mathcal{R}$ and $\mathcal{N}$ as well as the spaces $\mathcal{K}$ and $\mathcal{M}$ are Ω orthogonal complements.*

Proof By (S), the kernels of L and L^* are identical. It follows that the ranges are identical as well. If $v = Lu$ and if $Ly = 0$, then

$$\Omega(v, y) = \langle\!\langle Jv, y\rangle\!\rangle = - \langle\!\langle Lu, Jy\rangle\!\rangle = \langle\!\langle u, L^* Jy\rangle\!\rangle = 0$$

since, by hypothesis, Jy is in the kernel of L, that is, in the kernel of L^*. The proposition now follows. □

Hypothesis (S) implies that the mapping g maps $\mathcal{K} \to \mathcal{K}$. With the above assumption, it therefore makes sense to think of g as a vector field on the symplectic vector space $\mathcal{K}$. We now make a simple but crucial observation:

Theorem 6.2 *Under assumption (S), the map g is a Hamiltonian vector field with Hamiltonian $h(k) = H(k + w(k))$. Moreover, g and the function h have the same invariance properties as the given Hamiltonian H.*

Proof Let $k_1 \in \mathcal{K}$ and recall that $g(k) = (I - P)F(k + w(k))$. Use the fact that $Dw(k) \cdot k_1 \in \mathcal{M}$ and the fact that $PF(k + w(k)) = 0$ is the defining property of the implicit function w and Proposition 6.1 to verify

$$\begin{aligned} \Omega(g(k), k_1) &= \Omega(g(k), dw(k) \cdot k_1 + k_1) \\ &= \Omega(F(k + w(k)), dw(k) \cdot k_1 + k_1). \end{aligned}$$

Since F is Hamiltonian, we can use the chain rule to write the last expression as

$$dH(k + w(k)) \cdot (dw(k) \cdot k_1 + k_1) = dh(k) \cdot k_1.$$

Thus, g is a Hamiltonian vector field with Hamiltonian function h. Invariance properties follow in the usual way from uniqueness, assuming that the group action preserves the symplectic form and the complex structure. □

In particular, in this case we have reformulated the problem to one of searching for equilibria of a Hamiltonian vector field defined on the kernel. In what follows, we will show how to also use this result for finding periodic orbits by casting the problem into one of finding the equilibria of an infinite-dimensional Hamiltonian system on loop space. See Vanderbauwhede and van der Meer [1994] for an alternative discussion of the Hamiltonian structure of the bifurcation equations.

6.2 Periodic orbits and the loop space We start with a Hamiltonian system $\dot{z} = X_H(z)$ on a symplectic vector space (E, ω). As we have noted before, eventually H will also depend on a bifurcation parameter. (We will be assuming that the space E is finite-dimensional for simplicity, but in principle the methods that we outline here will apply to infinite-dimensional problems too; however, as is known, the resonance assumptions that are necessary in the infinite-dimensional case are severe using these methods.) We will be applying the above theorem not to this Hamiltonian system, but to an associated Hamiltonian system on loop space, to be described below. The point of view of using the loop space was used by Weinstein in one approach to the Weinstein-Moser theorem (see Weinstein [1978]) and it is also related to the method used by Bridges [1990].

We may assume that $E = \mathbb{C}^n$ and that ω is in canonical form, that is, ω is the skew-symmetric bilinear form given by

$$\omega(z_1, z_2) = -2\mathrm{Im}\,\langle z_1, z_2\rangle\,. \tag{6.1}$$

The factor of -2 is of course conventional and corresponds to the standard choice of symplectic structure. Also, the inner product here stands for the standard *Hermitian* inner product.

Introduce the loop space $\mathcal{C}^1_{2\pi}$ of C^1 maps $\mathbf{S}^1 \to E$ and let $\mathcal{C}^0_{2\pi}$ be the corresponding space of C^0 maps. As in Section 3, define

$$F : \mathcal{C}^1_{2\pi} \times \mathbb{R} \to \mathcal{C}^0_{2\pi}$$

by

$$F(u, \tau) = (1+\tau)\frac{du}{ds} - f(u),$$

where $f = X_H$. The Liapunov-Schmidt procedure is now applied to the map F.

Since the vector field $f = X_H$ is Hamiltonian, the identity

$$\omega(X_H(u), v) = dH(u)\cdot v$$

holds for all $v \in \mathbb{R}^{2n}$. Define the map

$$\Phi : \mathcal{C}^1_{2\pi} \times \mathbb{R} \times \mathcal{C}^0_{2\pi} \to [\mathcal{C}^1_{2\pi}]^*$$

by

$$\Phi(u, \tau, v)\cdot U = \int_0^{2\pi} \left\{\omega\left(v - (1+\tau)\frac{du}{ds}, U\right) + dH(u)\cdot U\right\} ds. \tag{6.2}$$

The implicit constraint

$$\Phi(u, \tau, F(u, \tau)) = 0 \tag{6.3}$$

is a restatement of the condition that the vector field f is Hamiltonian with Hamiltonian function H. This condition can also be rephrased as saying that the map

F, regarded as a (parameter-dependent) vector field on $\mathcal{C}^1_{2\pi}$, is Hamiltonian with respect to the weak symplectic form

$$\Omega(u,v) = \frac{1}{2\pi}\int_0^{2\pi} \omega(u(s),v(s))\,ds, \tag{6.4}$$

and with the Hamiltonian function

$$\mathcal{H}(u,\tau) = \frac{1}{2\pi}\int_0^{2\pi} \left\{\frac{1}{2}\omega\left((1+\tau)\frac{du}{ds},u\right) - H(u)\right\} ds. \tag{6.5}$$

We can also put a real inner product and a complex structure on loop space by averaging the corresponding objects on E around loops, as we did with the symplectic structure.

In this approach, we take the symplectic vector space to which the general theory is to be applied to be the loop space $V = \mathcal{C}^0_{2\pi}$. Of course, there is the usual derivative loss, so the vector field F is really only defined on the dense subspace $\mathcal{C}^1_{2\pi}$. Moreover, the real inner product with respect to which adjoints are taken is an L^2 inner product and so V is not complete. In addition, the symplectic form is only a weak symplectic form. However, as in Weinstein [1978] and Chernoff and Marsden [1974], while these are important technicalities, they are easily taken care of using ellipticity; they are similar to the technicalities that occur in this approach to the Hopf bifurcation (as we mentioned before, the reader can consult Golubitsky and Schaeffer [1985], p. 332 for additional information).

Theorem 6.2 shows that if the kernel of the linearized equations is symplectic and arranged to be invariant under the complex structure, then the bifurcation equation will be a Hamiltonian vector field. To be able to apply this, we need to do two things. First, we need to calculate the linearized equations and second, we need to calculate the induced symplectic structure on the kernel of the linearization and check that it is nondegenerate. In the remainder of this section we shall show how to do this for the Liapunov center theorem and later we shall show how to implement it for the case of resonance bifurcations.

6.3 The Hamiltonian Liapunov center theorem Now we apply the Hamiltonian version of the Liapunov-Schmidt procedure developed above to the case of the Hamiltonian Liapunov center theorem. As in (3.2) the linearization

$$L:\mathcal{C}^1_{2\pi} \to \mathcal{C}^0_{2\pi}$$

is

$$Lv = \frac{dv}{ds} - A_0 v,$$

where $A_0 = df(0)$. Since the linear map L is Fredholm of index zero we may use the L^2-orthogonal splittings

$$\begin{aligned} \mathcal{C}^1_{2\pi} &= \mathcal{K} \oplus \operatorname{range} L^* \\ \mathcal{C}^0_{2\pi} &= \operatorname{range} L \oplus \mathcal{N} \end{aligned}$$

where $\mathcal{K} = \ker L$ and $\mathcal{N} = \ker L^*$. The Liapunov-Schmidt reduced map obtained from F is

$$g:\mathcal{K} \times \mathbb{R} \to \mathcal{N}.$$

Next, we compute the kernel of L under the given eigenvalue assumptions. First of all, note that the real eigenspace of A_0 corresponding to eigenvalues $\pm i$ (with the frequency set equal to one, as before) is of real dimension two and is coincident with the generalized eigenspace since the eigenvalues are assumed to be simple. Thus, this space is symplectic. (As before, see, for example, Golubitsky and Stewart [1987] for a discussion of this Williamson type of result.) We can choose the complex structure on E so that this eigenspace is the one complex dimensional space corresponding to the eigenvalue i and that the operator A_0 is given by multiplication by i on this space. We can also assume that the other eigenspaces are ω-orthogonal to it. We can now realize E as $\mathbb{C}^n$ with this eigenspace given by the first factor. Thus, in this complex representation, the complex eigenspace with eigenvalue i (*i.e.*, the two-dimensional real eigenspace corresponding to eigenvalues $\pm i$) is given by the first component. Let $v_0 = (1, 0, \ldots, 0)$ denote this eigenvector.

The kernel of L was given previously in real terms, but in complex terms, it is simply the two-dimensional subspace of the loop space given by the set of all loops of the form

$$v(s) = e^{is} z v_0$$

for arbitrary complex z. This space is clearly invariant under the complex structure on loop space that was described earlier. Also, we see explicitly that this space is symplectic by noting that its symplectic structure is given, using self explanatory notation, by

$$\begin{aligned} \Omega(v_1, v_2) &= \frac{1}{2\pi} \int_0^{2\pi} \omega(e^{is} z_1 v_0, e^{is} z_2 v_0) \\ &= \frac{1}{2\pi} \int_0^{2\pi} -2\mathrm{Im}\, \left\langle e^{is} z_1 v_0, e^{is} z_2 v_0 \right\rangle \\ &= -2\mathrm{Im}\, z_1 \bar{z}_2. \end{aligned}$$

In other words, the symplectic structure on the kernel of L, after identification with $\mathbb{C}$, is the standard one. Thus, the general theory applies to show that the map g that occurs in the bifurcation equation $g = 0$ is indeed a Hamiltonian vector field.

Recall from (3.3) that due to $\mathbf{S}^1$-equivariance, the bifurcation map has the form

$$g : \mathbb{C} \times \mathbb{R} \to \mathbb{C}$$

where

$$g(z, \tau) = pz + qiz, \tag{6.6}$$

and p and q are real valued functions of $|z|^2$ and τ. In Section 3 we used (3.4) to show that p vanishes whenever q vanishes. However, in the present case, knowing that g is Hamiltonian implies that it has the form

$$g(z) = -2i \frac{\partial H}{\partial \bar{z}}$$

for a real valued function H of $(z, \bar{z})$. In particular, g is purely imaginary, and so we have the identity $p = 0$.

6.4 Conditions imposed on the bifurcation equation In this subsection we consider the situation when the hypothesis (S) might fail for the linearization. This will be needed in the next section when we search for periodic orbits in the Hamiltonian Hopf bifurcation. In this case, we still have an implicit constraint as is guaranteed by the general theory described earlier. To do this, we return to the implicit constraint on loop space, condition (6.3), but we do not necessarily assume hypothesis (S).

As in (3.2) the linearization

$$L: \mathcal{C}^1_{2\pi} \to \mathcal{C}^0_{2\pi}$$

is

$$Lv = \frac{dv}{ds} - A_0 v,$$

where $A_0 = df(0)$. Since the linear map L is Fredholm of index zero (we showed above that in general the kernel and the cokernel of L are isomorphic, and so have the same dimension), we may use the splittings

$$\begin{aligned} \mathcal{C}^1_{2\pi} &= \mathcal{K} \oplus \operatorname{range} L^* \\ \mathcal{C}^0_{2\pi} &= \operatorname{range} L \oplus \mathcal{N} \end{aligned}$$

where $\mathcal{K} = \ker L$ and $\mathcal{N} = \ker L^*$. The Liapunov-Schmidt reduced map obtained from F is

$$g: \mathcal{K} \times \mathbb{R} \to \mathcal{N}.$$

We use (6.3) to determine a condition that is independent of the implicit function w used to define g. For this purpose we choose the function U in (6.3) to be du/ds. With this choice it is clear that the last term in (6.3) vanishes, since it is the s-derivative of $H(u(s))$ and u is periodic. It is also obvious that the term

$$\omega((1+\tau)du/ds, du/ds)$$

is zero. Thus,

$$\Psi(u, \tau, v) = \int_0^{2\pi} \omega\left(\frac{du}{ds}, v\right) ds$$

is an implicit constraint for F. Setting $u = k + w(k, \tau)$ we obtain the identity

$$\psi(k, \tau) \equiv \int_0^{2\pi} \omega\left(\frac{dk}{ds} + \frac{dw}{ds}, g(k, \tau)\right) = 0$$

from the induced implicit constraint on g.

Lemma 6.3 *We have*

$$\int_0^{2\pi} \omega(L^* v, V) ds = -\int_0^{2\pi} \omega(v, L^* V) ds, \tag{6.7}$$

for all $v, V \in \mathcal{C}^1_{2\pi}$.

Proof Recall that

$$L^* V = -\frac{dV}{ds} - A_0^* V.$$

Since v and V are periodic, integration by parts yields

$$\begin{aligned}\int_0^{2\pi} \omega(v, L^*V)ds &= \int_0^{2\pi} \left\{\omega\left(v, -\frac{dV}{ds} - A_0^*V\right)\right\} ds \\ &= \int_0^{2\pi} \left\{\omega\left(\frac{dv}{ds}, V\right) + \omega(v, -A_0^*V)\right\} ds.\end{aligned}$$

Earlier we showed that both A_0 and A_0^* are Hamiltonian vector fields. This implies in particular that:

$$\omega(v, A_0^*V) = -\omega(A_0^*v, V).$$

This shows that

$$\begin{aligned}\int_0^{2\pi} \left\{\omega\left(\frac{dv}{ds}, V\right) + \omega(v, -A_0^*V)\right\} ds &= \int_0^{2\pi} \omega\left(\frac{dv}{ds} + A_0^*v, V\right) ds \\ &= -\int_0^{2\pi} \omega(L^*v, V)ds,\end{aligned}$$

as desired. □

Since $dw/ds \in \operatorname{range} L^*$ and $g \in \ker L^*$, this lemma establishes the symplectic orthogonality condition

$$\int_0^{2\pi} \omega\left(\frac{dw}{ds}, g(k, \tau)\right) = 0.$$

This proves the following:

Theorem 6.4 *The bifurcation mapping g satisfies the implicit constraint*

$$\psi(k, \tau) = \int_0^{2\pi} \omega\left(\frac{dk}{ds}, g(k, \tau)\right) = 0. \tag{6.8}$$

Part II Hamiltonian-Hopf Bifurcations

7 Introduction

A Hamiltonian-Hopf bifurcation occurs in a Hamiltonian system of ODEs when, as a parameter λ in the Hamiltonian is varied, two pairs of simple purely imaginary eigenvalues of the Jacobian at an equilibrium collide and form a quadruplet of complex eigenvalues. The splitting of eigenvalues into the complex plane forces the critical purely imaginary eigenvalues to be geometrically simple (though algebraically double). The Hamiltonian Hopf bifurcation has a long history that we shall not attempt to survey here. We refer to Meyer and Schmidt [1971], Abraham and Marsden [1978] and Arnol'd [1988] for discussions and some references. We discuss some of the relevant literature in the introduction to Part III.

Van der Meer [1985] studied this bifurcation and, through the application of normal form theory and singularity theory developed for the purpose, was able to classify the periodic solutions that are spawned by this resonance. Bridges [1990] showed that the periodic solutions in a Hamiltonian-Hopf bifurcation can be obtained using $\mathbb{Z}_2$ singularity theory with a distinguished parameter as developed in Golubitsky and Langford [1981]. In this Part we show that Bridges's results can be recovered as an example of the use of implicit constraints. We summarize the advantages of this approach.

- No discussion of the effects of the tails of Taylor series — such as occurs in all normal form analyses — need be given.
- The proof works equally easily for $2n$-dimensional systems as for four-dimensional systems. This fact may allow eventually the derivation of a formula describing the alternatives in the Hamiltonian-Hopf bifurcation such as now exists for Hopf bifurcation.

The disadvantages of this approach are summarized by:

- The issue of the linearized stability of the periodic solutions is not addressed (though a strengthening of the reduction technique such as occurs in the analysis of Hopf bifurcation, see Golubitsky, Stewart and Schaeffer [1988] Chapter XVI Section 6, might be able to recover this information).
- Dynamics other than periodic solutions — such as the existence of invariant tori — are not discussed. Again, we feel that this can be achieved with additional effort.

After rescaling we can assume that the critical eigenvalues are at $\pm i$. We use the Liapunov-Schmidt procedure for finding periodic solutions that are approximately 2π-periodic. As we will show, the assumption that the eigenvalues are geometrically simple permits a reduction of the problem of finding 2π-periodic solutions to one of solving an equation of the form

$$g(z, \lambda, \tau) = 0,$$

where

$$g: \mathbb{C} \times \mathbb{R} \times \mathbb{R} \to \mathbb{C}.$$

As in the proof of the Liapunov center theorem, $\mathbf{S}^1$ phase-shift symmetry on 2π-periodic functions implies that g is $\mathbf{S}^1$-equivariant. Equivariance in turn guarantees that

$$g(x, \lambda, \tau) = p(x^2, \lambda, \tau)x + q(x^2, \lambda, \tau)ix,$$

where p and q are real-valued functions. Solving $g = 0$ for nonzero solutions reduces to solving the pair of equations $p = q = 0$. As in Bridges [1990] we will show that the $\mathbf{S}^1$-invariant function q vanishes identically. (Note the difference with the Liapunov center theorem where p vanishes identically.) Thus the problem of finding periodic solutions in the original system of ODEs reduces, near a point where a Hamiltonian-Hopf bifurcation occurs, to finding solutions of

$$p(x^2, \lambda, \tau)x = 0. \tag{7.1}$$

As van der Meer notes, the solutions to (7.1) are most easily found by swapping the roles of λ and τ; that is, think of the perturbed period parameter τ as the bifurcation parameter and the system parameter λ as the unfolding parameter. The main technical result of this Part is the verification of the following:

Theorem 7.1
(a) $p(0,0,0) = 0, \quad p_\tau(0,0,0) = 0, \quad p_{\tau\tau}(0,0,0) > 0.$

(b) *Generically* $p_\lambda(0,0,0)$ *and* $p_{x^2}(0,0,0)$ *are nonzero.*

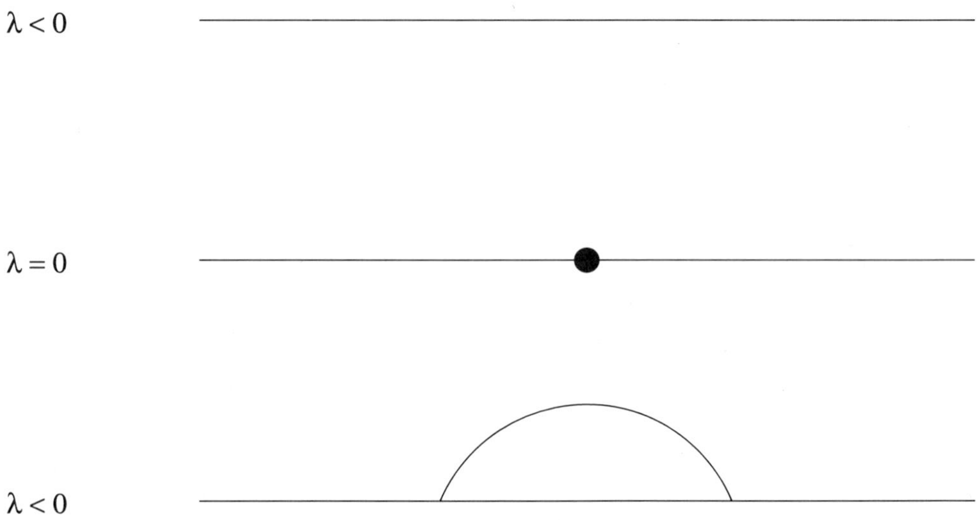

Figure 1 $(x^2 + \tau^2 + \lambda)x = 0$.

Theorem 7.1 shows that solving (7.1) is equivalent to solving the (singularity theory) normal form equation:

$$(\pm x^2 \pm \lambda + \tau^2)x = 0. \tag{7.2}$$

The bifurcation diagrams associated with these solutions are given in Figures 1 and 2. Note that when $\lambda < 0$ — which corresponds to the eigenvalues of the Jacobian being purely imaginary — there are two branches of periodic solutions emanating from the trivial equilibrium. These solutions are just the ones guaranteed by the Liapunov center theorem.

We end this introduction by discussing the structure of Part II. In the next section we discuss the linear algebra associated with a Hamiltonian-Hopf bifurcation. In Section 9 we introduce the generalities of the Liapunov-Schmidt reduction while in Section 10 we show how the Hamiltonian implicit constraint allow us to conclude that the function q is a multiple of p and that it even has to vanish identically. In Section 11 we prove Theorem 7.1 and in the last section we use singularity theory to derive (7.2).

8 The general setting

Let $H : \mathbb{R}^{2n} \times \mathbb{R} \to \mathbb{R}$ be a Hamiltonian parametrized by a system parameter λ, and let

$$\frac{dX}{ds} = f(X, \lambda) \tag{8.1}$$

be the corresponding Hamiltonian system of ODEs. We assume that at $\lambda = \lambda_0$ the system (8.1) undergoes a Hamiltonian-Hopf bifurcation at the equilibrium $X = X_0$. We simplify notation by setting $X_0 = 0$ and $\lambda_0 = 0$. Thus

$$f(0,0) = dH(0,0) = 0.$$

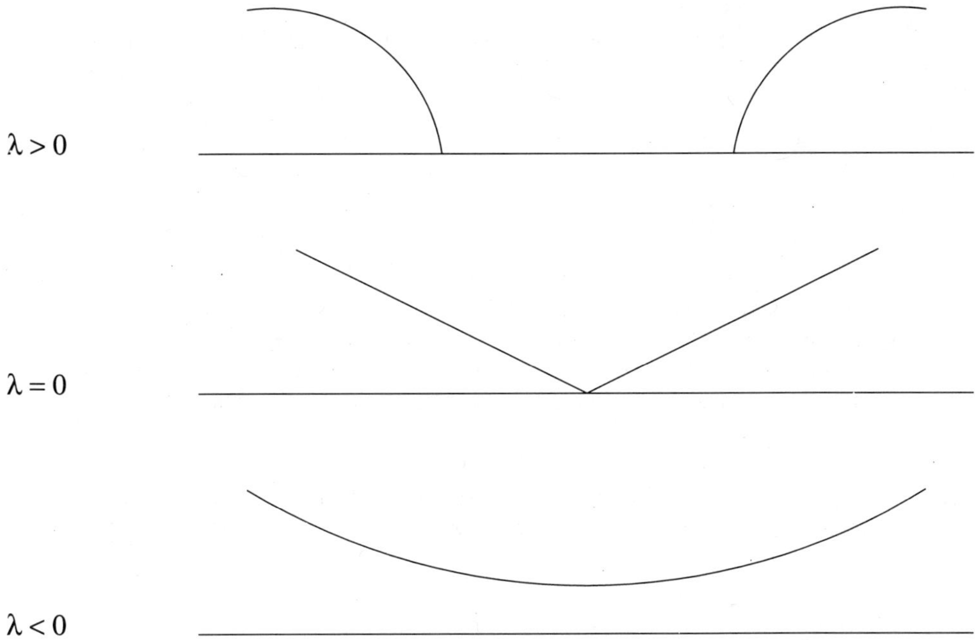

Figure 2 $(-x^2 + \tau^2 + \lambda)x = 0$.

In a Hamiltonian-Hopf bifurcation two pairs of simple purely imaginary eigenvalues collide on the imaginary axis as λ is varied. Generically, after the collision, the critical eigenvalues form a quadruplet of eigenvalues in the complex plane. Indeed, generically at criticality, the generalized eigenspace V corresponding to the critical eigenvalues is a four-dimensional subspace of $\mathbb{R}^{2n}$ on which

$$A_0 = df(0,0)$$

is nilpotent and has double purely imaginary eigenvalues $\pm i\omega$. We rescale time in (8.1) and assume that

$$\omega_0 = 1.$$

Nilpotency implies that A_0 has a two-dimensional invariant subspace

$$V_0 = \ker(A_0^2 + I),$$

and a complementary two-dimensional subspace V_1 mapped isomorphically onto V_0 by $A_0^2 + I$.

The linear map $A_0|V$ has a normal form that can be described by choosing an appropriate basis, as follows. Choose v_1, v_2 such that

$$V_0 = \mathbb{R}\{v_1, v_2\}$$

and

$$A_0(v_1) = v_2 \quad \text{and} \quad A_0(v_2) = -v_1.$$

Next, choose a basis

$$V_1 = \{v_3, v_4\}$$

where

$$A_0(v_3) = v_1 + v_4 \quad \text{and} \quad A_0(v_4) = v_2 - v_3. \tag{8.2}$$

In this basis, the matrix form of $A_0|V$ is

$$A_0|V = \begin{pmatrix} J & I \\ 0 & J \end{pmatrix},$$

where

$$J = \begin{pmatrix} 0 & -1 \\ 1 & 0 \end{pmatrix}.$$

At a generic Hamiltonian-Hopf bifurcation we assume *nonresonance* of the critical eigenvalues; that is, we assume that ki is not an eigenvalue of $df(0,0)$ when $k = 0, \pm 2, \pm 3, \ldots$. Since 0 is not an eigenvalue, the implicit function theorem implies that there is a smooth branch of equilibria for (8.1); that is,

$$f(u_0(\lambda), \lambda) = 0,$$

where $u_0(0) = 0$. We may change coordinates in u so that this family of equilibria is at 0; that is, $u_0(\lambda) \equiv 0$. Thus, we may assume

$$f(0, \lambda) = 0 \quad \text{for all } \lambda. \tag{8.3}$$

9 The reduced system

In the present context the operator F in (3.1) depends on the parameters τ and λ, and we define $F: \mathcal{C}^1_{2\pi} \times \mathbb{R} \times \mathbb{R} \to \mathcal{C}^0_{2\pi}$ by

$$F(u, \lambda, \tau) = (1 + \tau)\frac{du}{ds} - f(u, \lambda).$$

As above, let L be the linearization of F at 0,

$$Lv = \frac{dv}{ds} - A_0 v, \quad A_0 = df(0,0),$$

and let L^* be the adjoint operator

$$L^* v = -\frac{dv}{ds} - A_0^* v.$$

We view A_0 as an $n \times n$ matrix acting on $\mathbb{C}^{2n}$. Then we can choose complex eigenvectors $v_0, v_0^* \in \mathbb{C}^{2n}$ such that

$$A_0 v_0 = -i v_0 \quad \text{and} \quad A_0^* v_0^* = i v_0^*$$

by setting

$$v_0 = v_1 + i v_2 \quad \text{and} \quad v_0^* = v_3 + i v_4. \tag{9.1}$$

Let $\cdot$ denote the ordinary inner product on $\mathbb{C}^{2n}$. Then these choices satisfy

$$v_0^* \cdot \overline{v_0} = v_0^* \cdot v_0 = 0, \tag{9.2}$$

as a direct calculation using (9.1) shows.

Again we have the splittings (see Section 6)

$$\begin{aligned} \mathcal{C}^1_{2\pi} &= \mathcal{K} \oplus \text{range } L^* \\ \mathcal{C}^0_{2\pi} &= \text{range } L \oplus \mathcal{N} \end{aligned}$$

where $\mathcal{K} = \ker L$ and $\mathcal{N} = \ker L^*$. (Unlike in Hopf bifurcation, nilpotency of A_0 implies that we cannot choose $\mathcal{N} = \mathcal{K}$.) We compute

$$\begin{aligned} \mathcal{K} &= \{\mathrm{Re}(ze^{-is}v_0) \mid z \in \mathbb{C}\}, \\ \mathcal{N} &= \{\mathrm{Re}(ze^{-is}v_0^*) \mid z \in \mathbb{C}\}. \end{aligned}$$

In this way we have identified both $\mathcal{K}$ and $\mathcal{N}$ with $\mathbb{C}$, viewed as a two-dimensional real vector space. We denote elements of $\mathcal{K}$ by v and elements of $\mathcal{N}$ by v^*. Finally, let

$$E : \mathcal{C}_{2\pi} \to \mathrm{range}\, L$$

be the projection onto range L with kernel $\mathcal{N}$.

By the standard technique (implicit function theorem and $\mathbf{S}^1$-equivariance) we can reduce the problem of finding zeros of F to the solution of

$$g(x, \lambda, \tau) = p(x^2, \lambda, \tau)x + q(x^2, \lambda, \tau)ix = 0, \tag{9.3}$$

where $x \in \mathbb{R}$. Here we can identify x with the 2π-periodic function

$$\hat{x}(s) = x\mathrm{Re}(e^{-is}v_0).$$

10 Dependence of p and q

In the case of the Liapunov center theorem we have seen that the function p vanishes identically, and hence the solutions of the reduced system are determined by the equation $q = 0$ (see Sections 3 and 6). As we will show in this section, in the case of the Hamiltonian Hopf bifurcation q vanishes identically, and the structure of the set of periodic solutions is defined by $p = 0$.

We proceed in a way analogous to our treatment of the Liapunov center theorem. First we use the implicit constraint induced by Φ in (3.4) to show that q is a multiple of p. Then — using the implicit constraint of Section 6 — we prove that q vanishes identically.

Define

$$\Phi : C^1_{2\pi} \times C^0_{2\pi} \to \mathbb{R}$$

by

$$\Phi(u, v) = \int_0^{2\pi} dH(u(s)) \cdot v(s) ds.$$

As in Section 3, but with parameters suppressed,

$$\Phi(u, F(u)) \equiv 0.$$

This implies that

$$\varphi : \mathcal{K} \times \mathcal{N} \to \mathbb{R}$$

defined by

$$\varphi(u, v) = \Phi(u + w(u), v)$$

satisfies

$$\varphi(u, g(u)) \equiv 0,$$

where w is the implicitly defined function obtained from Liapunov-Schmidt reduction.

The general discussion in the previous paragraph implies that

$$\int_0^{2\pi} dH(u_x(s)) \cdot g(x, \lambda, \tau) ds = 0,$$

where $u_x(s)$ is the element of $\mathcal{C}^1_{2\pi}$ of the form $\hat{x}+O(x^2)$. Regarding $g(x,\lambda,\tau)$ as an element of $\mathcal{C}_{2\pi}$ by

$$z \mapsto \mathrm{Re}(ze^{-is}v_o^*)$$

and using the linearity of Φ and φ in the second variable we obtain

$$\begin{aligned} & p(x^2,\lambda,\tau)\int_0^{2\pi} dH(u_x(s))\cdot \mathrm{Re}(xe^{-is}v_0^*)ds \\ + \; & q(x^2,\lambda,\tau)\int_0^{2\pi} dH(u_x(s))\cdot \mathrm{Re}(ixe^{-is}v_0^*)ds \quad = \quad 0. \end{aligned}$$

We want to show that there exists a function $r = r(x,\lambda,\tau)$ such that

$$q(x^2,\lambda,\tau) = r(x,\lambda,\tau)p(x^2,\lambda,\tau) \tag{10.1}$$

in a neighborhood of $x = 0$, $x \neq 0$. For this to hold it is sufficient that the derivative of

$$\Psi(x) = \int_0^{2\pi} dH(u_x(s))\cdot \mathrm{Re}(ie^{-is}v_0^*)ds$$

does not vanish at $x = 0$. Since

$$\mathrm{Re}(e^{-is}v_0) = \cos(s)v_1 + \sin(s)v_2 \quad \text{and} \quad \mathrm{Re}(ie^{-is}v_0^*) = \sin(s)v_3 - \cos(s)v_4$$

and

$$d^2H(0,0) = \begin{pmatrix} 0 & J \\ -J & I \end{pmatrix}$$

we find that

$$\begin{aligned} \Psi'(0) &= \int_0^{2\pi} d^2H(0,0)(\mathrm{Re}(e^{-is}v_0), \mathrm{Re}(ie^{-is}v_0^*))ds \\ &= 2\pi. \end{aligned}$$

This proves (10.1).

We end this section by using the methods of Section 6 to show that $q \equiv 0$. First observe that all the considerations in that section leading to (6.8) are still valid if the operator F additionally depends on a parameter. Hence, for elements $k \in \mathcal{K} = \ker L$, we have

$$\psi(k,\lambda,\tau) = \int_0^{2\pi} \omega\left(\frac{dk}{ds}, g(k,\lambda,\tau)\right) = 0. \tag{10.2}$$

Substituting g and using the $\mathbf{S}^1$-invariance of p and q we obtain

$$p(x^2,\lambda,\tau)\int_0^{2\pi}\omega\left(\frac{dx}{ds},x\right)ds + q(x^2,\lambda,\tau)\int_0^{2\pi}\omega\left(\frac{dx}{ds},ix\right)ds = 0.$$

We claim that the first integral is zero whereas the second integral is nonzero, which shows that q vanishes identically. By the definitions of c and d in (9.1) we have $Jd = c$, where J is the matrix associated with the underlying symplectic form ω. Therefore

$$\begin{aligned} \int_0^{2\pi} \omega(\mathrm{Re}(-ie^{-is}v_0), \mathrm{Re}(e^{-is}v_0^*))ds &= \pi\mathrm{Re}(\omega(-iv_0,\overline{v_0^*})) = 0, \quad \textit{and} \\ \int_0^{2\pi} \omega(\mathrm{Re}(-ie^{-is}v_0), \mathrm{Re}(ie^{-is}v_0^*))ds &= \pi\mathrm{Re}(\omega(-v_0,\overline{v_0^*})) = -\pi|v_0|^2 \neq 0, \end{aligned}$$

which proves the claim.

11 Derivatives in the reduced system

By the results of the previous section we know that the set of periodic solutions is locally given by solutions of the equation

$$p(x^2, \lambda, \tau)x = 0.$$

Theorem 7.1, which allows us to recognize the structure of the solution set of this equation in a neighborhood of $(0,0,0)$, follows directly from the next four lemmas.

Lemma 11.1

$$p_\tau(0,0,0) = 0.$$

Proof Observe that

$$g_{x\tau}(0,0,0) = p_\tau(0,0,0) + iq_\tau(0,0,0).$$

Using the formulas in Golubitsky and Schaeffer [1985, p. 33], and the fact that $F_\tau = dF_\tau = \frac{d}{ds}$, we obtain at 0

$$\begin{aligned} g_{x\tau} &= \langle v_0^*, dF_\tau \cdot v_0 - d^2F(v_0, L^{-1}PF_\tau)\rangle \\ &= \langle \mathrm{Re}(e^{-is}v_0^*), \frac{d}{ds}(\mathrm{Re}(e^{-is}v_0))\rangle \\ &= 0 \quad \text{(by (9.2))}. \end{aligned}$$ □

Lemma 11.2

$$p_\lambda(0,0,0) = -\frac{1}{2}\mathrm{Re}(v_0^* \cdot (df_\lambda)v_0).$$

Proof It follows from (8.3) that $F_\lambda(0,0) = 0$. We compute

$$\begin{aligned} g_{x\lambda} &= \langle v_0^*, dF_\lambda \cdot v_0 - d^2F(v_0, L^{-1}PF_\lambda)\rangle \\ &= -\langle v_0^*, df_\lambda \cdot v_0\rangle \\ &= -\frac{1}{4}\langle v_0^*, df_\lambda \cdot \overline{v_0}\rangle - \frac{1}{4}\langle \overline{v_0^*}, df_\lambda \cdot v_0\rangle \\ &= -\frac{1}{2}\mathrm{Re}(v_0^* \cdot df_\lambda \cdot v_0). \end{aligned}$$ □

Lemma 11.3

$$p_{\tau\tau}(0,0,0) = \overline{v_0^*} \cdot v_0^*.$$

Proof This time we use formulas (3.37) and (3.38) in Golubitsky and Schaeffer [1985] and compute

$$\begin{aligned} g_{x\tau\tau} &= \langle v^*, dF_{\tau\tau} \cdot v + 2dF_\tau \cdot w\rangle \\ &= 2\left\langle v^*, \frac{d}{ds}w\right\rangle \\ &= -2\left\langle \frac{dv^*}{ds}, w\right\rangle, \end{aligned}$$

where w is defined by

$$w = -L^{-1}PdF_\tau \cdot v.$$

In particular, w satisfies the ODE

$$Lw = -P\frac{dv}{ds} = -\frac{dv}{ds}. \tag{11.1}$$

The second equality holds since $\frac{dv}{ds} \in \mathcal{K}$ and since the kernel $\mathcal{K}$ is contained in the range $\mathcal{R}$ by the assumptions on A_0. We note that

$$\frac{dv}{ds} = -\frac{i}{2}\left(e^{-is}v_0 - e^{is}\overline{v_0}\right)$$

and make the ansatz

$$w(s) = \mathrm{Re}(e^{-is}b_1).$$

Since $v = \mathrm{Re}(e^{-is}v_0)$ it follows with (11.1) that b_1 has to satisfy

$$(A_0 + iI)b_1 = iv_0.$$

Hence, by the definition of v_0 and v_0^* (9.1) and the action of A_0 on v_3 and v_4 (8.2), we conclude that $b_1 = iv_0^*$ which yields

$$\begin{aligned}
-2\langle\frac{dv^*}{ds}, w\rangle &= -\frac{1}{2}\langle -iv_0^*, iv_0^*\rangle - \frac{1}{2}\langle\overline{iv_0^*}, -\overline{iv_0^*}\rangle \\
&= \frac{1}{2}\left(\langle iv_0^*, iv_0^*\rangle + \langle -\overline{iv_0^*}, -\overline{iv_0^*}\rangle\right) \\
&= \mathrm{Re}(-\overline{iv_0^*} \cdot iv_0^*) \\
&= \overline{v_0^*} \cdot v_0^*.
\end{aligned}$$

Therefore, $p_{\tau\tau} = \overline{v_0^*} \cdot v_0^* > 0$, as desired. □

Lemma 11.4

$$p_{x^2}(0,0,0) = -\frac{3}{2}\mathrm{Re}\left[\overline{v_0^*} \cdot \left(\frac{1}{4}d^3f(v_0, v_0, \overline{v_0}) + d^2f(v_0, b_0) + d^2f(\overline{v_0}, b_2)\right)\right],$$

where

$$\begin{aligned}
A_0 b_0 &= \frac{1}{2}d^2f(v_0, \overline{v_0}) \\
(A_0 - 2iI)b_2 &= \frac{1}{4}d^2f(v_0, v_0).
\end{aligned}$$

Proof The proof of this lemma follows exactly the method of the derivation of the cubic term in the bifurcation equation at a Hopf bifurcation. We follow the proof of Proposition 3.3 in Golubitsky and Schaeffer [1985, pp. 352, 355–356].

Using Golubitsky and Schaeffer [1985] (3.26), we have

$$\frac{\partial p}{\partial(x^2)} = \langle v^*, d^3F(v,v,v) + 3d^2F(v,w)\rangle$$

where

$$w = -L^{-1}Pd^2F(v,v).$$

Observe that $d^kF = -d^kf$ when $k > 1$. Recall that

$$v = \frac{1}{2}(e^{-is}v_0 + e^{is}\overline{v_0}) \quad \text{and} \quad v^* = \frac{1}{2}(e^{-is}v_0^* + e^{is}\overline{v_0^*}).$$

Since $\langle e^{ims}, e^{ins}\rangle = 0$ if $m \neq n$ it is easy to compute

$$\langle v^*, -d^3f(v,v,v)\rangle = -\frac{3}{8}\mathrm{Re}(\overline{v_0^*}\cdot d^3f(v_0,v_0,\overline{v_0})).$$

Similarly, one can verify that

$$\langle v^*, d^2f(v,v)\rangle = 0,$$

from which it follows that $d^2f(v,v) \in$ range L and hence that $Pd^2f(v,v) = d^2f(v,v)$. It also follows that we may assume

$$w = b_0 + e^{-2is}b_2 + e^{2is}\overline{b_2},$$

and we may use this form to solve

$$Lw = d^2f(v,v)$$

for $b_0 \in \mathbb{R}^{2n}$ and $b_2 \in \mathbb{C}^{2n}$. Indeed,

$$d^2f(v,v) = \frac{1}{4}\{e^{-2is}d^2f(v_0,v_0) + 2d^2f(v_0,\overline{v_0}) + e^{2is}d^2f(\overline{v_0},\overline{v_0})\}.$$

Combining the various terms, we have that

$$\langle v^*, d^2F(v,w)\rangle = -\frac{1}{2}\mathrm{Re}[\overline{v_0^*}\cdot(d^2f(v_0,b_0) + d^2f(\overline{v_0},b_2))],$$

where

$$A_0b_0 = \frac{1}{2}d^2f(v_0,\overline{v_0})$$

and

$$(A_0 - 2iI)b_2 = \frac{1}{4}d^2f(v_0,v_0). \qquad \square$$

12 Singularity theory

Having proved Theorem 7.1 we follow van der Meer [1985] in viewing the bifurcation equation (7.1) as a bifurcation problem in the perturbed period parameter τ with imperfection parameter λ. We follow Bridges [1990] in solving $p = 0$ by using $\mathbb{Z}_2$-equivariant singularity theory. Setting $\lambda = 0$ in (7.1) and using the facts

$$p(0) = 0, \quad p_\tau(0) = 0, \quad p_{\tau\tau}(0) > 0 \quad \text{and} \quad p_{x^2}(0) \neq 0$$

proved in Theorem 7.1, we can apply the determinacy results from $\mathbb{Z}_2$-equivariant singularity theory with a distinguished parameter (Proposition 3.4(b) Golubitsky and Schaeffer [1985], p. 259) to show that (7.1) with $\lambda = 0$ is $\mathbb{Z}_2$-equivalent to

$$(\pm x^2 + \tau^2)x = 0.$$

Next we use the $\mathbb{Z}_2$ universal unfolding theorem (Theorem 3.3 of Golubitsky and Schaeffer [1985], p. 259) and the fact that $p_\lambda(0) \neq 0$ to prove that the reduced bifurcation equation (7.1) is $\mathbb{Z}_2$-equivalent to the normal form (7.2).

Part III Periodic Orbits near Resonances

In this part, we study two questions: how to find periodic solutions in reversible systems near equilibria in $k:\ell$ resonance and how to find periodic solutions in Hamiltonian systems near equilibria with $k:\ell$ resonances. The answer to the first question relies on using symmetry — both the $\mathbf{S}^1$-symmetry of phase shift and the reversible symmetry — and (a slightly nonstandard) singularity theory to solve the reduced equations obtained through Liapunov-Schmidt reduction. We use implicit constraints and symmetry to reduce the answer for the second question to the answer for the first question.

These subjects have been well studied and we now describe the relation of the present approach to $k:\ell$ resonances in reversible and Hamiltonian systems with that of other authors. The survey in Arnol'd [1988] applies directly only to systems that are in Birkhoff normal form. Although, as is remarked there, a great deal can subsequently be deduced for systems that are subject to higher order perturbations, these deductions are not made explicit. Duistermaat [1983], whose work is largely based on Schmidt [1974], obtains similar results to ours but the singularity-theoretic calculations are more delicate because he works in a different context with a different equivalence relation. Consequently, his normal forms have one extra modal parameter for $k:1$ resonances with $k \geq 3$. In compensation, he obtains more delicate information, such as stabilities. A reference for the reversible case is Sevryuk [1988], but again the singularity theory is treated differently.

There are many earlier papers, including Palmore [1969], Meyer and Palmore [1970], Meyer and Schmidt [1971], Roels [1971a, b], Henrard [1973], Schmidt and Sweet [1973], Sweet [1973] and Schmidt [1973, 1978]. They differ from the present work in method — most of them use perturbation expansions and prove convergence, although Schmidt and Sweet [1973] and Sweet [1973] use the 'alternative method' — which is essentially the same as Liapunov-Schmidt reduction. Moreover, the results of Sweet [1973] apply to any system with a first integral, not just a Hamiltonian system. A good general source for the related method of averaging and further references, is Sanders and Verhulst [1985].

The most important difference with other works and the present one, is that in this paper we obtain normal forms that include the effect of a 'detuning parameter'. Of the references cited, only Schmidt [1973], Duistermaat [1983], and Arnol'd [1988] do this: all others consider only the dynamics *at* resonance. Moreover, we find *all* k, ℓ, and 1-families of periodic solutions near equilibrium, and our results apply to reversible systems as well as to Hamiltonian ones. Our singularity-theoretic methods differ considerably from those of all authors except Duistermaat [1983] and Arnol'd [1988]. The latter gives no details about this method, but it is clear that singularity theory is involved.

Another more technical difference is that most of the above authors assume that the Hamiltonian is real analytic — a property that is essential for the series methods they employ. The exceptions are Arnol'd [1988], Duistermaat [1983] and Roels [1971a]; the latter assumes that the Hamiltonian is kinetic plus potential where the potential is C^1 and has locally Lipschitz gradient.

Most authors identify which coefficients in the Hamiltonian are relevant and which nondegeneracy conditions must hold. Some obtain stability information, which we do not attempt here, though the methods of Montaldi, Roberts, and Stewart [1990b] could in principle be used and the calculations would be very similar to those of Duistermaat [1983].

For examples of reversible systems in mechanics that may or may not be Hamiltonian, and their stability and bifurcation properties, we refer to O'Reilly, Malhotra and Namachchivaya [1993] and references therein.

Another useful tool in the study of resonances is that of pinched spheres and invariants. When a resonant system is put into normal form, the Hamiltonian acquires an $\mathbf{S}^1$ symmetry, which can be used to perform reduction, leading to an analysis of the normal form. The reduced spaces are spheres for the 1:1 resonance and are 'pinched spheres' for higher order resonances. These spheres carry an interesting Lie-Poisson type of Poisson structure. The 'pinches' come about because of the isotropy of special points that occurs for higher order resonances. The systems on these spheres can often be analyzed rather effectively. References for this approach are Kummer [1979], Cushman and Rod [1982], Iwai [1985], Kummer [1990], David, Holm, and Tratnick [1990] and references therein. Knobloch, Mahalov, and Marsden [1993] contains an exposition of the Poisson structure of this problem.

Of course, the methods above often can be extended to multiple resonances. We do not discuss this aspect here, except for the obvious remark that our methods are not restricted to simple resonances. Some of the relevant literature is Duistermaat [1984], Sanders and Verhulst [1985] (and references therein), Hoveijn and Verhulst [1990], Haller and Wiggins [1992], and Haller [1993] (and references therein).

13 Preliminaries

We assume that the state space is $\mathbb{R}^{2n}$. Many different 'time-reversal' symmetries are possible in a general system of ODEs; here we consider only the time-reversal symmetry that takes the form

$$R(x,y) = (x,-y), \tag{13.1}$$

where $x, y \in \mathbb{R}^n$. Abstractly, we write the system of ODEs as

$$\frac{dX}{ds} = f(X,\lambda), \tag{13.2}$$

where $X = (x,y) \in \mathbb{R}^{2n}$ and $\lambda \in \mathbb{R}$. We assume that (13.2) has an equilibrium at $\lambda = 0$ which is R-invariant in the time-reversible case, and which without loss of generality we take to be the origin; that is,

$$f(0,0) = 0.$$

Let $A_0 = df_{0,0}$ be the Jacobian matrix of this equilibrium and let $0 < k < \ell$ be coprime integers. Assume

(R1) $\pm ki$ and $\pm \ell i$ are simple eigenvalues of A_0 and $\pm mi$ is not an eigenvalue of A_0 where $m \neq k, \ell$ is any nonnegative integer.

Hypothesis (R1) implies that A_0 is invertible, so by the implicit function theorem there is a unique branch of equilibria $X(\lambda)$ with $X(0) = 0$. Therefore

$$f(X(\lambda),\lambda) \equiv 0.$$

For Hamiltonian systems we can change coordinates directly in X to assume $X(\lambda) \equiv 0$ and

$$f(0,\lambda) \equiv 0. \tag{13.3}$$

Slightly more care is needed to arrive at the same result in the time-reversible case. Recall that time-reversibility implies

$$f(RX,\lambda) = -Rf(X,\lambda) \tag{13.4}$$

$$A_0 R = -RA_0. \tag{13.5}$$

Hence

$$f(RX(\lambda),\lambda) \equiv 0,$$

and uniqueness of solutions obtained by the implicit function theorem implies that

$$RX(\lambda) = X(\lambda).$$

Now a change of coordinates that preserves time-reversibility leads to (13.3).

At resonance we may always factor out common denominators and rescale time so that the resonant eigenvalues are integer multiples of i, and we assume from now on that this has been done. In Hamiltonian systems it is well known that generically purely imaginary eigenvalues can pass through resonances as a parameter is varied. To verify the corresponding statement for time-reversible systems, we must show that eigenvalues in these systems are also locked onto the imaginary axis. This is also well known — but *less* well known — so we now give details.

In so doing, we discuss the reasonableness of the assumption (13.1) as well as its consequences. Let $V^+ = \{(x,0)\}$ be the $+1$ eigenspace of R and let $V^- = \{(0,y)\}$ be the -1 eigenspace of R. The identity (13.5) implies that

$$A_0 : V^+ \to V^- \quad \text{and} \quad A_0 : V^- \to V^+.$$

Observe that V^+ and V^- are invariant subspaces under A_0^2. Hypothesis (R1) implies that A_0 is invertible, so that $A_0^2|V^+$ and $A_0^2|V_-$ are similar matrices, that is,

$$A_0^2|V^+ = (A_0|V^+)^{-1}(A_0^2|V^-)(A_0|V^+).$$

Hence these restrictions have the same eigenvalues. It follows that the eigenvalues of A_0 are precisely the square roots of the eigenvalues of $A_0^2|V^+$, and the eigenvalues of A_0 fall into three classes:

(i) $\pm\sigma; \quad \sigma > 0,$

(ii) $\pm\mu i; \quad \mu > 0,$

(iii) $\pm\sigma \pm \mu i; \quad \sigma,\mu > 0,$

corresponding to whether the eigenvalues of $A_0^2|V^+$ are positive, negative or complex. It follows that the existence of purely imaginary eigenvalues of A_0 is generic.

We now show that the resonance hypothesis (R1) is a codimension one assumption in time-reversible systems. Let

$$A_\lambda = df_{0,\lambda}.$$

It follows that as λ is varied the purely imaginary eigenvalues can (generically be expected to) pass through $k:\ell$ resonances with eigenvalues $\pm ki$,$\pm \ell i$ when $\lambda = 0$. Generically these eigenvalues are simple. By rescaling time we can assume that the appropriate eigenvalues of A_λ, for all λ near zero, are $\pm ki$ and $\pm\omega(\lambda)i$ where ω is smooth in λ and $\omega(0) = \ell$. We assume that A_0 has no other resonant eigenvalues $\pm mi$, where $m \neq k,\ell$ is a non-negative integer; and that the critical eigenvalues pass through resonance with nonzero speed, that is,

(R2) $\omega'(0) \neq 0$.

It is well known that $\mathbf{S}^1$ has two families of nontrivial symplectic representations, in which

$$\theta \cdot z = e^{\ell i\theta} z$$

or

$$\theta \cdot z = e^{-\ell i\theta} z$$

where $\ell \geq 1$, $z \in \mathbb{C}$, and $\mathbb{C}$ carries its standard symplectic structure. Call these representations ρ_ℓ and $\rho_{-\ell}$. The map $z \mapsto \overline{z}$ shows that ρ_ℓ and $\rho_{-\ell}$ are isomorphic as abstract representations. However, this isomorphism does not preserve the symplectic structure — see Montaldi, Roberts and Stewart [1988] and Dellnitz and Melbourne [1993]. It is usual to distinguish between these two representations by referring to a $k:\ell$ resonance or a $k:-\ell$ resonance. More precisely, let (z_1, z_2) be complex coordinates on $\mathbb{C}^2$. Without loss of generality the $\mathbf{S}^1$-action on z_1 is by ρ_k where $k \geq 1$. Then the $\mathbf{S}^1$-action on z_2 is either ρ_ℓ or $\rho_{-\ell}$ where $\ell \geq 1$. The first case is a $k:\ell$ resonance and the second a $k:-\ell$ resonance.

14 Normal forms with $\mathbf{S}^1$-Symmetry

We now find periodic solutions to (13.2) with period approximately 2π using a Liapunov-Schmidt reduction. Let

$$F: C^1_{2\pi} \times \mathbb{R} \times \mathbb{R} \to C_{2\pi}$$

be defined by

$$F(u(s), \lambda, \tau) = (1+\tau)\frac{du}{ds} - f(u(s), \lambda),$$

and solve $F = 0$ by Liapunov-Schmidt reduction in the standard way. The resonance hypothesis $(R1)$ implies that the kernel of $dF_{0,0,0}$ is four-dimensional. Thus we can write the reduced equation as

$$g: \mathbb{C}^2 \times \mathbb{R} \times \mathbb{R} \to \mathbb{C}^2$$

and solving $g(z, \lambda, \tau) = 0$ yields all small-amplitude periodic solutions of (13.2) with period near 2π.

The $\mathbf{S}^1$ phase shift symmetry that acts on periodic functions in $\mathcal{C}_{2\pi}$ restricts the form of g considerably. Our eigenvalue assumptions imply that on the above kernel, $\theta \in \mathbf{S}^1$ acts by

$$\theta(z_1, z_2) = (e^{ki\theta} z_1, e^{\ell i\theta} z_2),$$

where $k \geq 1$ and ℓ (which may be positive or negative) are coprime and g commutes with this action. A calculation from invariant theory determines the general form of g. We state the result for the $k:\ell$ resonance in the following lemma, and discuss the $k:-\ell$ resonance afterwards.

Lemma 14.1 *Assume that $g: \mathbb{C}^2 \to \mathbb{C}^2$ is equivariant with respect to the $k:\ell$ action of $\mathbf{S}^1$. Then*

$$g(z_1, z_2) = (P_1 z_1 + Q_1 \overline{z_1}^{\ell-1} z_2^k, P_2 z_2 + Q_2 z_1^\ell \overline{z_2}^{k-1}) \tag{14.1}$$

where P_1, P_2, Q_1, Q_2 are complex-valued $\mathbf{S}^1$-invariant functions of the following three of the four $\mathbf{S}^1$-invariants

$$|z_1|^2, |z_2|^2, \mathrm{Re}(z_1^\ell \overline{z_2}^k).$$

Moreover, P_1, P_2, Q_1, Q_2 are uniquely determined by g.

Proof Write $g = (g_1, g_2)$ in coordinates. We derive the form of g_1; the derivation for g_2 is similar. Equivariance implies that

$$g_1(z_1, z_2) = e^{-ki\theta} g(e^{ki\theta} z_1, e^{\ell i\theta} z_2).$$

Write the Taylor series for g_1 as

$$g_1(z_1, z_2) = \sum A_{abcd} z_1^a \overline{z_1}^b z_2^c \overline{z_2}^d$$

and use the invariance condition to show that the only terms that survive are the ones for which

$$k(a - b - 1) + \ell(c - d) = 0.$$

Collect terms in the Taylor expansion of g_1 that have a common factor of $z_1\overline{z_1}$ or $z_2\overline{z_2}$ to write g_1 in the form

$$g_1(z_1, z_2) = \sum B_{abcd}(|z_1|^2, |z_2|^2) z_1^a \overline{z_1}^b z_2^c \overline{z_2}^d,$$

where two of the indices a, b, c, d are zero. Since k and ℓ are coprime and non-negative we conclude that

$$g_1(z_1, z_2) = A(|z_1|^2, |z_2|^2, z_1^\ell \overline{z_2}^k) z_1 + B(|z_1|^2, |z_2|^2, \overline{z_1}^\ell z_2^k) \overline{z_1}^{\ell-1} z_2^k.$$

We now assert that g_1 can be put into the form

$$g_1(z_1, z_2) = C(|z_1|^2, |z_2|^2, \mathrm{Re}(z_1^\ell \overline{z_2}^k)) z_1 + D(|z_1|^2, |z_2|^2, \mathrm{Re}(z_1^\ell \overline{z_2}^k)) \overline{z_1}^{\ell-1} z_2^k.$$

To verify this assertion observe that

$$\begin{aligned}(z_1^\ell \overline{z_2}^k)^m z_1 &= ((z_1^\ell \overline{z_2}^k + \overline{z_1}^\ell z_2^k) - \overline{z_1}^\ell z_2^k)^m z_1 \\ &= \sum_{i+j=m} k_{ij} [\mathrm{Re}(z_1^\ell \overline{z_2}^k)]^i (\overline{z_1}^\ell z_2^k)^j z_1.\end{aligned}$$

Then note that when $j > 0$

$$(\overline{z_1}^\ell z_2^k)^j z_1 = (\overline{z_1}^\ell z_2^k)^{j-1} |z_1|^2 \overline{z_1}^{\ell-1} z_2^k.$$

Similarly,

$$\begin{aligned}(\overline{z_1}^\ell z_2^k)^m \overline{z_1}^{\ell-1} z_2^k &= ((\overline{z_1}^\ell z_2^k + z_1^\ell \overline{z_2}^k) - z_1^\ell \overline{z_2}^k)^m \overline{z_1}^{\ell-1} z_2^k \\ &= \sum_{i+j=m} k_{ij} [\mathrm{Re}(z_1^\ell \overline{z_2}^k)]^i (z_1^\ell \overline{z_2}^k)^j \overline{z_1}^{\ell-1} z_2^k.\end{aligned}$$

Then note that when $j > 0$,

$$(z_1^\ell \overline{z_2}^k)^j \overline{z_1}^{\ell-1} z_2^k = (z_1^\ell \overline{z_2}^k)^{j-1} |z_1|^{2\ell-2} |z_2|^{2k} z_1.$$

A double induction completes the proof of the assertion.

To complete the proof of the lemma we verify that the invariant functions P_1, P_2, Q_1, Q_2 are unique. To see this assume that

$$P_1 z_1 + Q_1 \overline{z_1}^{\ell-1} z_2^k \equiv 0.$$

Using $\mathbf{S}^1$ symmetry we can assume that $z_1 = x_1$ is real and obtain

$$P_1 x_1 + Q_1 x_1^{\ell-1} z_2^k \equiv 0. \tag{14.2}$$

Since P_1 and Q_1 are clearly invariant under the action $z_2 \mapsto \overline{z_2}$ it follows that

$$P_1 x_1 + Q_1 x_1^{\ell-1} \overline{z_2}^k \equiv 0. \tag{14.3}$$

Subtracting (14.3) from (14.2) yields

$$Q_1 x_1^{\ell-1} \mathrm{Im}(z_2^k) \equiv 0,$$

so that $Q_1 \equiv 0$, and hence $P_1 \equiv 0$ as desired. □

For the $k:-\ell$ resonance the results are similar, but z_2 must be replaced by $\overline{z_2}$ and conversely. Clearly zeros of (14.1) are in one-to-one correspondence with zeros of the corresponding mapping with z_2 and $\overline{z_2}$ interchanged, so the zeros for the $k:-\ell$ resonance can be read off from those for the corresponding $k:\ell$ resonance. When seeking periodic orbits of the original Hamiltonian system, determined by these zeros, we may therefore confine attention to the $k:\ell$ resonance. We do so from now on.

15 Reduced equations in the time-reversible case

The form of g is further restricted by time-reversibility. It can be checked that the time-reversal symmetry R preserves the kernel and cokernel of dF and hence anticommutes with g, that is,

$$g(\overline{z_1}, \overline{z_2}) = -\overline{g(z_1, z_2)}.$$

This anticommutativity allows us prove a stronger version of Lemma 14.1.

Lemma 15.1 *Time-reversal symmetry implies that the invariant functions P_1, P_2, Q_1, Q_2 are purely imaginary.* □

Set $P_j = iP_j^i$ and $Q_j = iQ_j^i$. After dividing by i we see that finding periodic solutions to the differential equation (13.2) is equivalent to finding solutions to the following system of two complex equations:

$$\begin{aligned} P_1^i z_1 + Q_1^i \overline{z_1}^{\ell-1} z_2^k &= 0 && (15.1)\\ P_2^i z_2 + Q_2^i z_1^\ell \overline{z_2}^{k-1} &= 0, && (15.2) \end{aligned}$$

where $P_1^i, P_2^i, Q_1^i, Q_2^i$ are real-valued functions of $|z_1|^2, |z_2|^2, \mathrm{Re}(z_1^\ell \overline{z_2}^k), \lambda$ and τ.

We begin by finding the one-parameter family of periodic solutions that is always present by virtue of the Liapunov center theorem (for time-reversible systems). Suppose that $z_1 = 0$. Then (15.1) is satisfied and (15.2) reduces to $P_2^i z_2 = 0$. Thus either $z_2 = 0$ (the trivial equilibrium) or $P_2^i = 0$. Since $P_{2,\tau}^i(0) \neq 0$ we can use the implicit function theorem to solve $P_2^i = 0$ for τ and obtain a family of periodic solutions with period near $2\pi/\ell$. We refer to this family as the *ℓ-family* of solutions. Since $\ell > k$ the eigenvalues $\pm \ell i$ are nonresonant.

We may now assume that $z_1 \neq 0$. Applying the $\mathbf{S}^1$ action we further assume that $z_1 = x_1 > 0$ where $x_1 \in \mathbb{R}$. Dividing by x_1 in (15.1), shows that the remaining periodic solutions may be found by solving

$$\begin{aligned} P_1^i + Q_1^i x_1^{\ell-2} z_2^k &= 0 && (15.3)\\ P_2^i z_2 + Q_2^i x_1^\ell \overline{z_2}^{k-1} &= 0. && (15.4) \end{aligned}$$

Solving the imaginary part of (15.3) leads to the equation $Q_1^i \mathrm{Im}(z_2^k) = 0$. Since Q_1^i is the coefficient of a higher order term in the reduced equation (a term of order at least two) we can make the genericity hypothesis:

(R3) $Q_1^i(0) \neq 0$.

It follows from (R3) that

$$\mathrm{Im}(z_2^k) = 0 \tag{15.5}$$

at solutions.

Next, since $P_{1,\tau}^i(0) \neq 0$, we can use the implicit function theorem to solve the real part of (15.3) for τ and then substitute this solution for τ in (15.4). Thus finding the desired families of periodic solutions reduces to solving (15.4).

Equation (15.5) implies that $z_2^k \in \mathbb{R}$, that is,

$$z_2 = x_2 e^{mi\pi/k} \tag{15.6}$$

where m is an integer, $x_2 \in \mathbb{R}$ and $x_2 \geq 0$. Now the group generated by $e^{2\pi i/k} \in \mathbf{S}^1$ fixes x_1, so up to symmetry (15.6) reduces to two cases

$$\begin{aligned} z_2 &= x_2 \\ z_2 &= x_2 e^{i\pi/k}. \end{aligned}$$

It follows that $|z_2|^2 = x_2^2$ and $z_2^k = \epsilon x_2^k$ where $\epsilon = \pm 1$.

We divide our analysis into two cases: $k > 1$ and $k = 1$. However, note that when k is odd, we could have used the discrete symmetries to conjugate z_2 to $x_2 \in \mathbb{R}$; that is, we can permit x_2 to be negative as well as positive. This has the advantage of eliminating the parameter $\epsilon = \pm 1$ from the problem. We will use this observation when considering the case $k = 1$.

The case $k > 1$ When $k > 1$ we can set $z_2 = 0$ to obtain a family of solutions directly, because (15.4) is automatically satisfied. Equation (15.3) (which we solved previously for τ) now reduces to $P_1^i = 0$. We call this the *k-family* of solutions. It consists of solutions with period near $2\pi/k$ and is given by the nonresonant eigenvalues $\pm ki$ in the Liapunov center theorem.

Next we assume that $z_2 \neq 0$ and multiply (15.4) by $\overline{z_2}$ to get

$$P_2^i |z_2|^2 + Q_2^i x_1^\ell \overline{z_2}^k = 0. \tag{15.7}$$

The imaginary part of (15.7) is just

$$Q_2^i \mathrm{Im}(z_2^k) = 0$$

which holds by (15.5). Indeed, (15.7) may be rewritten as

$$P_2^i x_2^2 + \epsilon Q_2^i x_1^\ell x_2^k = 0.$$

Since $x_2 \neq 0$ and $k \geq 2$ we may divide by x_2^2 to obtain

$$P_2^i + \epsilon Q_2^i x_1^\ell x_2^{k-2} = 0, \tag{15.8}$$

where $\ell \geq 3$.

Since P_2^i and Q_2^i are functions of $x_1^2, x_2^2, \epsilon x_1^\ell x_2^k, \lambda, \tau$ and $\tau = \tau(x_1^2, x_2^2, \epsilon x_1^{\ell-2} x_2^k, \lambda)$, it follows that (15.8) has the form

$$p^\epsilon(x_1, x_2, \lambda) = r^\epsilon(x_1^2, x_2^2, \lambda) + s^\epsilon(x_1^2, x_2^2, \lambda) x_1^{\ell-2} x_2^{k-2} \tag{15.9}$$

where $s^\epsilon(0,0,\lambda) = 0$ and $\epsilon = \pm 1$ yields two distinct functions p^ϵ. It may be verified that the lowest order terms in r^ϵ do not depend on ϵ, that is,

$$r^\epsilon(x_1^2, x_2^2, \lambda) = a x_1^2 + b x_2^2 + c\lambda + \ldots$$

where a, b, c do not depend on ϵ. (The lowest order terms in s^ϵ do depend on ϵ.)

The case $k = 1$ When $k = 1$ the previous discussion implies that we must solve

$$P_2^i x_2 + Q_2^i x_1^\ell = 0 \tag{15.10}$$

where P_2^i, Q_2^i are functions of x_1^2, x_2^2, $x_1^\ell x_2$, λ, τ, and $\tau = \tau(x_1^2, x_2^2, x_1^{\ell-2} x_2, \lambda)$. Since $k = 1$ is odd, we may assume $x_2 \in \mathbb{R}, x_2 \neq 0$ rather than considering the two cases $\epsilon = \pm 1$. Also, note that generically (when $Q_2(0) \neq 0$) there is no 1-family when $x_2 = 0$ (since $x_2 = 0$ then implies that $x_1 = 0$). However, we will find 1-families of solutions when $x_2 \neq 0$.

As in the case $k > 1$ we can rewrite (15.10) uniquely as

$$p(x_1, x_2, \lambda) = r(x_1^2, x_2^2, \lambda) x_2 + s(x_1^2, x_2^2, \lambda) x_1^{\ell-2} = 0, \tag{15.11}$$

where $s(0,0,\lambda) = 0$.

16 Reduced equations in the Hamiltonian case

In this section we show that at a point of $k : \ell$ resonance in a Hamiltonian system we can also find time-periodic solutions by solving a system analogous to the one used to find time-periodic solutions in the reversible case. To establish this, we control the Liapunov-Schmidt reduced equations by using the extra structure imposed by the Hamiltonian character instead of using the time-reversal symmetry, as in Lemma 15.1.

We consider the equation $\dot{u} = f(u) = X_H(u)$ on a symplectic vector space isomorphic to $\mathbb{R}^{2n} = \mathbb{C}^n$, as before (later we will add parameters) and make assumptions similar to the reversible case:

(RH1) x_0 is a fixed point: $f(x_0) = 0$. Let $A_0 = df(x_0)$ be the linearization at x_0 and let k and ℓ be coprime positive integers.

(RH2) $\pm k\omega_0 i$ and $\pm \ell\omega_0 i$ are simple eigenvalues of A_0 and $p\omega_0 i$ is not an eigenvalue for p a nonnegative integer other than k or ℓ. Let v_k and v_ℓ be the eigenvectors of A_0 for the eigenvalues $k\omega_0 i$ and $\ell\omega_0 i$ respectively.

By (RH1) the function H has a critical point at the point x_0. By (RH2) the second derivative is nondegenerate on the four-dimensional eigenspace.

We can assume that $0 < k < \ell$, as in the reversible case, and that $\omega_0 = 1$ by rescaling time. Now apply the Liapunov-Schmidt procedure to the map F. Let $\mathcal{X} = \mathcal{C}^1_{2\pi} \times \mathbb{R}$ and $\mathcal{Y} = \mathcal{C}^0_{2\pi}$ (together with bifurcation parameters to be included later). Let $\mathbf{S}^1$ act on the first two components of $\mathcal{X}$ by $(\theta \cdot u)(s) = u(s + \theta)$ and similarly on $\mathcal{Y}$. Note that F is $\mathbf{S}^1$-equivariant.

The derivative of F at $(u, \tau) = (x_0, 0)$ is L, where

$$Lv \equiv dF(x_0, 0) \cdot v = \frac{dv}{ds} - A_0 v.$$

The kernel of $dF(x_0, 0)$ coincides in this case (unlike Hamiltonian Hopf) with the eigenspace of the four eigenvalues and therefore is symplectic and is even the symplectic sum of two two-dimensional symplectic subspaces. Thus, we will proceed by using the condition (S). By (RH2), $\ker dF(x_0, 0)$ is spanned by the four periodic functions

$$\mathrm{Re}(e^{iks} v_k),\ \mathrm{Im}(e^{iks} v_k),\ \mathrm{Re}(e^{i\ell s} v_\ell),\ \mathrm{Im}(e^{i\ell s} v_\ell).$$

Exactly as in our treatment of the Liapunov center theorem, we can use the complex structure already on phase space to rephrase this as the space spanned by the loops $e^{iks} z v_k$ and $e^{i\ell s} z v_\ell$. Thus we can identify the kernel of dF with $\mathbb{C}^2$; explicitly,

$$(z_1, z_2) \mapsto (z_1 e^{iks} v_k, z_2 e^{i\ell s} v_\ell).$$

The same calculation as in the case of the Liapunov center theorem shows that the symplectic structure on the kernel coincides with the standard symplectic structure on $\mathbb{C}^2$. We may choose coordinates on $\mathbb{C}^2$ so that the matrix A_0 has the block form

$$\begin{pmatrix} 0 & k & 0 & 0 \\ -k & 0 & 0 & 0 \\ 0 & 0 & 0 & \ell \\ 0 & 0 & -\ell & 0 \end{pmatrix}$$

and the second derivative $d^2H(0)$ is

$$\begin{pmatrix} \alpha & 0 & 0 & 0 \\ 0 & \alpha & 0 & 0 \\ 0 & 0 & \beta & 0 \\ 0 & 0 & 0 & \beta \end{pmatrix}$$

where α and β are nonzero constants. In these coordinates, the components of v_k are $(1, i, 0, 0, 0, \ldots, 0)$ and those of v_ℓ are $(0, 0, 1, i, 0, \ldots, 0)$.

Because condition (S) holds, the kernel of L^* equals the kernel of L and is isomorphic to $\mathbb{C}^2$ by means of

$$(z_1, z_2) \mapsto (z_1 e^{iks} v_k^*, z_2 e^{i\ell s} v_\ell^*),$$

where v_k^* is the eigenvector for A_0^* given by $A_0^* v_k^* = -ik^*$ and similarly for v_ℓ^*. The first four components of v_k^* and v_k are the same, as are those of v_ℓ^* and v_ℓ.

The Liapunov-Schmidt procedure in the context of assumption (S), may now be applied. Recall that we use the standard inner product on $\mathbb{C}^n$, and that $L^* v = -v' - A_0^* v$, $\mathcal{M} = \mathrm{range}\, L^*$, $\mathcal{K} = \ker L$, and $\mathcal{N} = \ker L^*$. The bifurcation equation for F has the form

$$g: \mathbb{C}^2 \times \mathbb{R} \times \mathbb{R} \to \mathbb{C}^2$$

and, being $\mathbf{S}^1$-equivariant, g has the form

$$\begin{aligned} g_1 &= P_1 z_1 + Q_1 \overline{z_1}^{\ell-1} z_2^k, \\ g_2 &= P_2 z_2 + Q_2 z_1^\ell \overline{z_2}^{k-1}. \end{aligned}$$

Of course, the bifurcation equations will be $g = 0$. By Theorem 6.2, the mapping g is a Hamiltonian vector field with an $\mathbf{S}^1$-invariant Hamiltonian H. Being invariant, the function H can be written as a function of the invariants which we shall denote $a = z_1\overline{z_1}$, $b = z_2\overline{z_2}$, $c = \frac{1}{2}(z_1^\ell\, \overline{z_2}^k + \overline{z_1}^\ell z_2^k)$, $d = -\frac{i}{2}(z_1^\ell \overline{z_2}^k - \overline{z_1}^\ell z_2^k)$, and the parameters λ and τ.

It is a standard and easily verified fact that a Hamiltonian vector field g with components g_j on $\mathbb{C}^2$ has the form

$$g_j = -2i\frac{\partial H}{\partial \overline{z_j}}$$

where we regard the Hamiltonian as a complex-valued (smooth) function of z_1 and $\overline{z_1}$, z_2, $\overline{z_2}$, λ and τ.

Putting these facts together, we find that

$$\begin{aligned} g_1 &= -2iH_a z_1 + \ell[H_d - H_c i]\overline{z_1}^{\ell-1} z_2^k \\ g_2 &= -2iH_b z_2 - k[H_d + H_c i] z_1^\ell \overline{z_2}^{k-1}. \end{aligned}$$

André Vanderbauwhede pointed out that these equations may be solved in a fashion similar to those obtained in the time-reversible case (15.8)–(15.11). Setting either $z_1 = 0$ or $z_2 = 0$ gives the $2\pi/k$ and $2\pi/\ell$ families of solutions as in Section 15. We assume that $z_1 \neq 0$ and $z_2 \neq 0$. Phase-shift $\mathbf{S}^1$ symmetry allows us to set $z_1 = x_1 > 0$. As in the time-reversible case, we solve for τ by using the implicit function theorem to solve $\mathrm{Im}(g_1) = 0$. Indeed,

$$\mathrm{Im}(g_1) = x_1[-2H_a + \ell x_1^{\ell-2}\mathrm{Im}([H_d - iH_c]z_2^k)] = 0$$

can be solved for τ since $H_{a,\tau} \neq 0$ (recall $P^i_{1,\tau}(0) \neq 0$).

Next note that $\mathrm{Re}(g_1) = 0$ and $\mathrm{Re}(g_2) = 0$ are identical equations. We can write the two remaining equations as $\mathrm{Re}(g_2\overline{z_2}) = 0$ and $\mathrm{Im}(g_2\overline{z_2}) = 0$. Explicitly

$$\begin{aligned} \mathrm{Re}(g_2\overline{z_2}) &= -kx_1^\ell \mathrm{Re}([H_d + iH_c]\overline{z_2}^k) = 0 && (16.1) \\ \mathrm{Im}(g_2\overline{z_2}) &= -2H_b|z_2|^2 - kx_1^\ell \mathrm{Im}([H_d + iH_c]\overline{z_2}^k) = 0. && (16.2) \end{aligned}$$

Equation (16.1) implies that

$$(H_d + iH_c)\overline{z_2}^k = \pm\sqrt{H_d^2 + H_c^2}|z_2|^k i.$$

Hence (16.2) may be rewritten as

$$2H_b|z_2|^2 \pm kx_1^\ell\sqrt{H_d^2 + H_c^2}|z_2|^k = 0. \tag{16.3}$$

Assuming the nondegeneracy assumption (recall (R3) in Section 15)

$$H_d(0) \neq 0 \quad \text{or} \quad H_c(0) \neq 0$$

implies that $\sqrt{H_d^2 + H_c^2}$ is smooth. Equation (16.3), upon division by $|z_2|$ or $|z_2|^2$ depending on whether $k = 1$ or $k > 1$, yields equations of the form (15.9) or (15.11). Once (16.3) is solved for $|z_2|$, the phase of z_2 is found using (16.1).

17 The case $k > 1$

We find solutions to (15.9) by using singularity theory to determine all small amplitude solutions. The singularity-theoretic results required in this context — such as the 'tangent space constant' theorem and the universal unfolding theorem — are a consequence of general singularity theory results proved in Damon [1984].

The singularity theory calculations are most easily handled using the 'invariant' functions r and s. (We omit the superscript ϵ in (15.9) in this discussion.) The changes of coordinates that we consider are given as follows. First, let

$$u = x_1^2 \quad \text{and} \quad v = x_2^2, \tag{17.1}$$

and consider changes of coordinates in the domain of p of the form

$$(x_1, x_2) \mapsto (x_1 X_1(u,v), x_2 X_2(u,v)), \tag{17.2}$$

where $X_1(0,0) > 0$ and $X_2(0,0) > 0$. These changes of coordinates preserve the x_1 and x_2 axes and the positive quadrant. We also allow multiplication of p by functions of the form

$$S(u,v) + R(u,v) x_1^{\ell-2} x_2^{k-2}$$

where $S(0,0) \neq 0$. It is a simple exercise to check that these transformations preserve the form of p and thus can be thought of as operations on the pair $(r(u,v), s(u,v))$, as is done for invariant functions in equivariant singularity theory.

In the context of the theory developed in Golubitsky and Schaeffer [1985], it is straightforward to compute the tangent space $T(p)$ for these coordinate changes. One finds that $T(p)$ is a module of function pairs in $(\mathcal{E}_{u,v}, \mathcal{M}_{u,v})$ where $\mathcal{E}_{u,v}$ is the ring of germs of smooth, real-valued functions in the variables u, v and $\mathcal{M}_{u,v} \subset \mathcal{E}_{u,v}$ is the maximal ideal generated by functions vanishing at the origin. This module has four generators:

$$(r,s), (u^{\ell-2} v^{k-2} s, r), (2ur_u, 2us_u + (\ell-2)s), (2vr_v, 2vs_v + (k-2)s). \tag{17.3}$$

Theorem 17.1 *Assume that $\ell > k \geq 2$. If*

$$r_u, r_v, r_u s_v - r_v s_u \tag{17.4}$$

are nonzero at the origin, then p is equivalent to the normal form

$$r(u,v) = u + \epsilon_1 v, \quad s(u,v) = \epsilon_2 u \tag{17.5}$$

where $\epsilon_1 = \operatorname{sgn}(r_u(0) r_v(0)) = \pm 1$ and $\epsilon_2 = -\epsilon_1 \operatorname{sgn}(r_u s_v - r_v s_u) = \pm 1$. This singularity has codimension one and a universal unfolding is

$$r(u,v,\lambda) = u + \epsilon_1 v + \lambda, \quad s(u,v) = \epsilon_2 u. \tag{17.6}$$

Proof Under the assumptions (17.4), $T(p) = (\mathcal{M}_{u,v}, \mathcal{M}_{u,v})$. Hence terms of degree two and above in r and s may be eliminated by coordinate changes. Moreover, the 'tangent space constant' theorem allows us to transform s to $s(u,v) = \pm u$. Rescaling gives the appropriate normalizations to r. □

We now discuss the zero sets of p^ϵ which has the form

$$p^\epsilon = X_1^2 + \epsilon_1 X_2^2 + \Lambda + \cdots,$$

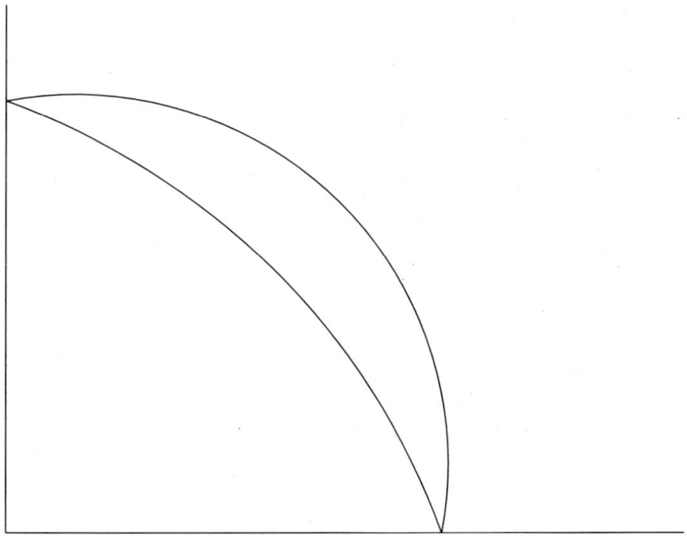

Figure 3 $\ell > k \geq 2$ and $\epsilon_1 = +1$. Equation $x_1^2 + x_2^2 + \lambda \pm x_1^\ell x_2^{k-2} = 0$ when $\lambda < 0$.

where $X(x_1, x_2, \lambda) = (X_1, X_2)$ and $\Lambda(\lambda)$ are invertible changes of coordinates. Note from (15.8) that $p^+ = 0$ and $p^- = 0$ have the same solutions when either $x_1 = 0$ or $x_2 = 0$. Note also that the sign ϵ_1 is the same for p^+ and p^- since ϵ affects only higher order terms. Hence the zero set, restricted to $x_1 > 0$, $x_2 > 0$, is as shown in Figure 3 for $\epsilon_1 = +1$ and $\lambda < 0$. The figures for $\lambda < 0$ and $\epsilon = -1$ may be drawn in a similar fashion.

18 The case $k = 1$

Using singularity theory we determine all small amplitude solutions of (15.11). Since we do not need to preserve the first quadrant, we can allow a more general system of coordinate changes than we did in the previous section. The coordinate changes that we allow are:

$$(x_1, x_2) \mapsto (x_1 X_1(u, v), x_2 X_2(u, v) + x_1^{\ell-2} X_3(u, v)) \tag{18.1}$$

where $X_2(0, 0) \neq 0$. These changes of coordinates still preserve the x_2 axis. Note that when $\ell = 2$ we must require that $X_3(0, 0) = 0$ since we always assume that $s(0, 0, 0) = 0$ in (15.11). (When $\ell = 3$ we will enlarge the system of coordinate changes even further, see (18.10).) We also allow multiplication of p by functions of the form

$$S(u, v) + R(u, v) x_1^{\ell-2} x_2$$

where $S(0, 0) \neq 0$. It is an exercise to check that these transformations preserve the form of p and hence can be thought of as operations on the pair $(r(u, v), s(u, v))$, as is done for invariant functions in equivariant singularity theory. However, it is not easy to write down the precise way that (r, s) is transformed by these changes of coordinates.

It is also a standard exercise to compute the tangent space $T(p)$ for these coordinate changes. One finds that $T(p)$ is a module of function pairs in $(\mathcal{E}_{u.v}, \mathcal{M}_{u.v})$ where $\mathcal{E}_{u,v}$ is the ring of germs of smooth, real-valued functions in the variables

u, v and $\mathcal{M}_{u,v} \subset \mathcal{E}_{u,v}$ is the maximal ideal generated by functions vanishing at the origin. This module has four generators:

$$(r,s), (u^{\ell-2}s, vr), (2ur_u, 2us_u + (\ell-2)s), (2vr_v + r, 2vs_v), (2u^{\ell-2}s_v, 2vr_v + r) \tag{18.2}$$

when $\ell > 3$. When $\ell = 2$ the last generator must be replaced by u and v times that generator.

As indicated, we distinguish three subcases: $\ell \geq 4$, $\ell = 2$ and $\ell = 3$. To complete the singularity theory for the case $\ell = 3$ we must modify the discussion slightly.

Theorem 18.1 *Assume that $\ell \geq 4$. If*

$$r_u, r_v, s_u, s_v, r_u s_v - 3r_v s_u \tag{18.3}$$

are nonzero at the origin, then p is equivalent to the normal form

$$r(u,v) = u + \epsilon_1 v, \quad s(u,v) = u \tag{18.4}$$

where $\epsilon_1 = \pm 1$. This singularity has codimension one and a universal unfolding is

$$r(u,v,\lambda) = u + \epsilon_1 v + \lambda, \quad s(u,v) = u. \tag{18.5}$$

Proof Under assumptions (18.3), $(\mathcal{M}^2_{u,v}, \mathcal{M}^2_{u,v}) \subset \mathcal{M}_{u,v} T(p)$. The second generator in (18.2) is not used in this calculation. The standard 'tangent space constant' theorem implies that p is 2-determined. Rescaling x_1 and x_2 and premultiplying p by $S(0,0)$ produces the normal form (18.4). Since $x_2 \neq 0$ when $k = 1$ we may allow $X_2(0,0)$ to be nonzero — not just positive. Finally, we can now compute $T(p)$ exactly and see that the codimension of (18.4) is one (the constant term in $r(0)$). □

Theorem 18.2 *Assume that $\ell = 2$. If*

$$s_u, s_v \tag{18.6}$$

are nonzero at the origin, then p is equivalent to the normal form

$$r(u,v) = 0, \quad s(u,v) = \epsilon_1 u + v \tag{18.7}$$

where $\epsilon = \pm 1$. This singularity has codimension one and a universal unfolding is

$$r(u,v,\lambda) = \lambda, \quad s(u,v) = \epsilon_1 u + v. \tag{18.8}$$

Proof Write $r = au + bv$ and $s = cu + dv$. Under assumptions (18.6),

$$T(p) = (\mathcal{M}_{u,v}, \mathcal{M}_{u,v}).$$

It follows that p is 2-determined and we can assume $b = 0$. Rescaling leads to the desired normal form. This singularity has codimension one with unfolding parameter λ as indicated in (18.8). □

Finally, we consider the case $\ell = 3$. Now

$$p(x_1, x_2) = r(u,v)x_2 + s(u,v)x_1 \tag{18.9}$$

where $s(0,0)=0$. These p's are just the odd functions, that is, they satisfy

$$p(-x_1,-x_2)=-p(x_1,x_2).$$

The appropriate coordinate changes are $X(x_1,x_2)$ where X is odd. In addition, these coordinate changes must preserve the half-plane $x_1 \geq 0$. The general such change of coordinates is:

$$X(x_1,x_2)=(Ax_1+Bux_2,Cx_1+Dx_2) \tag{18.10}$$

where A,B,C,D are functions of u,v, $A(0,0)>0$, and $D(0,0)\neq 0$. The contact equivalences in the range have the form

$$S(u,v)+R(u,v)x_1x_2,$$

where $S(0,0)\neq 0$.

We now compute the tangent space $T(p)$ and find that it has six generators

$$(2ur_u,2us_u+s),(2u^2s_u+us,2uvr_u),(2us_v,2vr_v+r),(2vr_v+r,2vs_v),(r,s),(us,vr)$$

corresponding to A,B,C,D,S and R.

We next put the lowest order terms of r,s into a normal form. Suppose that

$$r(u,v)=au+bv \quad \text{and} \quad s(u,v)=cu+dv.$$

Consider the change of coordinates

$$X(x_1,x_2)=(Ax_1,\mu Ax_1+Dx_2)$$

where A,D, and μ are constants with $A>0,D\neq 0$. Then

$$p(X(x_1,x_2))=\hat{r}(u,v)x_2+\hat{s}(u,v)x_1$$

where

$$\begin{aligned}\hat{r}(u,v) &= (a+2d\mu+3b\mu^2)A^2Du+bD^3v\\ \hat{s}(u,v) &= (c+a\mu+d\mu^2+b\mu^3)A^3u+(3b\mu+d)AD^2v.\end{aligned}$$

This calculation implies that

$$b\neq 0 \tag{18.11}$$

is required in a singularity of lowest codimension. Thus we assume that $r_v(0,0)\neq 0$ and set $\mu=-d/3b$. Next we assume that

$$c+a\mu+d\mu^2+b\mu^3\neq 0,$$

which is equivalent to

$$\alpha \equiv 2d^3-9abd+27b^2c\neq 0. \tag{18.12}$$

We now choose A, D and $S(0,0)$ so that

$$\hat{r}(u,v)=mu+v, \quad \hat{s}(u,v)=u, \tag{18.13}$$

where

$$m=3\mathrm{sgn}(\alpha)\frac{3ab-d^2}{\alpha^{2/3}}$$

is a modal parameter. No further simplification of (18.13) is possible.

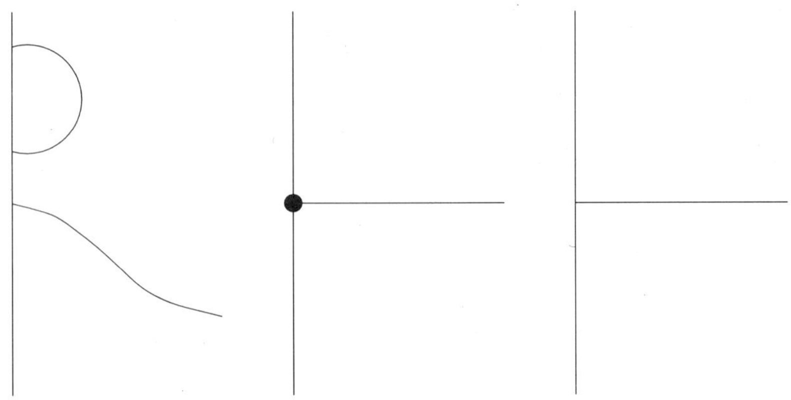

Figure 4 $k = 1$, $\ell \geq 4$. Equation $x_1^2x_2 + x_2^3 + \lambda x_2 + x_1^\ell = 0$.

Theorem 18.3 *Assume that $\ell = 3$ and that (18.11) and (18.12) are satisfied. Then p is equivalent to the normal form (18.13). This singularity has codimension two and a universal unfolding is*

$$r(u, v, \lambda) = mu + v + \lambda, \quad s(u, v) = u, \tag{18.14}$$

where m is a modal parameter.

Proof Using normal form (18.13) we can show that, independent of higher order terms,

$$T(p) = (\mathcal{M}_{u,v}^2, \mathcal{M}_{u,v}^2) \oplus \mathbb{R}\{(2mu, 3u), (0, mu + 3v), (mu + 3v, 0)\}.$$

The 'tangent space constant' theorem then says that we can eliminate higher order terms using a change of coordinates. The computation of the tangent space also shows that p has codimension two — but one of those unfolding parameters is the modal parameter m along which the germ p has constant codimension. □

19 Bifurcation diagrams when $k = 1$

We now consider the solutions that may be derived from the normal forms of the previous theorems.

The case $\ell \geq 4$ In this case normal form (18.5) is

$$x_1^2x_2 + \epsilon_1 x_2^3 + \lambda x_2 + x_1^\ell = 0$$

where $\epsilon_1 = \pm 1$. This equation yields the pictures in Figures 4 and 5.

The case $\ell = 3$ In this case normal form (18.14) is

$$x_1^3 + mx_1^2x_2 + x_2^3 + \lambda x_2 = 0$$

where $m \in \mathbb{R}$ is a modal parameter. The pictures are essentially the same as in the case $\ell \geq 4$.

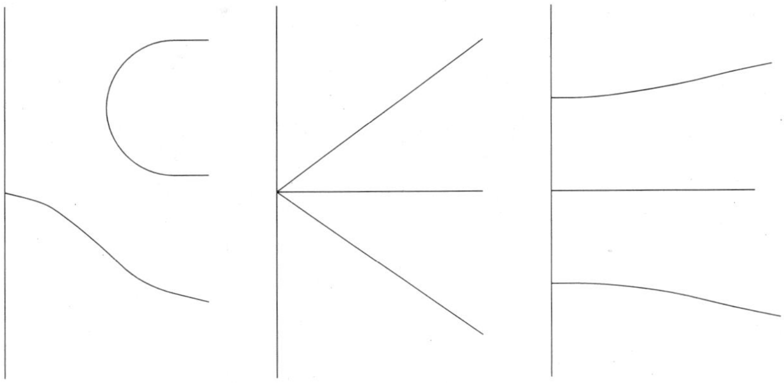

Figure 5 $k = 1$, $\ell \geq 4$. Equation $x_1^2 x_2 - x_2^3 + \lambda x_2 + x_1^\ell = 0$.

The case $\ell = 2$ In this case normal form (18.8) is

$$\epsilon_1 x_1^2 + x_2^2 + \lambda = 0$$

where $\epsilon_1 = \pm 1$. The case $\epsilon_1 = +1$ and $\lambda < 0$ is graphed in Figure 6.

We conclude with a remark about the relation between the reversible and the Hamiltonian cases. Our analysis shows that the resulting bifurcations for the reversible and the Hamiltonian $k:\ell$ resonant cases are closely related. Arnol'd and others have raised the issue of why it is that the reversible and Hamiltonian cases so often lead to the same bifurcation diagrams? When the Hamiltonian problem satisfies condition (S), our results show that we end up with a Hamiltonian system with a Hamiltonian that is $\mathbf{S}^1$-invariant, and that in these cases, one has similar bifurcation diagrams. However, as we saw, the nonsemisimple Hamiltonian-Hopf case was rather different in the reversible and Hamiltonian cases. Another interesting link that might be useful for this general question is that, subject to some mild conditions, any Hamiltonian system can be realized as the $\mathbf{S}^1$ reduction of a reversible Hamiltonian one. The proof of this, whose details we do not give here, uses the Kaluza-Klein trick along with the theory of reduction of Lagrangian systems (see Marsden [1992] and Marsden and Scheurle [1993]). Thus, for example, the bifurcation of an equilibrium of a Hamiltonian system can be directly realized as the reduction of the bifurcation of a relative equilibrium in a *reversible* Hamiltonian system with $\mathbf{S}^1$-symmetry.

Conclusions

In this paper we have shown how singularity theory and the Liapunov-Schmidt procedure for systems with implicit constraints leads directly and simply to specific results for the bifurcation of periodic orbits. This method was applied to the Liapunov center theorem in the cases of systems with an integral, Hamiltonian systems and reversible ones. It was also applied to the Hamiltonian Hopf bifurcation, and to resonance bifurcations in both the Hamiltonian and reversible cases. In this way, we were able to recover the corresponding bifurcation results of Duistermaat [1983],

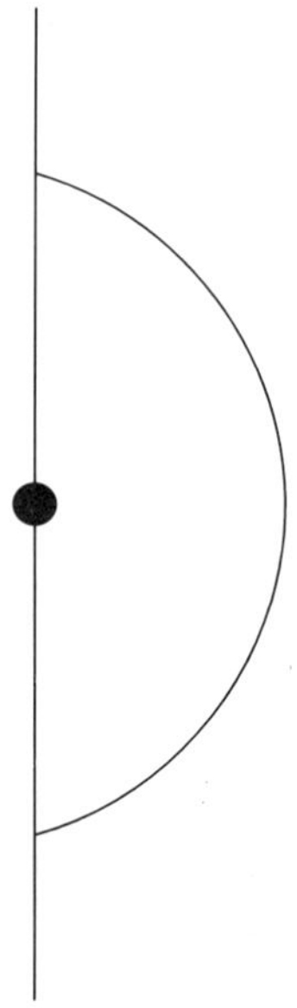

Figure 6 $k = 1$, $\ell = 2$. Equation $x_1^2 + x_2^2 + \lambda = 0$ when $\lambda < 0$.

Sevryuk [1986] and Bridges [1990]. The method here, which looks for periodic orbits by applying the Liapunov-Schmidt procedure directly on loop spaces, has a number of technical advantages over normal form methods. We showed that the mapping that occurs in the procedure is a Hamiltonian vector field if a semisimple type of condition (S) is satisfied, and this is used to show that the bifurcation analysis of the reversible and Hamiltonian resonances are closely related. This condition holds for the Hamiltonian Liapunov center theorem and resonance bifurcations, but not for the Hamiltonian Hopf bifurcation, which is analyzed directly using the implicit constraint. In addition to analyzing stability, which we believe can be done by enhancing the present context using the techniques used for the Hopf bifurcation (see Golubitsky, Stewart and Schaeffer [1988]) combined with those of Montaldi, Roberts and Stewart [1988], there are many other problems to which the methods of this paper could in principle, be applied. For example, it would be interesting to analyze the case of passing resonances for Hamiltonian systems with symmetry, as in Dellnitz, Melbourne and Marsden [1992].

References

Abraham, R. and Marsden, J.E. [1978] *Foundations of Mechanics*, Second Edition, Addison-Wesley Publishing Co., Reading, Mass.

Abraham, R., Marsden, J.E. and Ratiu, T.S. [1988] *Manifolds, Tensor Analysis, and Applications*, Second Edition, Springer-Verlag, New York.

Alexander, J.C. and York, J.A. [1978] *Global bifurcation of periodic orbits*, Am. J. Math., **100**, 263–292.

Arnol'd, V.I. (ed.) [1988] *Dynamical Systems III*, Encyclopaedia of Math. Sciences, **3**, Springer-Verlag, New York.

Bridges, T. [1990] *Bifurcation of periodic solutions near a collision of eigenvalues of opposite signature*, Math. Proc. Camb. Phil. Soc., **108**, 575–601.

Chernoff, P.R. and Marsden, J.E. [1974] *Properties of infinite dimensional Hamiltonian systems,* Lecture Notes in Math., **425**, Springer-Verlag, New York.

Cushman, R. and Rod, D.L. [1982] *Reduction of the semisimple* $1:1$ *resonance*, Physica D, **6**, 105–112.

Damon, J.N. [1984] *The unfolding and determinacy theorems for subgroups of* $\mathcal{A}$ *and* $\mathcal{K}$, Memoirs Vol. **50**, No. **306**, Amer. Math. Soc., Providence.

Dellnitz, M. and Melbourne, I. [1993] *The equivariant Darboux theorem,* Lectures in Appl. Math., **29**, 163–169.

Dellnitz, M., Melbourne, I. and Marsden, J.E. [1992] *Generic bifurcation of Hamiltonian vector fields with symmetry*, Nonlinearity, **5**, 979–996.

Duistermaat, H. [1983] *Bifurcations of periodic solutions near equilibrium points of Hamiltonian systems*, Lecture Notes in Math., **1057**, Springer-Verlag, New York.

Duistermaat, H. [1984] *Non-integrability of the* $1:2:2$ *resonance*, Ergodic Theory and Dyn. Sys., **4**, 553.

Golubitsky, M., Krupa, M. and Lim, C. [1991] *Time reversibility and particle sedimentation*, SIAM J. Appl. Math., **51**, 49–72.

Golubitsky, M. and Langford, W. F. [1981] *Classification of unfoldings of degenerate Hopf bifurcations*, J. Diff. Eqns., **41**, 375–415.

Golubitsky, M. and Schaeffer, D.G. [1985] *Singularities and Groups in Bifurcation Theory: Vol. I*, Applied Math. Sciences, **51**, Springer-Verlag, New York.

Golubitsky, M. and Stewart, I. [1987] *Generic bifurcation of Hamiltonian systems with symmetry*, Physica D, **24**, 391–405.

Golubitsky, M. and Stewart, I. [1993] *An algebraic criterion for symmetric Hopf bifurcation*, Proc. Roy. Soc. Lond. A, **440**, 727–732.

Golubitsky, M., Stewart, I. and Schaeffer, D. [1988] *Singularities and Groups in Bifurcation Theory: Vol. II*, Applied Math. Sciences, **69**, Springer-Verlag, New York.

Hale, J.K. [1969] *Ordinary Differential Equations*, Wiley, New York.

Haller, G. and Wiggins, S. [1992] *Orbits homoclinic to resonances: the Hamiltonian case*, Physica D, **66** 298–346.

Haller, G. [1993], *Multi-pulse Homoclinic Phenomena in Resonant Hamiltonian Systems,* Thesis, Caltech.

Henrard, J. [1973] *Lyapunov's center theorem for resonant equilibrium, J. Diff. Eq.* **14**, 431–441.

Hoveijn, I. and Verhulst, F. [1990] *Chaos in the* $1:2:3$ *Hamiltonian normal form*, Physica D, **44**, 397.

Iwai, T. [1985] *On reduction of two degrees of freedom Hamiltonian systems by an S^1 action, and $SO_0[1,2]$ as a dynamical group*, J. Math. Phys., **26**, 885–893.

Kelley, A. [1967] *The stable, center-stable, center, center-unstable, and unstable manifolds*, J. Diff. Eq., **3**, 546–570.

Knobloch, E., Mahalov, A. and Marsden, J.E. [1993] *Normal Forms for three-dimensional parametric instabilities in ideal hydrodynamics*, Physica D, to appear.

Kummer, M. [1979] *On resonant classical Hamiltonians with two degrees of freedom near an equilibrium point*, Lecture Notes in Physics, **93**, Springer-Verlag, 57–75.

Kummer, M. [1990] *On resonant classical Hamiltonians with n frequencies*, J. Diff. Eq., **83**, 220–243.

Liapunov, M.A. [1949] *Problème générale de la stabilité du mouvement*, Annals Math. Stud., **17**, Princeton University Press.

Marsden, J.E. [1992], *Lectures on Mechanics*, London Math. Society Lecture Note Series, **174**, Cambridge University Press, Cambridge.

Marsden, J.E. and McCracken, M. [1976] *The Hopf bifurcation and its Applications*, Applied Math. Sciences Series, **19**, Springer-Verlag, New York.

Marsden, J.E. and Scheurle, J. [1993] *The reduced Euler-Lagrange equations, Fields Institute Comm.*, **1**, 139–164.

Meyer, K.R. and Palmore, J. [1970] *A new class of periodic solutions in the restricted three-body problem*, J. Diff. Eq., **8**, 264–276.

Meyer, K.R. and Schmidt, D.S. [1971] *Periodic orbits near L_4 for mass ratios near the critical mass ratio of Routh*, Celest. Mech., **4**, 99–109.

Montaldi, J.A., Roberts, R.M. and Stewart, I.N [1988] *Periodic solutions near equilibria of symmetric Hamiltonian systems*, Phil. Trans. R. Soc. Lond., A **325**, 237–293.

O'Reilly, O., Malhotra, N.K. and Namachchivaya, N.S. [1993] *Destabilization of the equilibria of reversible dynamical systems*, preprint.

Palmore, J. [1969] *Bridges and natural centers in the restricted three-body problem*, Report, Univ. of Minnesota.

Rabinowitz, P.H. [1977] *A bifurcation theorem for potential operators*, J. of Funct. An., **25**, 412–424.

Roberts, J.A.G. and Quispel, G.R.W. [1992] *Chaos and time-reversal symmetry. Order and chaos in reversible dynamical systems*, Phys. Reports, **216** 63–177.

Roels, J. [1971a] *An extension to resonant cases of Liapunov's theorem concerning the periodic solutions near a Hamiltonian equilibrium*, J. Diff. Eq., **9**, 300–324.

Roels, J. [1971b] *Families of periodic solutions near a Hamiltonian equilibrium when the ratio of two eigenvalues is 3*, J. Diff. Eq., **10**, 431–447.

Sanders, J.A. and Verhulst, F. [1985] *Averaging Methods in Nonlinear Dynamical Systems*, Springer-Verlag, New York.

Schmidt, D.S. [1973] *Periodic solutions near a resonant equilibrium of a Hamiltonian system*, Celest. Mech., **9**, 81–103.

Schmidt, D.S. [1976a] *Hopf's bifurcation theorem and the center theorem of Liapunov*, Celest. Mech., **9**, 81–103.

Schmidt, D.S. [1976b] *Hopf's bifurcation theorem and the center theorem of Liapunov with resonance cases*, Section 3C of Marsden and McCracken [1976].

Schmidt, D.S. and Sweet, D. [1973] *A unifying theory in determining periodic families for Hamiltonian systems at resonance*, J. Diff. Eq., **14**, 597–609.

Sevryuk, M.B. [1986] *Reversible Systems*, Lecture Notes in Math., **1211**, Springer-Verlag, Berlin.

Sweet, D. [1973] *Periodic solutions for dynamical systems possessing a first integral in the resonance case*, J. Diff. Eq., **14**, 171–183.

Vanderbauwhede, A. [1982a] *Local Bifurcation and Symmetry*, Research Notes in Math., **75**, Pitman, Boston.

Vanderbauwhede, A. [1982b] *Families of periodic solutions for autonomous systems*; in Dynamical Systems II (A. Bednarek and L. Cesari, eds), Academic Press, 427–446.

Vanderbauwhede, A. [1990] *Hopf bifurcation for equivariant conservative and time-reversible systems*, Proc. Roy. Soc., Edinburgh, **116A**, 103–128.

Vanderbauwhede, A. and van der Meer, J.C. [1994] *A general reduction method for periodic solutions near equilibria in Hamiltonian systems*, Fields Institute Comm., to appear.

van der Meer, J.C. [1985] *The Hamiltonian-Hopf Bifurcation*, Lecture Notes in Math., **1160**, Springer-Verlag, Berlin.

Weinstein, A. [1978] *Bifurcations and Hamilton's principle*, Math. Zeit., **159**, 235–248.

Fields Institute Communications
Volume 4, 1995

Whiskered tori and chaos in resonant Hamiltonian normal forms

G. Haller
Division of Applied Mathematics
Brown University
Providence, RI, USA 02912

S. Wiggins
Department of Applied Mechanics
Caltech 104-44
Pasadena, CA, USA 91125

Abstract. In this paper we study the dynamics near resonant elliptic equilibria in certain three-degree-of-freedom Hamiltonian systems. We assume that the frequencies of the linearized vectorfield are in strong resonance and the corresponding Birkhoff normal form is integrable at cubic order. Using a new global perturbation method we show that on each energy surface of the quartic normal form there exist two families of 2-tori which are connected through heteroclinic intersections of their stable and unstable manifolds. These intersections are transversal when restricted to a given level surface of the invariant corresponding to the S^1 symmetry of the normal form. We also obtain the generic existence of another family of whiskered tori with *double-pulse* homoclinic manifolds. The splitting of the whiskers of the tori in both cases is of the order of the square of the distance from the equilibrium, in particular, it is *not exponentially small.*

1 Introduction

The basic nature of dynamics near resonant equilibria of multi-degree-of-freedom Hamiltonian systems is still an open question (see Arnold [1964]). It appears that instead of trying to solve this problem in general, one has to look at specific resonances in detail (Sanders and Verhulst [1985]). This approach is motivated by the diversity of different resonant normal forms arising in the simplification of the underlying Hamiltonian near the equilibrium. While in some cases the truncations

---

1991 *Mathematics Subject Classification.* Primary 70H33; Secondary 58F27, 58F36.
This research was supported by an NSF Presidential Young Investigator Award and an ONR Young Investigator Award.

of these normal forms turn out to be integrable (see, e.g., van der Aa and Verhulst [1984]), in other cases they can be shown non-integrable (see Duistermaat [1984] and Hoveijn [1992]). Even in the cases of proven non-integrability, the global geometry of the truncated normal forms is not fully understood and the implications of the dynamics of the normal forms for the full system are not explored. A thorough study of particular normal forms for higher multiplicity resonances would be very important in identifying the basic structures which are responsible for the irregularity of motions near resonant equilibria.

This paper intends to contribute to this program by a geometric analysis of non-integrability and chaotic behavior in a broad class of resonances in three-degree-of-freedom (3DOF) Hamiltonian systems. We study a class of multiplicity-two 3DOF resonances which admit integrable normal forms at cubic order. Our basic goal is to understand the integrable geometry in detail and then treat higher order normal form terms as perturbations on this structure. In the discussion we mainly restrict ourself to an exposition of the results and the methods we use. A more detailed account with motivation, proofs and examples will appear in Haller and Wiggins [1994], where we also show how the results for the finite jet of the normal form are related to the full Hamiltonian. A concrete example and a discussion of the physical implications of the results presented here can be found in Haller and Wiggins (in preparation).

The exposition of this paper is as follows. In Section 2 we identify the class of resonances we are concerned with. This class involves resonances like the $1: 2: 2$, $1: 2: \omega_3$ with $|\omega_3| > 5$ and other resonances. All these resonances admit the same type of integrable cubic normal form which we analyze in Section 3. We identify an invariant 3-sphere on every nonzero energy level of the cubic normal form and also find four-dimensional invariant manifolds homoclinic to these spheres. The 3-spheres are foliated by periodic orbits of the same period. After a symplectic reduction, these periodic orbits are all mapped into a conical singularity of the reduced phase space. In order to study them via smooth perturbation methods, we blow up the singularity of the reduced phase space in Section 4 by introducing a singular change of coordinates. In the new coordinates we consider the partially reduced problem which admits two *hyperbolic two-manifolds of relative equilibria* connected through heteroclinic three-manifolds. To study the fate of this degenerate structure under perturbation, we use a special global perturbation method described in Section 5. We apply this method in Section 6 to our problem and find certain invariant two-tori with two different time scales in the quartic normal form. The stable and unstable manifolds of these *whiskered tori* intersect transversally within an invariant subset of the energy surface and give rise to the existence of nearby Smale-horseshoes. This indicates strong non-integrability and homoclinic chaos in the normal form truncated at fourth or higher order, as discussed in Section 7 together with other implications of our results.

2 Set-up and assumptions

We will be concerned with Hamiltonians of the form

$$H(q,p) = H_2(q,p) + \tilde{H}(q,p), \tag{2.1}$$

where (q,p) are canonical coordinates on the phase space $(\mathbb{R}^6, \omega)$ with $\omega = dq \wedge dp$. In (2.1) H_2 is a quadratic polynomial and $\tilde{H}$ is a C^4 function with $D\tilde{H}(0) = 0$ and

$D^2\tilde{H}(0) = 0$. With the notation $x = (q, p)$ the Hamiltonian equations associated with (2.1) take the form

$$\dot{x} = J_4 DH(x), \quad J_4 = \begin{pmatrix} 0 & Id_2 \\ -Id_2 & 0 \end{pmatrix}, \tag{2.2}$$

where $Id_2 \in \mathbb{R}^{3\times 3}$ is the identity matrix. We assume that $x = 0$ is an elliptic fixed point of system (2.2) and $J_4 D^2 H_2|_0$ is semisimple, in which case a linear canonical transformation puts H_2 to the form

$$H_2(q,p) = \frac{1}{2}\sum_{k=1}^{3} \omega_k (q_k^2 + p_k^2), \tag{2.3}$$

where $\Omega = (\omega_1, \omega_2, \omega_3) \in \mathbb{R}^3 - \{0\}$ is the frequency vector of the linear flow generated by H_2. We now introduce the canonical change of variables

$$z = q - ip, \quad \bar{z} = q + ip, \quad z, \bar{z} \in \mathbb{C}^3, \tag{2.4}$$

with its inverse defined as $T_1 \colon (z, \bar{z}) \mapsto (q, p)$, and consider the transformed system on the phase space $(\mathbb{C}^3, \tilde{\omega})$ with $\omega = \frac{1}{2}\mathrm{Im}(d\bar{z} \wedge dz)$. Letting $z \longrightarrow \varepsilon z$, $\bar{z} \longrightarrow \varepsilon\bar{z}$ with $\varepsilon > 0$ small, and dividing (2.1) by ε^2, we arrive at the Hamiltonian

$$H(z, \bar{z}; \varepsilon) = H_2(z, \bar{z}) + \varepsilon\tilde{H}(z, \bar{z}; \varepsilon), \tag{2.5}$$

with

$$H_2(z, \bar{z}) = \frac{1}{2}\sum_{k=1}^{3} \omega_k |z_k|^2, \quad \tilde{H}(z, \bar{z}; \varepsilon) = \sum_{|l|+|m|\geq 3} \varepsilon^{|l|+|m|-3} h_{lm} z^l \bar{z}^m, \tag{2.6}$$

where $h_{lm} = \bar{h}_{ml} \in \mathbb{C}$ with $(l, m) \in \mathbb{N}^3 \times \mathbb{N}^3$ and $|l| = l_1 + l_2 + l_3$. In (2.6) we used the usual notation $z^p \colon = z_1^{p_1} z_2^{p_2} z_3^{p_3}$. To describe the resonances we will study, let us introduce the resonant module

$$M = \{ n \in \mathbb{Z}^3 \ | \langle \omega, n \rangle = 0 \ \}. \tag{2.7}$$

We then assume that, after a possible reindexing of the variables $z_k, \bar{z}_k$,

(H1) Either $(2,1,0) \in M$ or $(2,-1,0) \in M$ holds *and* $\dim M = 2$.

(H2) If $s \in M$ and $s \notin \mathrm{span}\{(2,1,0), (2,-1,0)\}$ then *one of the following* is satisfied:

(i) $|s| > 4$,

(ii) For any $l, m \in \mathbb{N}^3$ with $|l| + |m| = 3$, $l - m = s$, we have $h_{lm} = 0$.

Here (H1) means that the first two frequencies satisfy the strong resonance relationship $|\omega_1| \colon |\omega_2| = 1 \colon 2$ *and* there is one more independent resonance relationship among the three frequencies. If (H2)(i) holds then $(2, \pm 1, 0)$ is the unique third order generator of the resonant module M. This is satisfied by all resonances of the form $1 \colon \pm 2 \colon \omega_3$ with $|\omega_3| \geq 5$, and by a number of other resonances, like the $1\colon 2\colon 6$, $2\colon 4\colon 3$, $3\colon 6\colon 1$, $3\colon 6\colon 2$, $3\colon 6\colon 10$, etc. If, alternatively, (H2)(ii) holds then $(2, \pm 1, 0)$ is not the only third order generator of M. We then require certain coefficients of $\tilde{H}$ to vanish, which usually translates to the assumption that system (2.6) is close to having a discrete symmetry at cubic order (one can also assume $h_{lm} = \nu\varepsilon$ which requires the symmetry breaking terms to be less in order than the symmetric cubic terms). We note that all genuine first order resonances in three-degree-of-freedom Hamiltonians (i.e., $1\colon \pm 2\colon \omega_3$, $|\omega_3| = 1, 2, 3, 4$) with appropriate weakly broken discrete symmetries satisfy hypotheses (H1) and (H2)(u) (see Sanders and Verhulst [1985] for the definition of genuine resonances and a discussion on the effect of discrete symmetries).

Based on hypothesis (H1) we can rescale the frequencies by letting

$$\omega_1 = 1, \quad \omega_2 = 2, \quad \omega_3 \longrightarrow \frac{\omega_3}{\omega_1}, \tag{2.8}$$

where we have set the sign of ω_2 positive for convenience. Note that assumption (H1) implies the rescaled frequency ω_3 to be rational. In order to simplify the Hamiltonian (2.6) further, we apply a near-identity canonical change of variables which puts our system in Birkhoff normal form up to some order $r \geq 4$ (see Arnold [1988] or Sanders and Verhulst [1985], etc.). Using assumptions (H1)–(H2), the rescaling (2.8), and allowing some slight detuning from the exact resonance, we can write the normalized part of the Hamiltonian in the form

$$H = H_2 + \varepsilon H_3 + \varepsilon^2 \tilde{H}_4 + \varepsilon^3 H_5 + \cdots + \varepsilon^{r-2} H_r,$$

$$H_2 = \frac{1}{2}|z_1|^2 + |z_2|^2 + \frac{1}{2}\omega_3 |z_3|^2, \qquad H_3 = \frac{a}{2}\mathrm{Re}\,(z_1^2 \bar{z}_2), \quad a \in \mathbb{R}, \qquad \tilde{H}_4 = H_4 + \hat{H}_4, \tag{2.9}$$

with

$$\{H_2, H_j\} = 0, \quad 3 \leq j \leq r, \qquad \{H_2, \hat{H}_4\} = 0, \tag{2.10}$$

where $\{\ ,\ \}$ is the canonical Poisson bracket. In (2.9) H_j contains resonant terms of the form $z^l \bar{z}^m$ with $|l| + |m| = j$, $l - m \in M$. In general, the normalization procedure leads to the cubic term $H_3 = \mathrm{Re}\,(A z_1^2 \bar{z}_2)$ with $A \in \mathbb{C}$, and we have to apply an additional symplectic change of coordinates $z_1 \to \exp(-i\arg(A)/2) z_1$, $\bar{z}_1 \to \exp(-i\arg(A)/2)\bar{z}_1$, to put H_3 in the form of (2.9) with $a = 2|A|$ (see Kummer [1990]). From this point we will assume that the nondegeneracy condition $a \neq 0$ holds. The expression $\hat{H}_4$ contains some possible unnormalized perturbation of system (2.1) Poisson commuting with H_2, as well as cubic terms $h_{lm} z^l \bar{z}^m$ of the normal form with $h_{lm} = \mathcal{O}(\varepsilon)$, which may arise in the case of hypothesis (H2). We finally remark that for $\Omega = (1\colon \pm 2\colon \pm 2)$ system (2.6) can be put in the form (2.9) even if (H2) is not satisfied (see Van der Aa and Verhulst [1984], Kummer [1986], and Cushman [1985]).

3 Integrability and geometry of the cubic normal form

In this section we briefly discuss the geometry of the normal form (2.9) for $r = 3$. Related results can be found in Kummer [1979], Churchill *et al.* [1983], and Knobloch *et al.* [1992] for two-degree-of-freedom resonances, Kummer [1975], Cushman [1985], and Holm [1989] for certain three-degree-of-freedom resonances, and Kummer [1990] for n-degree-of-freedom systems. For more details the reader is referred to Haller and Wiggins [1994], while the standard terminology of symplectic reduction can be found in Abraham and Marsden [1978].

First, let us consider the action of the Lie group $G = \mathbb{T}^2 = S^1 \times S^1$ on $(P = \mathbb{C}^3, \omega)$ given by

$$\begin{aligned} \phi\colon G \times P &\rightarrow P, \\ (g_1, g_2, z_1, z_2, z_3) &\mapsto (e^{-ig_1} z_1, e^{-2ig_1} z_2, e^{-ig_2} z_3). \end{aligned} \tag{3.1}$$

A momentum map for this action can be shown to be

$$J\colon P \rightarrow \mathfrak{g}^* \simeq \mathbb{R}^2, \tag{3.2}$$

$$(z_1, z_2, z_3) \mapsto (\frac{1}{2}|z_1|^2 + |z_2|^2, \frac{1}{2}|z_3|^2), \tag{3.3}$$

where $\mathfrak{g}^*$ denotes the dual of the Lie algebra $\mathfrak{g}$ of G. Note that J is Ad^* equivariant (G is Abelian) hence the reduced phase space $P_\mu = J^{-1}(\mu)/G_\mu$ for the action (3.1) of G is well defined. Since G is compact, the action of the isotropy subgroup $G_\mu = \{g \in G | Ad^*_{g^{-1}}\mu = \mu\} = G$ is proper on $J^{-1}(\mu)$, but it is not free (G has nonidentity elements leaving $(0, z_2, z_3) \in J^{-1}(\mu)$ fixed); hence the Marsden-Weinstein reduction theorem does not apply. In order to carry out the reduction and find a Poisson bracket on the reduced phase space, one could use singular reduction techniques (see Lerman and Sjamaar [1991] and Arms *et al.* [1991]). Here we use the work of Churchill *et al.* [1983] (which is more closely related to symmetries in resonant normal forms) to obtain a representation of P_μ. Following Churchill *et al.* [1983] we introduce the Euler-variables

$$W_1 = \mathrm{Re}(z_1^2 \bar{z}_2), \quad W_2 = \mathrm{Im}(z_1^2 \bar{z}_2), \quad W_3 = \frac{1}{2}|z_1|^2, \tag{3.4}$$

and note that

$$W_1^2 + W_2^2 = 4W_3^2(J_1(z) - W_3). \tag{3.5}$$

Let $Z_\mu \subset \mathbb{R}^3$ denote the subset of $\mathbb{R}^3$ defined by (3.5). As is shown in Churchill *et al.* [1983], the reduced phase space P_μ can be identified with Z_μ.

We now turn to the study of the cubic truncation of the normal form (2.9) given by

$$H_0 = \frac{1}{2}|z_1|^2 + |z_2|^2 + \frac{1}{2}\omega_3|z_3|^2 + \varepsilon\frac{a}{2}\mathrm{Re}\,(z_1^2 \bar{z}_2). \tag{3.6}$$

It is easy to verify that $\phi_g \circ H_0 = H_0$ holds hence, based on the construction above, H_0 defines an *integrable Hamiltonian system* with H_0, J_1, and J_2 being three independent integrals. As we have found, the reduced phase space for H_0 can be realized as $P_\mu \simeq Z_\mu$, embedded in $\mathbb{R}^3$. It is not hard to see from (3.5) that P_μ is homeomorphic to S^2. In fact, Z_μ is a pinched sphere with its pinch p_μ at the origin of the W-space, and its axis of rotation is just the W_3 axis. As noted in the references cited at the beginning of the section, the reduced flow of (3.6) on P_μ is given by the Euler-like equations

$$\dot{W} = \nabla H_0 \times \nabla f_\mu, \tag{3.7}$$

with

$$f_\mu = \frac{1}{2}\big(W_1^2 + W_2^2 - 8W_3^2(\mu_1 - W_3)\big).$$

Using the fact for any $\mu \in \mathfrak{g}^*$ a subset of $J^{-1}(\mu)$ is a 2-torus bundle over $P_\mu \backslash p_\mu$, we can reconstruct certain invariant structures in the phase space of (3.6) based on our knowledge of the orbits of the reduced system (3.7). To identify the reduced orbits on P_μ one can use the fact that the Euler-variable W_1 is constant on these orbits (being a first integral for the cubic normal form (3.6)). Hence, following the idea of Kummer [1979], we can visualize the orbit structure of (3.7) by intersecting the pinched sphere Z_μ with planes $W_1 = const.$ Each such intersection is either a point, or a smooth closed curve, or a continuous closed curve through the pinch of Z_μ at $W = 0$. This latter curve contains a relative equilibrium p_μ at $W = 0$ which is connected to itself by a homoclinic orbit in the reduced phase space satisfying

$$W_1 = 0, \quad W_2 = 2\sqrt{2} W_3 \sqrt{\mu_1 - W_3}, \quad W_3 > 0. \tag{3.8}$$

As we pointed out, a relative equilibrium or a periodic solution of (3.7) indicates the presence of an invariant 2-torus or an invariant 3-torus, respectively, in $J^{-1}(\mu)$. One can ask what happens to these invariant sets under perturbation and, using an appropriate version of the KAM theory (see Haller and Wiggins [1994]), one can show that most of the invariant 3-tori persist in the phase space of (2.9). However, this is a phenomenon which one also finds in nonresonant normal forms. In particular, this result does not bring us any closer to an understanding of the source of irregularity in strongly resonant systems. To this end, we will study what happens to the homoclinic structure described by (3.8) above.

We start by defining the invariant set $M^h \subset P$ with $h > 0$ as

$$M^h = \cup_{\mu_1 + \mu_2 = h} \pi_W^{-1}(p_\mu), \tag{3.9}$$

where $\pi_W : P \to \mathbb{R}^3$, $z \mapsto W(z)$ is the Hopf-map defined in (3.5). Clearly, M^h is the set of orbits in a given energy-surface $H_0 = h$ which are mapped into relative equilibria at $W = 0$ in P_μ under the quotient projection $P \to P_\mu$. Using (3.4) and (2.4), one can immediately see that the image of M^h under the diffeomorhism T_1 satisfies

$$T_1(M^h) = \{ (q,p) \in \mathbb{R}^6 \,|\, q_1 = p_1 = 0, \; q_2^2 + p_2^2 + \frac{1}{2}\omega_3 (q_3^2 + p_3^2) = h \}. \tag{3.10}$$

This shows that for $\omega_3 > 0$ M^h is diffeomorphic to S^3, while for $\omega_3 < 0$ it is a three-dimensional hyperbolic surface of revolution. In either case, M^h is connected to itself by a four dimensional homoclinic manifold W^h which lies in the preimage of the set defined in (3.8) under π_W.

One can check that for any $h > 0$ M^h is entirely filled with periodic orbits. Two of these closed orbits are distinguished: they are the (nonlinear) *normal modes* of H_0 given by

$$N_2^h = \{ z \in \mathbb{C}^3 \,|\, z_1 = z_3 = 0, \; |z_2| = \sqrt{h} \}, \tag{3.11}$$

$$N_3^h = \{ z \in \mathbb{C}^3 \,|\, z_1 = z_2 = 0, \; |z_3| = \sqrt{\frac{2h}{\omega_3}} \}. \tag{3.12}$$

These two normal modes survive from the quadratic Hamiltonian H_2 under the effect of the cubic terms in (3.6). With the exception of N_3^h, all the periodic orbits in $\mathcal{M}^h$ have two-dimensional stable and unstable manifolds which foliate W^h into a two-parameter family of cylindrical surfaces.

Since M^h is a manifold of periodic orbits, the flow of the Hamiltonian (3.6) restricted to M^h can be considered as the action of the group S^1 on M^h. We can define $F^h = M^h/S^1$, the quotient space corresponding to this action, with the usual quotient projection $\pi_F\colon M^h \to F^h$. In other words, F^h is the orbit space of the periodic solutions contained in M^h. From (3.10) we see that the set $T_1(M^h) \simeq M^h$ can be considered as the three-dimensional energy surface for a Hamiltonian system of two linear oscillators which, by assumption (H1), are in $2\colon\omega_3$ resonance. Accordingly, F^h can be viewed as the reduced phase space for these oscillators with respect to the $2\colon\omega_3$ resonant action of S^1. As is shown in e.g. Churchill *et al.* [1983], this reduced phase space is homeomorphic to S^2. Furthermore, if κ_1 and κ_2 are relatively prime positive integers with

$$\frac{2}{|\omega_3|} = \frac{\kappa_1}{\kappa_2}, \tag{3.13}$$

then F^h will be a pinched sphere with a κ_1-order singularity at its north pole, and with a κ_2-order singularity at its south pole (e.g. $\kappa_1 = 1, 2, 3$ would mean no singularity, conical singularity, and cusp singularity, respectively). We summarize the observations of this section in the following.

Proposition 3.1 *Suppose that $\omega_3 > 0$ holds. Then on any energy surface $H_0 = h$ (with $h > 0$) of the integrable Hamiltonian system defined by H_0 there exists an invariant set M^h defined by $z_1 = 0$, which is diffeomorphic to S^3 and entirely filled with periodic orbits. Furthermore,*

(*i*) *Any invariant subset $M_0^h \subset M^h$ not containing the third normal mode N_3^h is normally hyperbolic and connected to itself by a four-dimensional homoclinic manifold W^h.*

(*ii*) *The orbit space $F^h = M^h/S^1$ of periodic solutions in M^h is homeomorphic to a 2-sphere with a κ_1-order singularity at its north pole, and with a κ_2 order singularity at its south pole (see (3.13)).*

We note that invariant spheres similar to M^h have been found recently in a large class of Hamiltonian resonances by Hoveijn [1992]. The methods we use in the following sections can be used to study perturbations of those spheres as well if they admit a homoclinic structure similar to W^h.

4 The "blown-up" normal form and the formulation of the perturbation problem

The goal of this section is to introduce coordinates which are suited to the study of what happens to the manifolds M^h and W^h in the normal form (2.9) under the perturbative effects of the terms $\varepsilon^2\hat{H}_4 + \cdots \varepsilon^{r-2}H_r$. The new coordinates are needed to eliminate the singularity of the reduced phase space and make it possible to view a subset of F^h as a normally hyperbolic invariant manifold.

Let us first introduce action-angle variables for the quadratic part H_2 of (2.9) letting

$$z_k = \sqrt{2I_k}e^{i\phi_k}, \quad \bar{z}_k = \sqrt{2I_k}e^{-i\phi_k}, \qquad k = 1, 2, 3, \tag{4.1}$$

with the inverse change of variables $T_2\colon (I,\phi) \mapsto (z,\bar{z})$. We apply a further canonical change of variables

$$\begin{aligned} \psi_1 &= \phi_1, & K_1 &= I_1 + 2I_2 + \omega_3 I_3, \\ \psi_2 &= \phi_3 - \omega_3\phi_1, & K_2 &= I_3, \\ x_1 &= \sqrt{2I_2}\sin(\phi_2 - 2\phi_1), & x_2 &= \sqrt{2I_2}\cos(\phi_2 - 2\phi_1). \end{aligned} \tag{4.2}$$

Note that the inverse coordinate transformation $T_3\colon (K,\psi,x) \mapsto (I,\phi)$ is defined for any $I_j \in \mathbb{R}^+$ and $\phi_j \in S^1$ $(j = 1,2,3)$. In this final set of variables our Hamiltonian (3.6) takes the form

$$H_0(x,K,\psi) = K_1 + \varepsilon a(K_1 - \omega_3 K_2 - |x|^2)x_2. \tag{4.3}$$

The corresponding Hamiltonian vectorfield is now *smooth* and defined on the set

$$\mathcal{P} = \{(x,K_1,\psi_1,K_2,\psi_2)|\; x \in \mathbb{R}^2, K \in \mathbb{R}^2, \psi \in \mathbb{T}^2\} \tag{4.4}$$

through the symplectic form $\omega = dx_1 \wedge dx_2 + d\psi \wedge dK$, but it is related to system (3.6) only in the domain

$$\bar{\mathcal{P}} = \{(x,K_1,\psi_1,K_2,\psi_2) \in \mathcal{P}|\; K_1 - \omega_3 K_2 \geq 0,\; K_2 \geq 0\}. \tag{4.5}$$

Our main goal with the sequence of transformations (4.1)–(4.2) is to "blow up" the singularity of the reduced phase space. This is achieved, as we will see next, by extending the transformations (4.1)–(4.2) to the domain $I_1 = 0$, in which case (4.1) is not a diffeomorhism any longer and ϕ_1 is not well defined. Since M^h is characterized by $I_1 = 0$, we have to study the effect of this singular transformation if we want to relate our later results on the perturbation of the blown-up normal form (4.3) back to the original normal form (2.9).

Proposition 4.1 *Suppose that $\omega_3 > 0$ is satisfied. Then the following hold:*
(*i*) *$T_3^{-1} \circ T_2^{-1}(M^h) = \mathcal{D}^h \cup \mathcal{M}^h$ where the set $\mathcal{D}^h$ satisfies*

$$\mathcal{D}^h = \{\,(x,K_1,\psi_1,K_2,\psi_2) \in \mathcal{P}\,\big|\, |x|^2 + \omega_3 K_2 = h,\; |x_2| > 0,\; K_1 = h,\; \psi \in \mathbb{T}^2\,\},$$

hence it is diffeomorphic to the disjoint union of two five dimensional open discs. The set $\mathcal{M}^h$ satisfies

$$\mathcal{M}^h = \{\,(x,K_1,\psi_1,K_2,\psi_2) \in \mathcal{P}\,\big|\, x_2 = 0,\; x_1^2 + \omega_3 K_2 = h,\; K_1 = h,\; \psi \in \mathbb{T}^2\,\},$$

hence it is diffeomorphic to $S^2 \times S^1$. Furthermore, $\mathcal{M}^h$ is filled with periodic orbits of (4.3), while $\mathcal{D}^h$ consists of orbits of (4.3) positively and negatively asymptotic to periodic orbits in $\mathcal{M}^h$.
(*ii*) *The manifold $\mathcal{W}^h = T_3^{-1} \circ T_2^{-1}(W^h)$ satisfies*

$$\mathcal{W}^h = \{\,(x,K_1,\psi_1,K_2,\psi_2) \in \mathcal{P}\,\big|\, x_2 = 0,\; x_1^2 + \omega_3 K_2 < h,\; K_1 = h,\; \psi \in \mathbb{T}^2\,\},$$

thus it is diffeomorphic to $B^3 \times S^1$ where B^3 is the open unit ball in $\mathbb{R}^3$. Further, $\mathcal{W}^h$ is filled with a three parameter family of orbits positively and negatively asymptotic to periodic orbits in $\mathcal{M}^h$.

(iii) Let $\pi_{\mathcal{F}}\colon \mathcal{M}^h \to \mathcal{F}^h = \mathcal{M}^h/S^1$ be the quotient projection from $\mathcal{M}^h$ to its orbit space $\mathcal{F}^h$, and let Q^h be defined through the diagram:

$$\begin{array}{ccc} \mathcal{M}^h & \xrightarrow{T_2 \circ T_3} & M^h \\ \pi_{\mathcal{F}} \downarrow & & \downarrow \pi_F . \\ \mathcal{F}^h & \xrightarrow{Q^h} & F^h \end{array}$$

In other words, let Q^h be the map between the quotient spaces induced by $T_2 \circ T_3$. Then $\mathcal{F}^h$ is diffeomorphic to S^2. Moreover, if M_0^h is a compact subset of M^h which does not contain the periodic orbits N_2^h and N_3^h, then $\mathcal{F}^h$ has a compact subset $\mathcal{F}_0^h$ not containing $\mathcal{N}_2^h = (Q^h)^{-1} \circ \pi_F(N_2^h)$ and $\mathcal{N}_3^h = (Q^h)^{-1} \circ \pi_F(N_3^h)$, such that Q^h restricts to a κ_1-fold smooth covering map onto $\pi_F(M_0^h)$ on any of the two connected components of $\mathcal{F}_0^h = \pi_{\mathcal{F}}(\mathcal{M}_0^h)$.

Proof The proof of statements (*i*)–(*ii*) is a direct computation based on the definition of T_2 and T_3, which we omit. To prove (*iii*) we first note that the change of variables (4.1) puts the vectorfield corresponding to (3.6) to the form

$$\begin{aligned} \dot{I}_1 &= \varepsilon 2aI_1\sqrt{2I_2}\sin(2\phi_1 - \phi_2), \\ \dot{I}_2 &= -\varepsilon aI_1\sqrt{2I_2}\sin(2\phi_1 - \phi_2), \\ \dot{I}_3 &= 0, \\ \dot{\phi}_1 &= 1 + \varepsilon a\sqrt{2I_2}\cos(2\phi_1 - \phi_2), \\ \dot{\phi}_2 &= 2 + \varepsilon a\frac{I_1}{\sqrt{2I_2}}\cos(2\phi_1 - \phi_2), \\ \dot{\phi}_3 &= \omega_3. \end{aligned} \tag{4.6}$$

Straightforward calculations show that the periodic orbits in $T_2^{-1}(M^h)$ (described by $I_1 = 0$) can be labelled by the two parameters

$$\alpha = I_2 = \frac{1}{2}(q_2^2 + p_2^2) \in [0, \frac{h}{2}], \quad \beta = \omega_3\phi_2 - 2\phi_3 \bmod \omega_3\frac{2\pi}{\kappa_2}, \tag{4.7}$$

which, combined with the definition of T_2, gives a parametrization of F^h. The parametrization is singular on $\pi_F(N_3^h)$ and $\pi_F(N_2^h)$: all points of F^h with $\alpha = 0$ should be identified with $\pi_F(N_3^h)$ and all points with $\alpha = h/2$ should be identified with $\pi_F(N_2^h)$. Using (4.2) and (4.7) we can see that on the set $\mathcal{M}^h$ the coordinates (x, K, ψ) and the parameters (α, β) satisfy the following relationships:

$$\begin{array}{ll} \dot{\psi}_1 = 1, & K_1 = h, \\ \psi_2 = -\beta/2 \pm \omega_3\pi/4, & K_2 = (h - 2\alpha)/\omega_3, \\ x_1 = \pm\sqrt{2\alpha}, & x_2 = 0. \end{array} \tag{4.8}$$

Factoring out the uniform rotation in the ψ_1 coordinate in the above equations is equivalent to passing to the quotient space $\mathcal{F}^h$. Performing this, we find from (4.8) that $\mathcal{F}^h$ can be identified with the set defined by the equation

$$x_1^2 + \omega_3 K_2 = h. \tag{4.9}$$

But (4.9) represents a 2-sphere in the (x_1, K_2, ψ_2) space which proves the first statement of (*iii*). Choosing a compact set M_0^h as in statement (*iii*) of the proposition,

(4.7) and (4.8) show that $\mathcal{F}_0^h = \pi_{\mathcal{F}}(\mathcal{M}_0^h)$ has two components: one on the northern hemisphere ($x_1 > 0$) and one on the southern hemisphere ($x_1 < 0$) of $\mathcal{F}^h$. We denote these two components by $\mathcal{A}_0^1$ and $\mathcal{A}_0^2$, respectively, and note that both can be globally parametrized by the variables (K_2, ψ_2). From (4.2) we in fact obtain a coordinate representation of $Q^h|\mathcal{A}_0^1 \cup \mathcal{A}_0^2$ in the form

$$\begin{array}{rcl} Q_0^h\colon \mathcal{F}_0^h = \mathcal{A}_0^1 \cup \mathcal{A}_0^2 & \longrightarrow & \pi_F(M_0^h) \subset F^h, \\ (K_2, \psi_2) & \mapsto & (\frac{1}{2}(h - K_2\omega_3), 2\psi_2 - \omega_3\frac{\pi}{2}\operatorname{sign} x_1), \end{array} \tag{4.10}$$

which, together with (3.13) and (4.7) implies the second part of statement (iii) of the proposition. □

We now rewrite the full normalized Hamiltonian (2.9) in terms of the (x, K, ψ) coordinates to obtain

$$H(x, K, \psi_2; \varepsilon) = H_0(x, K) + \varepsilon^2 \tilde{H}_4(x, K, \psi_2) + \cdots + \varepsilon^{r-2} H_r(x, K, \psi_2). \tag{4.11}$$

Note that the quantities listed in (4.11) do not depend on ψ_1, which follows from the fact that the symplectic transformations defined in (4.1)–(4.2) preserve the bracket relations in (2.10). The underlying idea of our study will be the following: the Hamiltonian (4.11) has an unbroken S^1 symmetry corresponding to rotations in the ψ_1 coordinate which allows a reduction to a two-degree-of-freedom system. Due to the proper choice of our coordinates, this reduction can be carried out by fixing the value of the action K_1 and considering the Hamiltonian system which derives from (4.11) through the symplectic form

$$\omega_R = dx_1 \wedge dx_2 + d\psi_2 \wedge dK_2, \tag{4.12}$$

on the phase space

$$\mathcal{P}^h = \{\, (x, K_2, \psi_2) \,|\, x \in \mathbb{R}^2,\ K_2 \in \mathbb{R}^+,\ \phi_2 \in S^1 \,\}. \tag{4.13}$$

For the invariant manifolds $\mathcal{M}^h$ and $\mathcal{W}^h$ this reduction means a passage to the quotient spaces $\mathcal{F}^h$ and $\mathcal{W}^h/S^1$, respectively. Since $\mathcal{M}^h$ is entirely filled with periodic orbits, the two-degree-of-freedom reduced system will have a 2-manifold of equilibria which we identify with $\mathcal{F}^h$. This invariant 2-sphere will be connected to itself by a three-dimensional homoclinic manifold which we identify with $\mathcal{W}^h/S^1$. Any subset of $\mathcal{F}^h$ not containing the poles and the equator will therefore appear as a normally hyperbolic 2-manifold of equilibria which is connected to some other subset of $\mathcal{F}^h$ through the appropriate subset of the homoclinic 3-manifold $\mathcal{W}^h/S^1$. Based on (4.10), two relative equilibria $p_1 \neq p_2$ in $\mathcal{F}^h \backslash (\pi_F(\mathcal{N}_2^h) \cup \pi_F(\mathcal{N}_3^h))$ represent the same periodic solution in M^h if and only if $Q^h(p_1) = Q^h(p_2)$, i.e. one of the following holds:

1. p_1 and p_2 lie on the same hemisphere of $\mathcal{F}^h$, their K_2 coordinates are the same, and their ψ_2 coordinates differ by some integer multiple of

$$\tilde{\psi}_2 = \omega_3 \frac{\pi}{\kappa_2} = \frac{2\pi}{\kappa_1}. \tag{4.14}$$

2. p_1 and p_2 lie on different hemispheres of $\mathcal{F}^h$, their K_2 coordinates are the same, and their ψ_2 coordinates differ by

$$\bar{\psi}_2 = \omega_3 \pi (\frac{k}{\kappa_2} \pm \frac{1}{2}) \tag{4.15}$$

for some integer k.

We would like to find out what happens to $\mathcal{F}^h$ and $\mathcal{W}^h/S^1$ under the effect of higher order normal form terms in (4.11). Due to the presence of a whole manifold of equilibria, one cannot answer this question by a direct application of some version of the Melnikov-method (see Wiggins [1988] for a discussion of various Melnikov-type methods). Combining the Melnikov method with ideas from geometric singular perturbation theory of Fenichel [1979], one can obtain a partial picture of the perturbed dynamics near perturbed hyperbolic manifolds of equilibria (see Haller and Wiggins [1992]). To gain a more comprehensive understanding of the geometry of the Hamiltonian (4.11) we will use a new global perturbation technique, the *energy-phase method.* This method is generally applicable to perturbations of hyperbolic manifolds of equilibria in two-degree-of-freedom Hamiltonian systems. In contrast to the Melnikov method, it is capable to predict the existence of multiple pulse orbits which are negatively and positively asymptotic to invariant sets arising from the perturbation of manifolds of equilibria. For a detailed and general exposition of the theory the reader is referred to Haller and Wiggins [1995], while a review of some of the results related to our set-up can be found in the next section.

5 Perturbation of hyperbolic manifolds of equilibria: the energy-phase method

We consider two-degree-of-freedom Hamiltonian systems of the form

$$\dot{x} = J_2 D_x H_0(x, I) + \varepsilon J_2 D_x H_1(x, I, \phi; \varepsilon), \tag{5.1}$$

$$\dot{I} = -\varepsilon D_\phi H_1(x, I, \phi; \varepsilon), \tag{5.2}$$

$$\dot{\phi} = D_I H_0(x, I) + \varepsilon D_I H_1(x, I, \phi; \varepsilon), \tag{5.3}$$

on the phase space $\mathcal{L} \subset \mathbb{R}^2 \times \mathbb{R} \times S^1$ equipped with the symplectic form $\omega = dx_1 \wedge dx_2 + d\phi \wedge dI$. In (5.1) we used the notation

$$J_2 = \begin{pmatrix} 0 & 1 \\ -1 & 0 \end{pmatrix}.$$

We assume that the underlying Hamiltonian $H = H_0 + \varepsilon H_1$ is C^ρ smooth in its arguments with $\rho \geq 4$, and

(A1) There exist $I_1, I_2 \in \mathbb{R}$, $I_1 < I_2$ such that for $\varepsilon = 0$ any $I \in [I_1, I_2]$ (5.1) has two hyperbolic fixed points, $\bar{x}^1(I)$ and $\bar{x}^2(I)$, connected by a cycle of heteroclinic trajectories, $x^{h1}(t, I)$ and $x^{h2}(t, I)$, with

$$\lim_{t \to -\infty} x^{h1}(t, I) = \lim_{t \to +\infty} x^{h2}(t, I) = \bar{x}^1(I),$$
$$\lim_{t \to +\infty} x^{h1}(t, I) = \lim_{t \to -\infty} x^{h2}(t, I) = \bar{x}^2(I).$$

(A2) For every $I \in [I_1, I_2]$

$$D_I H_0\big(\bar{x}^j(I), I\big) = 0, \quad j = 1, 2. \tag{5.4}$$

It follows from assumption (A1) that system (5.1)–(5.3) possesses two two-dimensional invariant manifolds (with boundary) defined as

$$\mathcal{A}_0^j = \{(x, I, \phi) \in \mathcal{L} |\ x = \bar{x}^j(I), I \in [I_1, I_2], \phi \in S^1\}, \quad j = 1, 2\ .$$

These sets are the images of the annulus $A = [I_1, I_2] \times S^1$ under the embeddings

$$\begin{array}{rcl} g_0^j : A & \rightarrow & \mathcal{L}, \\ (I, \phi) & \mapsto & (\bar{x}^j(I), I, \phi). \end{array} \tag{5.5}$$

Note that $\mathcal{A}_0^1$ has a three-dimensional unstable manifold $W^u(\mathcal{A}_0^1)$ which coincides with the three-dimensional stable manifold $W^s(\mathcal{A}_0^2)$ to form a heteroclinic manifold $\mathcal{W}_0^1$. Assumption (A1) also shows that there exists another similar heteroclinic manifold in the phase space, defined as $\mathcal{W}_0^2 = W^s(\mathcal{A}_0^1) \cap W^u(\mathcal{A}_0^2)$.

As a consequence of assumption (A2), the manifolds $\mathcal{A}_0^j$ are entirely filled with equilibria. Solutions of (5.1)–(5.3) falling in $\mathcal{W}_0^j$ are heteroclinic connections between these equilibria. An important quantity associated with these heteroclinic connections is the net change of the coordinate ϕ along them. This change will be referred to as *phase shift*, and denoted by $\Delta\phi^j$ for orbits in $\mathcal{W}_0^j$. From (5.3) we easily find that

$$\Delta\phi^j(I) = \int_{-\infty}^{+\infty} D_I H_0(x^{hj}(t, I), I)\, dt, \quad j = 1, 2. \tag{5.6}$$

Since, for $j = 1, 2$, $\mathcal{A}_0^j$ is a compact normally hyperbolic invariant manifold, for small $\varepsilon > 0$ system (5.1)–(5.3) has a two-dimensional invariant manifold $\mathcal{A}_\varepsilon^j$, which is C^r ε-close to $\mathcal{A}_0^j$, and is still a C^r embedding of the annulus A through the map

$$g_\varepsilon^j : A \rightarrow \mathcal{L},$$

$$(I, \phi) \mapsto (\bar{x}_\varepsilon^j(I, \phi), I, \phi) = (\bar{x}^j(I) + \varepsilon\tilde{x}^j(I, \phi; \varepsilon), I, \phi). \tag{5.7}$$

Let $i_\varepsilon^j : \mathcal{A}_\varepsilon^j \hookrightarrow \mathcal{P}$ be the inclusion map of $\mathcal{A}_\varepsilon^j$ with $\varepsilon \geq 0$. Then it can be shown (see e.g. Haller and Wiggins [1992]) that for small $\varepsilon \geq 0$ and $j = 1, 2$ $(\mathcal{A}_\varepsilon^j, (i_\varepsilon^j)^*\omega)$ is a symplectic 2-manifold with

$$(i_\varepsilon^j)^*\omega = (1 + \mathcal{O}(\varepsilon)) d\phi \wedge dI,$$

on which the vector field (5.1)–(5.3) derives from the *restricted Hamiltonian*

$$\mathcal{H}_\varepsilon^j = H|\mathcal{A}_\varepsilon^j = (i_\varepsilon^j)^* H = h_0 + \varepsilon\mathcal{H}^j + \mathcal{O}(\varepsilon^2), \tag{5.8}$$

with

$$\begin{array}{rcl} h_0 & = & H_0|\mathcal{A}_0^j = const., \\ \mathcal{H}^j(I, \phi) & = & H_1(\bar{x}^j(I), I, \phi; 0). \end{array} \tag{5.9}$$

Equation (5.8) justifies the usual terminology *slow manifold* for a manifold perturbing from a set of equilibria. We call $\mathcal{H}^j(I, \phi)$ the *reduced Hamiltonian* corresponding to the manifold $\mathcal{A}_0^j$, and consider it defined on the annulus A. We say that an orbit $\gamma \subset A$ of some Hamiltonian system defined on A is an *internal orbit* if it is either a periodic orbit or an orbit homoclinic to a hyperbolic fixed point, *and* it is bounded away from ∂A. Similarly, an orbit $\gamma_\varepsilon \in \mathcal{A}^j_\varepsilon$ of the restricted Hamiltonian $\mathcal{H}^j_\varepsilon$ is called an *internal orbit* if $(g^j_\varepsilon)^{-1}(\gamma_\varepsilon)$ is an internal orbit of the Hamiltonian $(g^j_\varepsilon)^*\mathcal{H}^j_\varepsilon$ on $(A, (g^j_\varepsilon)^*\omega)$. By definition, internal orbits are structurally stable, hence for small ε the internal orbits of the reduced Hamiltonian $\mathcal{H}^j$ give rise to C^r ε-close internal orbits of $(g^j_\varepsilon)^*\mathcal{H}_\varepsilon$ (see (5.8)). We also note that for small nonzero ε we also have persisting locally invariant 3-manifolds $W^s_{loc}(\mathcal{A}^j_\varepsilon)$ and $W^u_{loc}(\mathcal{A}^j_\varepsilon)$ C^r ε-close to $W^s_{loc}(\mathcal{A}^j_0)$ and $W^u_{loc}(\mathcal{A}^j_0)$, respectively (see Fenichel [1971]). As usual, $W^s_{loc}(\mathcal{A}^j_\varepsilon)$ and $W^u_{loc}(\mathcal{A}^j_\varepsilon)$ can be extended to globally defined invariant sets $W^s(\mathcal{A}^j_\varepsilon)$ and $W^u(\mathcal{A}^j_\varepsilon)$.

In the following we will give conditions for the existence of *N-pulse heteroclinic orbits* connecting the two manifolds $\mathcal{A}^1_\varepsilon$ and $\mathcal{A}^2_\varepsilon$ to one another. An N-pulse orbit is negatively asymptotic to e.g. $\mathcal{A}^1_\varepsilon$, leaves a neighborhood of $\mathcal{A}^1_\varepsilon$ to enter a neighborhood of the slow manifold $\mathcal{A}^2_\varepsilon$, then passes back to a neighborhood of $\mathcal{A}^1_\varepsilon$, etc. Finally, after making $N-1$ excursions, the orbit tends asymptotically to some invariant set in $\mathcal{A}^2_\varepsilon$. We will discuss the existence of such orbits under the simplifying assumption:

> **(A3)** For any $\varepsilon > 0$ $W^s(\mathcal{A}^1_\varepsilon) = W^u(\mathcal{A}^2_\varepsilon)$, i.e. $\mathcal{W}^2_0$ possibly deforms but does not break under the perturbation considered.

This symmetry assumption is not necessary for the general theory, but greatly simplifies the statement of the results. Moreover, it will be shown to hold for the perturbation problem outlined in the previous section.

The program of the energy-phase method described below is to predict the existence of multiple-pulse connections between the slow manifolds $\mathcal{A}^1_\varepsilon$ and $\mathcal{A}^2_\varepsilon$ using the *n-th order energy-difference function* $\Delta^n\mathcal{H}\colon A \to \mathbb{R}$ defined as

$$\Delta^n\mathcal{H}(I,\phi) = \mathcal{H}^2(I,\phi) - \mathcal{H}^1(I,\phi - n\Delta\phi^1(I) - (n-1)\Delta\phi^2(I))$$

$$= H_1(\bar{x}^2(I), I, \phi; 0) - H_1(\bar{x}^1(I), I, \phi - n\Delta\phi^1(I) - (n-1)\Delta\phi^2(I); 0). \quad (5.10)$$

This function carries *energy*-type information from the perturbed problem and *phase*-type information from the unperturbed problem. To verify the conditions of the energy-phase method we will need to study the relation of the zero set of (5.10) to the orbits of the reduced Hamiltonian (5.9). To this end, we introduce the zero set V^n_+ of $\Delta^n\mathcal{H}$:

$$V^n_+ = \{\ (I,\phi) \in A \mid \Delta^n\mathcal{H}(I,\phi) = 0\ \}, \quad (5.11)$$

and the transversal zero set

$$Z^n_+ = \{\ (I,\phi) \in V^n_+ \mid D\Delta^n\mathcal{H}(I,\phi) \neq (0,0)\ \}. \quad (5.12)$$

We will also need the $-\big(n\Delta\phi^1(I) + (n-1)\Delta\phi^2(I)\big)$ translate of these sets defined as

$$V^n_- = \mathcal{R}^{-1}_n(V^n_+), \qquad Z^n_- = \mathcal{R}^{-1}_n(Z^n_+), \quad (5.13)$$

with the rotation map

$$\begin{aligned} \mathcal{R}_j\colon A &\to A, \\ (I,\phi) &\mapsto (I, \phi + j\Delta\phi^1(I) + (j-1)\Delta\phi^2(I)). \end{aligned} \quad (5.14)$$

For any internal orbit $\gamma^1 \subset A$ of $\mathcal{H}^1$ we introduce the *pulse number*

$$N(\gamma^1) = \min\{\, n \geq 1 \,|\, V_-^k \cap \gamma^1 = \emptyset,\ k = 1, ..., n-1,\ Z_-^n \pitchfork \gamma^1 \,\}, \tag{5.15}$$

where the symbol $\pitchfork$ refers to *nonempty transversal intersection.* We will also need the *resonant pulse number*

$$N_R(\gamma^1) = \min\{\, n \geq 1 \,|\, V_-^k \cap \gamma^1 = \emptyset,\ k = 1, ..., n-1,\ \gamma^1 \subset V_-^n \,\}. \tag{5.16}$$

One can easily verify that if $N \equiv N(\gamma^1) < \infty$ then $\mathcal{R}_N(\gamma^1)$ intersects an orbit γ^2 of $\mathcal{H}^2$ transversally in the annulus A, such that the two orbits are "isoenergetic", i.e. $\mathcal{H}^1|\gamma^1 = \mathcal{H}^2|\gamma_2$. Moreover, $j = N$ is the minimal index for $\mathcal{R}_j$ such that such a transverse isoenergetic intersection occurs. Also note that $N_R \equiv N_R(\gamma^1) < \infty$ implies $N(\gamma^1) = \infty$. In that case $j = N_R$ is the minimum index for which $\mathcal{R}_j$ maps γ^1 *onto* an orbit γ^2 of $\mathcal{H}^2$ with the same energy. The energy-phase method predicts intersections of stable and unstable manifolds of internal orbits of the restricted Hamiltonians $\mathcal{H}_\varepsilon^1$ and $\mathcal{H}_\varepsilon^2$, based on the intersection of $\mathcal{R}_N(\gamma^1)$ or $\mathcal{R}_{N_R}(\gamma^1)$ with γ^2 or γ^1, respectively.

Theorem 5.1 *Let us assume that (A1)–(A3) hold. Suppose that for an internal orbit $\gamma_0^1 \subset A$ of the reduced Hamiltonian $\mathcal{H}^1$*

(A4) $N \equiv N(\gamma_0^1) < \infty$.

(A5) *Let $b_1 \in Z_-^N \cap \gamma_0^1$ and $b_2 = \mathcal{R}_N(b_1)$. Assume that the orbit $\gamma_0^2 \subset A$ of the reduced Hamiltonian $\mathcal{H}^2$ which contains b_2 is an internal orbit with $Z_+^N \pitchfork \gamma_0^2$.*

(A6) *If $D_x H_0$ points outwards on the heteroclinic cycle $\mathcal{W}^1 \cup \mathcal{W}^2$ of manifolds, we assume that*

$$h_0^1 = \mathcal{H}^1|\gamma_0^1 < \mathcal{H}^2|\mathcal{R}_k(\gamma_0^1), \quad k = 1, ..., N-1.$$

If $D_x H_0$ points inwards on $\mathcal{W}^1 \cup \mathcal{W}^2$ we assume that

$$h_0^1 = \mathcal{H}^1|\gamma_0^1 > \mathcal{H}^2|\mathcal{R}_k(\gamma_0^1), \quad k = 1, ..., N-1.$$

Then there exists $\varepsilon_0 > 0$ such that for $0 < \varepsilon < \varepsilon_0$ the following are satisfied:

(i) There exists an N-pulse heteroclinic orbit y_ε^N which is negatively asymptotic to an internal orbit $\gamma_\varepsilon^1 \in \mathcal{A}_\varepsilon^1$ and positively asymptotic to an internal orbit $\gamma_\varepsilon^2 \in \mathcal{A}_\varepsilon^2$. Moreover, $(g_\varepsilon^1)^{-1}(\gamma_\varepsilon^1)$ and γ_0^1 are $C^{\rho-1}$ ε-close, and $(g_\varepsilon^2)^{-1}(\gamma_\varepsilon^2)$ and γ_0^2 are C^0 ε-close. If γ_0^2 is periodic, this latter statement can be strengthened to "$C^{\rho-1}$ ε-close".

(ii) y_ε^N lies in the intersection of $W^u(\gamma_\varepsilon^1)$ and $W^s(\gamma_\varepsilon^2)$ which is transversal within the energy surface $\{H = h\}$ with $h = H|\gamma_\varepsilon^1 = H|\gamma_\varepsilon^2$.

If, alternatively, we assume that

(A4′) $N \equiv N_R(\gamma_0^1) < \infty$,

and $A1 - A3$ and $A6$ still hold, then for small $\varepsilon > 0$ we have:

(i′) There exist two N-pulse heteroclinic orbits, y_ε^N and $\tilde{y}_\varepsilon^N$, which are asymptotic in negative time to an internal orbit $\gamma_\varepsilon^1 \in \mathcal{A}_\varepsilon^1$ and is positive time to an internal orbit $\gamma_\varepsilon^2 \in \mathcal{A}_\varepsilon^2$. The orbits γ_0^1 and $(g_\varepsilon^1)^{-1}(\gamma_\varepsilon^1)$ are $C^{\rho-1}$ ε-close in A, while the orbits $\gamma_0^2 = \mathcal{R}_N(\gamma_0^1)$ and $(g_\varepsilon^2)^{-1}(\gamma_\varepsilon^2)$ are C^0 ε-close in A. If γ_0^1 is periodic, this latter statement can be strengthened to "$C^{\rho-1}$ ε-close".

(ii') y_ε^N and $\tilde{y}_\varepsilon^N$ lie in the intersection of $W^u(\gamma_\varepsilon^1)$ and $W^s(\gamma_\varepsilon^2)$ within the energy surface $H = h$ with $h = H|\gamma_\varepsilon^1 = H|\gamma_\varepsilon^2$.

Proof See Haller and Wiggins [1993] for a proof, discussion of the related geometry, and more results. □

For a detailed description of the usage of the energy-phase method for resonance bands the reader is referred to Haller and Wiggins [1995]. More details regarding the application of the method to 3DOF resonances will appear in Haller and Wiggins (in preparation).

6 Whiskered tori and chaos in the normal form

In this section we apply the method described above to study the effects of the higher order terms on the cubic truncation of the normalized Hamiltonian (4.11). In accordance with our discussion in Section 2, we will first analyze the two-degree-of-freedom subsystem

$$\dot{x} = J_2 D_x H_0(x, K_2) + \varepsilon J_2 D_x H_1(x, K_2, \psi_2; \varepsilon), \tag{6.1}$$

$$\dot{K}_2 = -\varepsilon D_{\psi_2} H_1(x, K_2, \psi_2; \varepsilon), \tag{6.2}$$

$$\dot{\psi}_2 = D_{K_2} H_0(x, K_2) + \varepsilon D_I H_1(x, I, \phi; \varepsilon), \tag{6.3}$$

with

$$H_0(x, K_2) = a(K_1 - \omega_3 K_2 - |x|^2)x_2, \tag{6.4}$$

$$H_1(x, K_2, \psi_2; \varepsilon) = \tilde{H}_4(x, h, K_2, \psi_2) + \sum_{j=5}^{r} \varepsilon^{j-2} H_j(x, h, K_2, \psi_2), \tag{6.5}$$

which derives from (4.11) on $(\mathcal{P}^h, \omega_R)$ (see (4.12)–(4.13) and note that in (6.1)–(6.3) we rescaled the time by ε). As we noted in the proof of Proposition 3.1, the reduction with respect to the S^1 symmetry in the ψ_1 coordinate reduces $\mathcal{M}^h$ (which is diffeomorphic to $S^2 \times S^1$) to its orbit space $\mathcal{F}^h$ which is diffeomorphic to S^2. Thus system (6.1)–(6.3) possesses a two-dimensional manifold of equilibria which is a 2-sphere and has a three-dimensional homoclinic manifold $\mathcal{W}^h$ attached to it. We will consider two subsets of the southern and northern hemispheres of $\mathcal{F}^h$ which will play the same role as the manifolds of equilibria $\mathcal{A}_0^1$ and $\mathcal{A}_0^2$, respectively, of the previous section. This will enable us to apply the energy phase method and obtain information for $\varepsilon > 0$ small on the dynamics of the system defined by the normal form Hamiltonian (4.11).

Fixing some small but arbitrary $\lambda > 0$, one can easily verify that for $\varepsilon = 0$ and $K_2 \in [\lambda, h/\omega_3 - -\lambda]$ (6.1) has two hyperbolic equilibria

$$\bar{x}^1(K_2) = (-\sqrt{h - K_2\omega_3}, 0), \quad \bar{x}^2(K_2) = (+\sqrt{h - K_2\omega_3}, 0). \tag{6.6}$$

Moreover, these equilibria are connected by three heteroclinic trajectories, $x^{h1}(t;K_2)$, $x^{h2+}(t;K_2)$, and $x^{h2-}(t;K_2)$ with

$$\begin{aligned}
\lim_{t\to-\infty} x^{h1}(t,K_2) &= \lim_{t\to+\infty} x^{h2+}(t,K_2) = \lim_{t\to+\infty} x^{h2-}(t,K_2) = \bar{x}^1(K_2),\\
\lim_{t\to+\infty} x^{h1}(t,K_2) &= \lim_{t\to-\infty} x^{h2-}(t,K_2) = \lim_{t\to-\infty} x^{h2+}(t,K_2) = \bar{x}^2(K_2),\\
\{x^{h1}(t,K_2)\}_{t=-\infty}^{\infty} &= \{x|\,|x_1| < \sqrt{h-K_2\omega_3},\, x_2=0\},\\
\{x^{h2+}(t,K_2)\}_{t=-\infty}^{\infty} &= \{x|\,x_2>0,\, x_1^2+x_2^2 = h-K_2\omega_3\},\\
\{x^{h2-}(t,K_2)\}_{t=-\infty}^{\infty} &= \{x|\,x_2<0,\, x_1^2+x_2^2 = h-K_2\omega_3\}.
\end{aligned}$$

Hence (6.1) satisfies assumptions (A1)–(A2) (see also (6.3) above) with two possible choices for the heteroclinic trajectory $x^h(I,\phi)$. Consequently, all the sets and quantities of Section 3 can be defined for system (6.1)–(6.3) with the substitution $(I,\phi)\to(K_2,\psi_2)$, and $(\mathcal{L},\omega)\to(\mathcal{P}^h,\omega_R)$. In particular, introducing $A=[\lambda, h/\omega_3-\lambda]\times S^1$, we have two normally hyperbolic 2-manifolds

$$\begin{aligned}
\mathcal{A}_0^1 &= \{(x,K_2,\psi_2)\in\mathcal{P}_h\,|\,x_1 = -\sqrt{h-\omega_3K_2},\, x_2=0,\,(K_2,\psi_2)\in A\},\\
\mathcal{A}_0^2 &= \{(x,K_2,\psi_2)\in\mathcal{P}_h\,|\,x_1 = +\sqrt{h-\omega_3K_2},\, x_2=0,\,(K_2,\psi_2)\in A\},
\end{aligned}$$

connected through the heteroclinic manifolds

$$\begin{aligned}
\mathcal{W}_0^1 &= \{(x,K_2,\psi_2)\in\mathcal{P}^h\,|\,|x_1| < \sqrt{h-\omega_3K_2},\, x_2=0,\,(K_2,\psi_2)\in A\},\\
\mathcal{W}_0^{2+} &= \{(x,K_2,\psi_2)\in\mathcal{P}^h\,|\,x_1^2+x^2 = h-\omega_3K_2,\, x_2>0\,(K_2,\psi_2)\in A\},\\
\mathcal{W}_0^{2-} &= \{(x,K_2,\psi_2)\in\mathcal{P}^h\,|\,x_1^2+x^2 = h-\omega_3K_2,\, x_2<0\,(K_2,\psi_2)\in A\}.
\end{aligned}$$

The phase shifts defined in (5.6) can be computed to equal

$$\Delta\psi_2^1(K_2)=0,\quad \Delta\psi_2^{2\pm}(K_2) = \pm\tilde{\psi}_2 = \pm\omega_3\frac{\pi}{\kappa_2} = \pm\frac{2\pi}{\kappa_1}, \tag{6.7}$$

where the $+$ sign is valid for the phase shift on $\mathcal{W}_0^{2+}$, while the $-$ sign is valid on $\mathcal{W}_0^{2-}$. The *restricted Hamiltonians* $\mathcal{H}_\varepsilon^j$ defined in (5.8) take the form

$$\mathcal{H}_\varepsilon^j(K_2,\psi_2) = \big(H_0(x,K_2)+\varepsilon H_1(x,K_2,\psi_2;\varepsilon)\big)|\mathcal{A}_\varepsilon^j = h_0+\varepsilon\mathcal{H}^j(K_2,\psi_2)+\mathcal{O}(\varepsilon^2), \tag{6.8}$$

with the *reduced Hamiltonians* being specifically

$$\mathcal{H}^j(K_2,\psi_2) = \tilde{H}_4((-1)^j\sqrt{h-K_2\omega_3},0,h,K_2,\psi_2),\quad j=1,2. \tag{6.9}$$

Note that by our discussion after Proposition 3.1 (see (4.14)–(4.15)) for any $k\in\mathbb{Z}$ we have

$$\begin{aligned}
\mathcal{H}^j(K_2,\psi_2+k\tilde{\psi}_2) &= \mathcal{H}^j(K_2,\psi_2),\quad j=1,2, &(6.10)\\
\mathcal{H}^1(K_2,\psi_2) &= \mathcal{H}^2(K_2,\psi_2+\bar{\psi}_2). &(6.11)
\end{aligned}$$

To satisfy assumption (A3) of Section 4 we require the following

(A3′) The resonant module M has no element of the form $(1,n_2,n_3)$ with $|n_2|+|n_3|=3$.

An easy calculation shows that assumption (A3$'$) implies the quartic truncation $H^* = H_0 + \varepsilon^2 \tilde{H}_4$ of (2.9) to have an invariant 3-sphere M_ε^h which lies in the transversal intersection of the energy surface $H^* = h$ and the subspace $z_1 = 0$. Applying the transformation $T_3^{-1} \circ T_2^{-1}$ to this set one obtains two sets, $\mathcal{M}_\varepsilon^h$ and $\mathcal{D}_\varepsilon^h$, with the same properties as $\mathcal{M}^h$ and $\mathcal{D}^h$ in Proposition 3.1. But $\mathcal{W}^{2+} \cup \mathcal{W}^{2-} = \mathcal{D}^h/S^1$, implying that both $\mathcal{W}^{2+}$ and $\mathcal{W}^{2-}$ persist unbroken under the effect of terms deriving from $\tilde{H}_4 = H_1|_{\varepsilon=0}$, hence (A3) holds for system (6.1)–(6.3). Note that (A3') is always satisfied if κ_1 is odd.

To apply the energy-phase method we need to compute the energy difference function as defined in (5.10). In the case of system (6.1)–(6.3), using (4.14), (4.15) with $k = 0$, (6.10)–(6.11), and (6.7), we can write

$$\begin{aligned} \Delta^1 \mathcal{H}(K_2, \psi_2) &= \tilde{H}_4(\sqrt{h - K_2\omega_3}, 0, h, K_2, \psi_2) \\ &\quad -\tilde{H}_4(\sqrt{h - K_2\omega_3}, 0, h, K_2, \psi_2 + \tfrac{\pi\kappa_2}{\kappa_1}), \end{aligned} \tag{6.12}$$

$$\Delta^2 \mathcal{H}(K_2, \psi_2) \equiv 0. \tag{6.13}$$

These equations and (5.11)–(5.13) immediately imply

$$V_+^1 = V_-^1, \quad Z_+^1 = Z_-^1, \quad V_+^2 = V_-^2 = A, \quad Z_+^2 = Z_-^2 = \emptyset. \tag{6.14}$$

Then (5.15), (5.16), and (6.14) show that for the pulse numbers N and N_R of any internal orbit $\gamma^1 \subset A$ of $\mathcal{H}^1$ the following hold:

Fact 1 $N(\gamma_1) = 1$ and $N_R(\gamma_1) = \infty$ if $\gamma^1 \pitchfork Z_-^1$,

Fact 2 $N(\gamma_1) = \infty$ and $N_R(\gamma_1) = 2$ if $\gamma^1 \cap Z_-^1 = \emptyset$.

Based on the results of Section 4, this shows that if $Z_-^1 = Z_+^1 \neq \emptyset$ then almost all internal orbits in $\mathcal{A}_\varepsilon^1$ and $\mathcal{A}_\varepsilon^2$ will participate either in transverse single-pulse heteroclinic connections or (not necessarily transverse) double-pulse connections.

We are now ready to apply Theorem 4.1 to (6.1)–(6.3). However, we will state the results immediately in terms of the full blown-up normal form Hamiltonian (4.11) defined on $(\mathcal{P}, \omega)$. This can be done using the fact that for $\varepsilon > 0$ any invariant set $\mathcal{S}$ in the phase space $(\mathcal{P}^h, \omega_R)$ indicates the existence of an invariant set $\mathcal{S}^*$ in the $K_1 = h$ hypersurface of the phase space $(\mathcal{P}, \omega)$, such that $\mathcal{S}^*$ is diffeomorphic to $\mathcal{S}^* \times S^1$. We will address the question of existence of these structures for the original normal form (2.9) after Theorem 5.1.

Note that for small fixed $\varepsilon > 0$ the normally hyperbolic manifolds $\mathcal{A}_\varepsilon^j \subset \mathcal{P}^h$ of (6.1)–(6.3) give rise to normally hyperbolic invariant 4-manifolds $\mathcal{B}_\varepsilon^j \subset \mathcal{P}$ of the full normal form (4.11), which are given by the two $C^{\rho-1}$ smooth embeddings

$$\begin{aligned} G_\varepsilon^j \colon \mathbb{R}^+ \times S^1 \times A &\to \mathcal{P}, \\ (K_1, \psi_1, K_2, \psi_2) &\mapsto (K_1, \psi_1, g_\varepsilon^j(K_2, \psi_2)). \quad j = 1, 2. \end{aligned} \tag{6.15}$$

Although it is not hyperbolic, $\mathcal{B}_0^j$ still exists as a smooth limit of manifolds, with the corresponding embedding $G_0^j : \mathbb{R}^+ \times S^1 \times A \to \mathcal{P}$, $j = 1, 2$. To facilitate the statement of the results, we introduce the set

$$E_\varepsilon(h_1, h_2) = \{\, (x, K_1, \psi_1, K_2, \psi_2) \in \mathcal{P} \,\big|\, K_1 = h_1,\ H(x, K, \psi; \varepsilon) = h_2 \,\}.$$

We then have the following theorem as an immediate application of Theorem 4.1 to system (6.1)–(6.3) with the results interpreted for the full normal form (4.11).

Theorem 6.1 *Suppose that assumption (A3′) is satisfied and for some* $\bar{h} \in \mathbb{R}^+$

(α) $Z^1_+ = Z^1_- \neq \emptyset$, *where* $Z^1_+ = Z^1_-$ *is the transverse zero set of* $\Delta^1\mathcal{H}$ *defined in (6.12).*

(β) γ^1 *and* γ^2 *are internal orbits of the reduced Hamiltonians* $\mathcal{H}^1$ *and* $\mathcal{H}^2$ *(see (6.9)), respectively, and both intersect* Z^1_+ *transversally at* $p = (\bar{K}_2, \bar{\psi}_2) \in A$.

Then there exists $\varepsilon_0 > 0$ *and an open set* $U \subset \mathbb{R}^+$ *with* $\bar{h} \in U$ *such that for* $0 < \varepsilon < \varepsilon_0$ *and* $K_1 = h \in U$ *the following are satisfied:*

(i) *Let* $I^j_\varepsilon \colon \mathcal{B}^j_\varepsilon \to \mathcal{P}$ *be the inclusion map of* $\mathcal{B}^j_\varepsilon$. *Then* $(\mathcal{B}^j_\varepsilon, (I^j_\varepsilon)^*\omega)$ *is a symplectic 4-manifold for* $j = 1, 2$, *on which the (integrable) dynamics is generated by the restricted Hamiltonian* $\mathcal{H}^j_\varepsilon$ *defined in (6.8).*

(ii) *The Hamiltonian (4.11) has two two-dimensional invariant manifolds,* $\mathcal{T}^1_\varepsilon$ *and* $\mathcal{T}^2_\varepsilon$. *If* γ^j *is periodic then* $\mathcal{T}^j_\varepsilon$ *is* $C^{\rho-1}$ *diffeomorphic to* $\mathbb{T}^2$, *and is* $C^{\rho-1}$ ε-*close to the set* $G^j_0(\{h\} \times S^1 \times \gamma^j)$. *If* γ^j *is not periodic, one has to replace "*$C^{\rho-1}$ *diffeomorphic" with "homeomorphic" and "*$C^{\rho-1}$ ε-*close" with "*C^0 ε-*close" in the above statement.*

(iii) $\mathcal{T}^1_\varepsilon$ *has a three-dimensional unstable manifold* $W^u(\mathcal{T}^1_\varepsilon)$ *and* $\mathcal{T}^2_\varepsilon$ *has a three-dimensional stable manifold* $W^s(\mathcal{T}^2_\varepsilon)$. $W^u(\mathcal{T}^1_\varepsilon)$ *and* $W^s(\mathcal{T}^2_\varepsilon)$ *are injective immersions of* $\mathbb{R}^+ \times S^1 \times \gamma^1$ *and* $\mathbb{R}^+ \times S^1 \times \gamma^2$, *respectively. Furthermore, they intersect transversally within* $E(h, H|\mathcal{T}^1_\varepsilon)$ *in a two-dimensional heteroclinic manifold* Y_ε.

Suppose that assumption (α) *above holds together with*

(β') γ^1 *and* γ^2 *are internal orbits of the reduced Hamiltonians* $\mathcal{H}^1$ *and* $\mathcal{H}^2$ *respectively, with* $\mathcal{R}_2(\gamma^1) = \gamma^2$ *for some choice of the sign in (6.7).*

Then statements (i)–(ii) *still hold. However, in contrast to* (iii), *there exist two double-pulse heteroclinic manifolds,* Y^1_ε *and* Y^2_ε *lying in the (not necessarily transversal) intersection of* $W^u(\mathcal{T}^1_\varepsilon)$ *and* $W^s(\mathcal{T}^1_\varepsilon)$.

Proof The theorem is an immediate application of Theorem 4.1 but see Haller and Wiggins [1994] for more details. Here we only note that assumption (A6) of Theorem 4.1 is always satisfied for one of the two possible choices for the sign in (6.7), i.e. for one of the two possible definitions of the rotation map $\mathcal{R}_k$ in (5.14). □

Theorem 5.1 gives us information about structures which exist in the blown-up normal form. Since the blown-up normal form (4.11) is related to the original normal form (2.9) through a singular transformation, we have to argue that the orbits and invariant manifolds we found are indeed present in the original normal form. Simple calculations show that the coordinate expressions for the map $T = T_1 \circ T_2 \circ T_3$, $(x, K, \psi) \mapsto (q, p)$ (which relates (4.11) to (2.9)) can be written as

$$\begin{aligned}
q_1 &= \sqrt{2(K_1 - \omega_3 K_2 - |x|^2)}\cos\psi, & p_1 &= -\sqrt{2(K_1 - \omega_3 K_2 - |x|^2)}\sin\psi,\\
q_2 &= x_2\cos 2\phi_1 - x_1\sin 2\phi_1, & p_2 &= x_1\cos 2\phi_1 + x_2\sin 2\phi_1,\\
q_3 &= \sqrt{2K_2}\cos(\psi_2 + \omega_3\psi_1), & p_3 &= -\sqrt{2K_2}\sin(\psi_2 + \omega_3\psi_1).
\end{aligned}$$

This implies that T is a C^r diffeomorphism away from the sets defined by $K_2 = 0$ and $K_1 - \omega_3 K_2 - |x|^2 = 0$. Since $W^u_{loc}(\mathcal{T}^i_\epsilon)$ and $W^s_{loc}(\mathcal{T}^i_\epsilon)$ do not intersect these sets, it follows that invariant manifolds diffeomorphic to $W^u(\mathcal{T}^i_\epsilon)$ and $W^s(\mathcal{T}^i_\epsilon)$ will also exist for the Hamiltonian (2.9). We also know that subsets M^h of the type M^h_0 (see

Proposition 3.1) are normally hyperbolic; hence they persist as C^r-close manifolds, M_ε^h, in the original normal form under the effect of quartic terms. The closure of, e.g., $W_{loc}^u(\mathcal{T}_\epsilon^i)$ intersects M_ϵ^h in an object diffeomorphic to $\mathcal{T}_\epsilon^i$. Using the results of Fenichel [1979] on the foliation of unstable manifolds, one can conclude that this object is indeed an invariant subset of M_ϵ^h and it carries the same types of motions as $\mathcal{T}_\epsilon^i$ (see Haller and Wiggins [1994] for details). Consequently, all statements of Theorem 5.1 carry over to the original normalized Hamiltonian (2.9).

In most cases the internal orbits γ^1 and γ^2 in assumption (α) of Theorem 5.1 are members of families of periodic orbits in A. It may also happen that they map to the same set of points in the orbit space $\mathcal{F}^h$ under the map $T_2 \circ T_3$ (see e.g. Haller and Wiggins (in preparation) for the case of the 1: 2: 2 resonance). Thus transverse heteroclinic connections between their perturbed counterparts actually yield transverse homoclinic connections in system (2.9). For this case, using Theorem 5.1, the Smale-Birkhoff homoclinic theorem (Smale [1963]), and a theorem of Moser [1973] we obtain the more specific

Corollary 6.2 *Let us suppose that the assumptions (α)–(β) of Theorem 5.1 hold and γ^1 and γ^2 are periodic orbits with $\mathcal{R}_2(\gamma_1) = \gamma_2$. Then for $\varepsilon > 0$ small enough*

(*i*) *$\mathcal{T}_\varepsilon^1$ and $\mathcal{T}_\varepsilon^2$ are members of two-parameter families of whiskered 2-tori. The whiskers $W^u(\mathcal{T}_\varepsilon^1)$ and $W^s(\mathcal{T}_\varepsilon^2)$ intersect transversally within $E_\varepsilon(h, H|\mathcal{T}_\varepsilon^1)$.*

(*ii*) *System (2.9) does not possess any nontrivial analytic integral other than H and K_1.*

(*ii*) *On energy surfaces $\mathcal{O}(\varepsilon)$ close to the surface $H = \bar{h} \in U$ an appropriately defined Poincare-map of system (2.9) has invariant Cantor-sets on which it is homeomorphic to a full shift on N symbols.*

7 Discussion

In this note we sketched how the energy-phase method developed in Haller and Wiggins [1995] can be used to prove the existence of whiskered tori in a class of three-degree-of-freedom Hamiltonian resonances. A more detailed exposition and proofs can be found in Haller and Wiggins [1994], where the results for the normal form are also related to the full unnormalized system. This is done by an application of a version of the KAM theory to the whiskered tori we found, together with a weak hyperbolicity-type argument about the persistence of their whiskers.

Corollary 6.2 above implies that the normal form of the resonances we studied is generically non-integrable when truncated at (at least) quartic order. This non-integrability is the consequence of the presence of the transverse intersection of stable and unstable whiskers of the above 2-tori *within* the surface $E_\varepsilon(h, h + \mathcal{O}(\varepsilon))$. The splitting distance and the splitting angles corresponding to these transversal intersections are of the order $\mathcal{O}(\varepsilon^2)$, in particular, they are not exponentially small. Such phenomenon has also been noted by Duistermaat [1984] and Hoveijn [1992] for specific perturbations of the cubic normal forms of the 1: 1: 2 and the 1: 2: 3 resonances, respectively. This indicates much stronger chaotic behavior in three degrees of freedom than in two degrees of freedom, where the normal form truncated at any order is integrable.

For an illustration of our results on a concrete example the reader is referred to Haller and Wiggins (in preparation) where the 1: 2: 2 resonance with detuning (first considered by van der Aa and Verhulst [1984]) is analyzed. In that reference we show how one obtains a "cross section" of a stochastic layer in the normal form based on the phase portrait of the reduced Hamiltonian (6.9). We also confirm the existence of the invariant sets found above by the direct simulation of a related four dimensional symplectic map. We show numerically that the double pulse heteroclinic manifolds lie in the *transversal* intersection of stable and unstable manifolds of 2-tori, giving rise to additional Smale-horseshoes in the quartic normal form. Finally, we discuss how the whiskered tori we found form building blocks for Arnold-diffusion (see Arnold [1988]) near the origin in the full unnormalized system.

References

Abraham, R. and Marsden, J.E. [1978] *Foundations of Mechanics*, The Benjamin/Cummings Publishing Co. Inc., Reading.

Arms, J., Cushman, R. and Gotay, M. [1991] *A universal reduction procedure for Hamiltonian group actions* in: The Geometry of Hamiltonian Systems, (Ratiu, T., ed.), **33**, Springer-Verlag, New York.

Arnold,V.I. [1964] *Instability of dynamical systems with many degrees of freedom*, Dokl. Akad. Nauk., USSR, **156**, 9.

Arnold,V.I., Kozlov,V.V. and Neishtadt, A.I. [1988] *Mathematical Aspects of Classical and Celestial Mechanics* in: Dynamical Systems III, (V.I. Arnold, ed.), Springer-Verlag, New York.

Churchill, R.C., Kummer, M. and Rod, D.L. [1983] *On averaging, reduction and symmetry in Hamiltonian systems*, J. Diff. Eqs., **49**, 359.

Cushman, R. [1985] *Notes on the* 1: 2: 2 *resonance*, unpublished.

Cushman, R. and Rod, D.L. [1982] *The reduction of the* 1: 1 *resonance*, Physica D, **6**, 105.

Duistermaat, J.J. [1984] *Non-integrability of the* 1: 2: 1 *resonance*, Ergod. Theory and Dyn. Sys., **4**, 553.

Fenichel, N. [1971] *Persistence and smoothness of invariant manifolds for flows*, Ind. Univ. Math. J., **21**, 193.

Fenichel, N. [1979] *Geometric singular perturbation theory for ordinary differential equations*, J. Diff. Eqs., **31**, 53.

Haller, G. and Wiggins, S. [1992] *Orbits homoclinic to resonances: the Hamiltonian case*, Physica D, **66**, 298.

Haller, G. and Wiggins, S. [1995] *N-pulse homoclinic orbits in perturbations of resonant Hamiltonian systems*, to appear in: Arch. Rat. Mech. and Anal.

Haller, G. and Wiggins, S. [1994] *Geometry and chaos near resonant equilibria of 3DOF Hamiltonian systems: I. Theory*, submitted to Physica D.

Haller, G. and Wiggins, S. *Geometry and chaos near resonant equilibria of 3DOF Hamiltonian systems: II. Applications*, in preparation.

Haller, G. and Wiggins, S. [1994] *Multi-pulse jumping orbits and homoclinic trees in a two-mode truncation of the damped-forced nonlinear Schrödinger equation*, to appear in Physica D.

Holm, D. [1989] *Notes on reduction of integrable resonant normal forms*, unpublished.

Hoveijn, I. [1992] *Aspects of Resonance in Dynamical Systems*, Ph.D. thesis, Rijksuniversiteit Utrecht.

Knobloch, E., Mahalov, A. and Marsden, J. [1992] *Normal forms for three-dimensional parametric instabilities in ideal hydrodynamics*, preprint.

Kummer, M. [1975] *An interaction of three resonant modes in a nonlinear lattice*, J. Math. Anal. and Apps., **52**, 64.

Kummer, M. [1979] *On resonant classical Hamiltonians with two degrees of freedom near an equilibrium point*, in: Stochastic Behavior in Classical and Quantum Hamiltonian Systems, Lect. Notes in Phys., **93**, Spriger-Verlag, New-York.

Kummer, M. [1986] *On resonant Hamiltonian systems with finitely many degrees of freedom*, in: Local and Global Methods of Nonlinear Dynamics, Lect. Notes in Phys., **252**, Springer-Verlag, New York.

Kummer, M. [1990] *On resonant classical Hamiltonians with n frequencies*, J. Diff. Eqs., **83**, 220.

Lerman, E. and Sjamaar, R. [1991] *Stratified symplectic spaces and reduction*, Ann. Math., **134**, 375.

Moser, J. [1973] *Stable and Random Motions in Dynamical Systems*, Princeton University Press, Princeton.

Sanders, J.A. and Verhulst, F. [1985] *Averaging Methods in Nonlinear Dynamical Systems*, Springer-Verlag, New York.

Smale, S. [1963] *Diffeomorphisms with many periodic points*, in: Differential and Combinatorial Topology, Princeton Univ. Press, Princeton.

van der Aa, E. and Verhulst, F. [1984] *Asymptotic integrability and periodic solutions of a Hamiltonian system in* 1: 2: 2 *resonance*, SIAM J. Math. Anal., **15**, 890.

Wiggins, S. [1988] *Global Bifurcations and Chaos—Analytical Methods*, Springer-Verlag, New York.

Fields Institute Communications
Volume **4**, 1995

Normal Forms for Perturbations of the Euler Top

Heinz Hanßmann
Department of Mathematics
University of Groningen
P.O. Box 800
9700 AV Groningen
The Netherlands

Abstract. A method for studying the behaviour of a rigid body in a general force field is proposed. The study is restricted to the case of a perturbation of the Euler top by a small external force. A normal form approach yields a formal 2-torus symmetry, which in turn leads to the reduction to a one-degree-of-freedom system. The behaviour of this system is used to identify quasi-periodic motions of the rigid body with two or three independent frequencies.

The method is tested in the integrable cases of Lagrange and Kirchhoff. Furthermore, it is applied to other cases in which two of the principle moments of inertia of the rigid body are equal.

Introduction

Of all integrable cases of the rigid body the free one, named after Euler, deserves our special attention. On the one hand it is the limit of all other cases, integrable or not, when the potential governing the external forces goes to zero. Therefore, one can view the general problem of a rigid body in an arbitrary weak potential as a perturbation of the Euler case.

On the other hand the Euler case is even *superintegrable*. Although the number of degrees of freedom is three, it has four independent integrals: the energy and the three components of the angular momentum. Hence, the general motion is restricted to a 2-torus. Let us digress on the terminology used here. To this end we first replace one of the components of the angular momentum by its length. Then two of the integrals commute with all four, in particular there are three commuting integrals. Therefore our superintegrable system is a "degenerate" integrable system as discussed in Nehorošev [1972]. One also observes that the Lie algebra spanned by the energy and the three components of the angular momentum has rank 2. Hence, the present case has also the extra structure implied by noncommutative integrability in the sense of Mishchenko and Fomenko [1978].

1991 *Mathematics Subject Classification*. Primary 70E20; Secondary 58F05.

In an integrable case with three degrees of freedom and (only) three (commuting) integrals, a typical motion fills a 3-torus densely. In this case the behaviour of a slightly perturbed system is quite well understood by standard KAM-theory (see Arnol'd [1963,1974] or Pöschel [1982]). Most of the 3-tori survive the perturbation, the resonant ones break up.

However, the 2-tori of the Euler case generally will not lead to invariant 2-tori for the perturbed problem, *i.e.* for the motion of a rigid body subject to a weak external conservative force. Quite on the contrary, we expect only 'isolated' (in a sense that will become clear later) 2-tori to be present after the perturbation. If we know how these are distributed in phase space, how the flow behaves in their neighbourhood and what are the homoclinic and heteroclinic connections between them, we will get at least a partial understanding of the global behaviour.

In order to identify the surviving 2-tori, we express the perturbed Hamilton function in action-angle variables of the Eulerian problem. Because of its superintegrability the kinetic energy will only depend on two of the actions. By means of a normal form theorem we make the perturbed Hamilton function independent of the two conjugate angles — up to higher order terms. Then we truncate the Hamilton function. The two actions become integrals, and we can treat them as (distinguished) parameters. In this fashion we may reduce to a one-degree-of-freedom problem.

Now the program is clear. First we study this one-degree-of-freedom problem. At this point we see whether we truncated the Hamilton function at a sufficiently high order. We require the one-degree-of-freedom problem to be structurally stable as a family, parametrised by the two actions. In the examples treated in this paper these 'distinguished' parameters will indeed give a versal unfolding.

In more involved examples it may be necessary to vary the external force field. Then one can use also the parameters the potential energy of the rigid body depends on to get a versal unfolding. Here one has to be careful about the type of transformation in parameter space one allows. As a rule, a bifurcation of *e.g.* codimension 3 requires three distinguished and three external parameters for a versal unfolding. Having only two distinguished parameters in the one-degree-of-freedom problem, one would look for a versal unfolding of this 2-parameter family, again by three external parameters. See the Appendix of Broer et al. [1993] for a theory of unfoldings of families with distinguished parameters.

The second part of the program is to interpret the results in terms of the original problem. Naïvely one would relate fixed points to 2-tori and closed orbits to 3-tori. This gives the flow for the truncated Hamilton function. We will use a generalization of KAM-theory, as developed in Moser [1967] or Broer, Huitema and Takens [1990], to see how these tori are perturbed when the higher order terms are added.

This paper is organised as follows. In Section 1 we give a description of the possible motions of the free rigid body and introduce the action-angle variables. Section 2 contains the formulation and proof of the normal form theorem. In the remaining Sections 3–5 we apply the above program to some special cases. We will assume the body to be *dynamically symmetric*, by this we mean that two of its principal moments of inertia are equal. In this case it is possible to calculate the action-angle variables by means of elementary functions. The potentials we consider will be linear or quadratic. This renders the necessary computations feasible and includes the classical cases of Lagrange and Kirchhoff. In these cases we can compare the results gained by our approach with the known behaviour of the top.

1 Action-angle variables for the free rigid body

We will perturb the free rigid body, fixed at one point. Therefore, we now recapitulate the main facts about it. The reader may find a comprehensive introduction in Abraham and Marsden [1978], Arnol'd [1974], Cushman and Bates [1994] or Goldstein [1950]. We choose a set of axes $\vec{e}_x, \vec{e}_y, \vec{e}_z$ fixed in space and a body set of axes $\vec{e}_1, \vec{e}_2, \vec{e}_3$ along the principal axes of inertia (with the same origin and orientation). The configuration space is the group $SO(3)$ of orientation preserving orthogonal three-by-three matrices. The elements of $SO(3)$ specify how to transform $\vec{e}_x, \vec{e}_y, \vec{e}_z$ into $\vec{e}_1, \vec{e}_2, \vec{e}_3$. The phase space is the cotangent bundle $T^*SO(3)$ (the space of positions and angular momenta), which we may identify with $TSO(3)$ (the space of positions and angular velocities) by means of the left invariant metric defining the kinetic energy.

There are two different trivializations of this bundle. For two elements g,h of $SO(3)$ let $R_g(h) := h \cdot g$ and define

$$\begin{array}{rccc} \varrho: & T^*SO(3) & \longrightarrow & SO(3) \times I\!R^3 \\ & \alpha_g & \mapsto & (g, T_e^* R_g(\alpha_g)) \end{array}$$

(where $\alpha_g \in T_g^* SO(3)$). This corresponds to measuring the angular momentum in a set of axes fixed in space. Accordingly, the angular momentum vector $\mu := T_e^* R_g(\alpha_g)$ is a constant of motion. Hence, the dynamics takes place in $SO(3) \times \{\mu\}$. For $\mu \neq 0$ we can divide out the subgroup $SO(2)$ consisting of those elements of $SO(3)$ that leave μ fixed. (In case $\mu = 0$ the rigid body does not move at all.) These are the rotations around μ, dynamically speaking they correspond to the precession of the body around the angular momentum. We end up with the reduced phase space $(SO(3) \times \{\mu\})/_{SO(2)}$.

For the left trivialization let $L_g(h) := g \cdot h$ and define

$$\begin{array}{rccc} \lambda: & T^*SO(3) & \longrightarrow & SO(3) \times I\!R^3 \\ & \alpha_g & \mapsto & (g, T_e^* L_g(\alpha_g)) \end{array}$$

(again $\alpha_g \in T_g^* SO(3)$). The vector $\ell := T_e^* L_g(\alpha_g) = T_e^* L_g(T_e^* R_g)^{-1}(\mu) = g^{-1}(\mu)$ consists of the components of the angular momentum, now measured in the body set of axes $\vec{e}_1, \vec{e}_2, \vec{e}_3$. The mass distribution remains fixed in this set of axes (the body is rigid), so it is in these *body co-ordinates* that we define our Riemannian metric by means of an inner product $<..\,|\,..>$. Since our basis $\vec{e}_1, \vec{e}_2, \vec{e}_3$ lies along the principal axes of inertia, the symmetric matrix $I = (<\vec{e}_i\,|\,\vec{e}_j>)^{-1}$ is already diagonalised — the eigenvalues being the principal moments of inertia I_1, I_2 and I_3. The kinetic energy is

$$T = \frac{1}{2} <\ell\,|\,\ell> \;=\; \frac{\ell_1^2}{2I_1} + \frac{\ell_2^2}{2I_2} + \frac{\ell_3^2}{2I_3}\,.$$

The isomorphism

$$\begin{array}{ccccc} (SO(3) \times \{\mu\})/_{SO(2)} & \longrightarrow & S^2_{|\mu|} & = & \{\, \ell \in I\!R^3 \;/\; |\ell| = |\mu| \,\} \\ {[g,\mu]}^{\sim} & \mapsto & g^{-1}(\mu) & & \end{array}$$

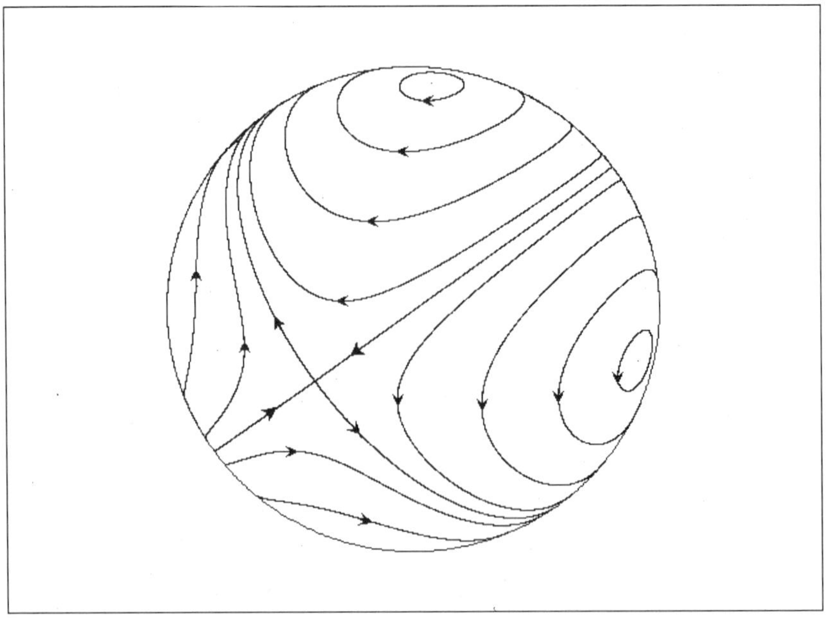

Figure 1 The flow of Euler's equations on the reduced phase space $S^2_{|\mu|}$.

reveals a nice interpretation of the reduced phase space. It is the space of all angular momenta of length $|\mu|$, measured in body co-ordinates. The reduced Hamilton function is given by

$$H_\mu(\ell) = \frac{\ell_1^2}{2I_1} + \frac{\ell_2^2}{2I_2} + \frac{\ell_3^2}{2I_3},$$

and the symplectic structure is the negative of the surface element on $S^2_{|\mu|}$, divided by the radius $|\mu|$. The phase curves are obtained as intersections of $S^2_{|\mu|}$ with the various energy ellipsoids $H_\mu^{-1}(h)$. (See *Figure 1.*)

The governing equations $\dot{\ell} = \ell \times I^{-1}(\ell)$ are Euler's equations. Usually they are written down for the angular velocities $\omega = I(\ell)$.

In *Figure 1* we see four stable equilibria (centres) and two unstable ones (saddles). The four centres correspond to steady rotations around the long and the short axis of inertia, and the two saddles correspond to steady rotations around the middle principal axis. The motion that gets reduced to a separatrix corresponds to what in Ashbaugh, Chicone and Cushman [1991] is called a half twist of the body: the body nearly rotates around the middle axis $\vec{e}_2$. Then this axis suddenly reverses, after which the body continues to rotate nearly around $\vec{e}_2$, but now in the opposite direction. (With respect to the axis $\vec{\mu}$, fixed in space, the rotation will again be in the same direction.) The general motion is reduced to $H_\mu^{-1}(h) \cap S^2_{|\mu|}$. This circle surrounds the principal axis of inertia corresponding to the smallest or the biggest principal moment of inertia. We call this axis the *figure axis*. In the dynamically symmetric case $I_1 = I_2$ this axis is always the one along $\vec{e}_3$. The general motion itself takes place on a trivial S^1-bundle over a circle $H_\mu^{-1}(h) \cap S^2_{|\mu|}$, hence, a 2-torus. Let us identify the angles. The fibre angle ξ measures the precession (in space) of the figure axis around the angular momentum axis $\vec{\mu}$. The angle ρ along the circle $H_\mu^{-1}(h) \cap S^2_{|\mu|}$ measures the rotation of the body around the figure axis.

Recall the Poinsot description of the motion of a free rigid body around a fixed point. The ellipsoid of inertia $\{ (I\omega|\omega) = 2h \}$ is rolling without sliding on the invariable plane $\{ (\mu|\Omega) = 2h \}$, with $g(\omega) = \Omega$. Here and elsewhere in this paper $(..|..)$ denotes the standard inner product. The point of contact traces out the polhode $\omega(t)$ on the ellipsoid of inertia and the herpolhode $\Omega(t)$ on the invariable plane. Under I the polhode gets mapped to $H_\mu^{-1}(h) \cap S^2_{|\mu|}$.

The fibres of the reduction mapping (that factors out $SO(2)$, see above) are easily described. They correspond to a *pure precession*, with one fixed point ω_0 on the ellipsoid of inertia sliding on the invariable plane. If $I_1 = I_2$ there is an obvious dualization of this concept; letting the ellipsoid of inertia rotate about the figure axis, with fixed point of contact Ω_0 on the invariable plane, defines in this case the angle ρ of rotation. Note that a counter-clockwise rotation about the figure axis leads to a clockwise movement of the point of contact $\omega(\rho)$ on the ellipsoid of inertia. In the general case we define an angle φ of rotation in a similar way, keeping the point of contact $\Omega(\varphi)$ on a straight line through the origin $\frac{2h}{|\mu|^2}\mu$ of the invariable plane. We require $\dot{\rho}$ to be constant, so we may define ρ as a monotonic function of φ. Note that this function $\rho(\varphi)$ involves the normal elliptic integral of the first kind. A zero-section $\{ \xi = 0 \}$ as determined by the above $\Omega(\varphi)$ would make ξ the polar angle of the invariable plane. Since we want $\dot{\xi}$ to be constant, we choose an appropriate homotopic zero-section $\{ \xi = 0 \}$.

We already found two angles of our action-angle co-ordinates, namely ξ and ρ. The action conjugate to the precession angle ξ is $|\mu|$ and the action conjugate to the angle ρ of rotation is given by $\Im := \frac{1}{2\pi}\oint_\gamma \vartheta$. Here ϑ is the canonical one-form on $T^*SO(3)$ and γ is given as a section over $H_\mu^{-1}(h) \cap S^2_{|\mu|}$, with ξ fixed. For the computation of $\Im$ see Sadov [1970] or Tantalo [1993]. In the dynamically symmetric case $I_1 = I_2$ the calculation of this integral is trivial. It is the component ℓ_3 of the angular momentum along the figure axis $\vec{e}_3$. One can read this directly off the phase portrait. (See *Figure 2.*) This corresponds to the simplification of Euler's equations to

$$\dot{\ell}_1 = -\frac{I_3 - I_1}{I_1 I_3}\ell_2\ell_3 , \quad \dot{\ell}_2 = \frac{I_3 - I_1}{I_1 I_3}\ell_1\ell_3 , \quad \dot{\ell}_3 = 0.$$

The remaining two action-angle variables stand for the possible directions of the angular momentum in space. The set of all these directions is a sphere. In order to get action-angle variables we want the quotient of the surface element of this sphere divided by its radius to be $-\mathrm{d}\vartheta - \mathrm{d}\xi \wedge \mathrm{d}|\mu| - \mathrm{d}\rho \wedge \mathrm{d}\Im$. Therefore, we fix the radius of this sphere to be $|\mu|$. In fact this is again the sphere of all angular momenta of length $|\mu|$, but now measured in a set of axes fixed in space. For every single point on this sphere we can choose the zero-section $\{\xi = 0\}$, but the Hairy Ball Theorem prevents us from doing so simultaneously on the whole sphere. Hence, we take out the poles $\pm|\mu| \cdot \vec{e}_z$ and choose as co-ordinates for the remaining directions the height $p = (\mu \,|\, \vec{e}_z)$ and the angle q around $\vec{e}_z$. This completes our construction of the action-angle variables, indeed $\mathrm{d}\xi \wedge \mathrm{d}|\mu| + \mathrm{d}\rho \wedge \mathrm{d}\Im + \mathrm{d}q \wedge \mathrm{d}p = -\mathrm{d}\vartheta$.

For $I_1 = I_2$ these action-angle variables are called Andoyer or Deprit co-ordinates and can be obtained quite similar to the Euler angles. We take $q \in S^1$ as the oriented angle between $\vec{e}_y$ and $\vec{e}_z \times \frac{\vec{\mu}}{|\mu|}$, $\xi \in S^1$ as the oriented angle between

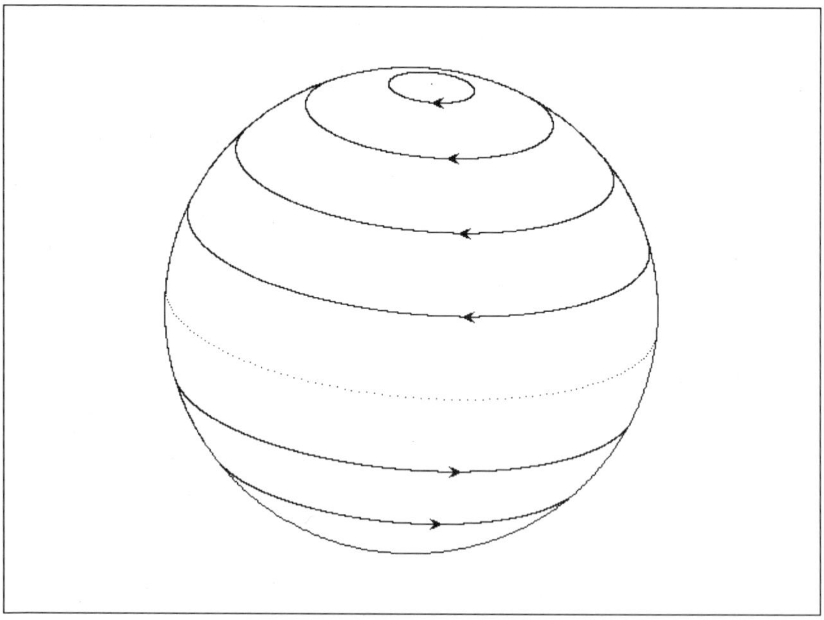

Figure 2 The reduced flow on $S^2_{|\mu|}$ in the dynamically symmetric case $I_1 = I_2$.

$\vec{e}_z \times \frac{\vec{\mu}}{|\mu|}$ and $\frac{\vec{\mu}}{|\mu|} \times \vec{e}_3$, and $\rho \in S^1$ as the oriented angle between $\frac{\vec{\mu}}{|\mu|} \times \vec{e}_3$ and $\vec{e}_2$. Note that the conventions chosen here differ slightly from the ones in Deprit [1967] and Arnol'd, Kozlov and Neishtadt [1985]. They work with $l = \rho - \frac{\pi}{2}$, $L = \Im$, $g = \xi$, $G = |\mu|$, $h = q + \frac{\pi}{2}$ and $H = p$.

The kinetic energy of the free rigid body depends only on $|\mu|$ and $\Im$, and the phase space is foliated by the invariant 2-tori on which we have a parallel flow with frequency vector

$$\omega(|\mu|, \Im) = \begin{pmatrix} \dfrac{\partial T(|\mu|, \Im)}{\partial |\mu|} \\ \dfrac{\partial T(|\mu|, \Im)}{\partial \Im} \end{pmatrix} .$$

In the dynamically symmetric case $I_1 = I_2$ the kinetic energy is easily calculated as

$$T(\xi, \rho, q, |\mu|, \Im, p) = \frac{|\mu|^2}{2I_1} - \frac{I_3 - I_1}{I_1 I_3} \cdot \frac{\Im^2}{2} .$$

2 A normal form theorem

Now we add a small Hamiltonian perturbation $V_\lambda(\xi, \rho, q, |\mu|, \Im, p)$ to the system. In our applications in Sections 3–5, V_λ will be the potential energy of the rigid body in some external force field. We let this perturbation depend on a parameter $\lambda \in \mathbb{R}^p$, and by small we mean that it vanishes for $\lambda = 0$. We want to apply normal form theory as can be found in Arnol'd [1978] or Broer [1981]. The object of the game is to simplify the perturbing Hamilton function by means of a series of co-ordinate transformations; we want to get rid of the dependence on the angles ξ and ρ. A Hamilton function which is totally independent of ξ and ρ possesses a 2-torus symmetry. We can not expect to gain the full 2-torus symmetry from the approximating Euler case. Instead we will push the 2-torus symmetry through the Taylor series. In each step the effect of the co-ordinate transformation is averaging along a 2-torus. Since averaging on resonant tori is hopeless, compare Arnol'd [1978], we single out a $(|\mu|_0, \Im_0)$ with (even) a *diophantine* frequency vector $\omega(|\mu|_0, \Im_0)$ and concentrate on the situation around the associated torus.

Recall that $\omega = (\omega_1, \omega_2)$ is called diophantine if there are constants $\gamma, \varepsilon > 0$ with

$$|(\omega|k)| \geq \frac{\gamma}{|k|^{2+\varepsilon}} \quad \forall\, k \in \mathbb{Z}^2 \backslash \{0\}.$$

The diophantine frequency vectors satisfying this condition for fixed γ and ε form a Cantor set of half lines. As $\gamma \to 0$, these half lines extend to the origin. The complement in any compact set of frequency vectors satisfying a diophantine condition with fixed γ and ε has a measure of order $O(\gamma)$ as $\gamma \to 0$.

We expand our Hamilton function $H = T + V_\lambda$ as a Taylor series in $(|\mu|, \Im) - (|\mu|_0, \Im_0)$ and proceed step by step by first averaging first order terms, then second order terms and so on. We also want to average the terms of the perturbation constant in $(|\mu|, \Im) - (|\mu|_0, \Im_0)$. Therefore, we expand in λ as well. Since the perturbation vanishes for $\lambda = 0$, this makes every term of the perturbation a term of at least order one in $(|\mu| - |\mu|_0, \Im - \Im_0, \lambda)$.

We simplify and unify the notation by letting $y_1 := |\mu| - |\mu|_0$, $y_2 := \Im - \Im_0$, $x_1 := \xi$, $x_2 := \rho$ and $z := (p, q)$. Recall that $x = (x_1, x_2)$ takes its values on a 2-torus $\mathbf{T}$, $y = (y_1, y_2)$ takes its values in an open part $\mathbf{D}$ of $\mathbb{R}^2$ and z takes its values in the punctured sphere

$$\mathbf{S} := S^2_{|\mu|_0} \backslash^{\{\pm |\mu|_0 \cdot \vec{e}_z\}}.$$

Note that we do not expand our Hamilton function in q and p as well, our construction is global with respect to $\mathbf{S}$. In the above co-ordinates we may write our Hamilton function as

$$H(x, y, z, \lambda) \;=\; t + (\omega\,|\,y) + \tau(y) + V_\lambda(x, y, z)$$

where $t := T(|\mu|_0, \Im_0)$ is the constant part of the kinetic energy, $(\omega\,|\,y) = \omega_1 y_1 + \omega_2 y_2$ the linear part and where $\tau(y) := T(|\mu|_0 + y_1, \Im_0 + y_2) - t - (\omega\,|\,y)$ contains the higher order terms of the kinetic energy.

In the dynamically symmetric case $I_1 = I_2$ we have

$$t = \frac{|\mu|_0^2}{2I_1} - \frac{I_3 - I_1}{I_1 I_3} \cdot \frac{\Im_0^2}{2},$$

$$\omega_1 = \frac{|\mu|_0}{I_1},$$

$$\omega_2 = -\frac{I_3 - I_1}{I_1 I_3} \cdot \Im_0,$$

and

$$\tau(y) = \frac{y_1^2}{2I_1} - \frac{I_3 - I_1}{I_1 I_3} \cdot \frac{y_2^2}{2}$$

is exactly the quadratic part of the kinetic energy.

Finally we introduce the following *gradation* of the ring of all (=formal) power series in y and λ with coefficients in $C^\infty(\mathbf{T} \times \mathbf{S})$ that vanish for $\lambda = 0$,

$$\mathcal{A}_n \;:=\; \{\, P \in C^\infty(\mathbf{T} \times \mathbf{S})[y, \lambda] \;/\; D^k P(0,0) = 0 \;\; \forall\, k \neq n\,,\; P(y, 0) \equiv 0 \,\}$$

together with the *filtration* $\mathcal{F}_n := \prod_{k\geq n} \mathcal{A}_k$. This terminology replaces expressions like " $+\, O(\lambda\cdot(y+\lambda)^n)$ uniformly as $y, \lambda \to 0$ " by " mod $\mathcal{F}_{n+1}$ ", compare Broer [1981]. With $\pi_{\mathcal{A}_n}$ we denote the projection of $\prod_{k\geq 1} \mathcal{A}_k$ onto $\mathcal{A}_n$, and $\pi_{\mathbf{D}}$ denotes the projection of $\mathbf{T} \times \mathbf{D} \times \mathbf{S}$ onto the second component.

Normal Form Theorem *Let* $H(x,y,z,\lambda) = T(y) + V_\lambda(x,y,z)$ *be a perturbed Hamilton function on* $\mathbf{T} \times \mathbf{D} \times \mathbf{S}$. *Suppose that the frequency vector* $\omega = \frac{\partial T}{\partial y}(0)$ *is diophantine, and fix an integer* $n \in \mathbb{N}$. *Then there is a symplectic co-ordinate transformation* ψ *such that*

$$(H \circ \psi)(x,y,z,\lambda) \;=\; T(y) \,+\, W_\lambda(y,z) \;\; (\text{mod}\; \mathcal{F}_{n+1}).$$

The terms up to order n in $W_\lambda(y,z)$ are called the normal form (of order n). In the new co-ordinates, H becomes independent of x up to the prescribed order n. The proof of this theorem is standard, compare Section 2 of Braaksma, Broer and Huitema [1990] or Section 5 of Arnol'd [1978]. It is constructive, and the algorithm presented in the proof will be used to compute the normal form in the examples.

Proof We will construct our symplectic co-ordinate transformation inductively, and in each step, as the time one flow ψ_1 of a Hamiltonian vector field $Y = X_K$. Note that for $K \in \mathcal{A}_n$ we have $\pi_{\mathbf{D}}\psi_1 = y - \frac{\partial K}{\partial x}$ (mod $\mathcal{F}_{2n-1}$). This leads to

$$H \circ \psi_1 \;=\; t \,+\, (\omega\,|\,y) \,+\, \tau(y) \,-\, (\omega\,|\,\frac{\partial K}{\partial x}) \,+\, V_\lambda(x,y,z) \;\; (\text{mod}\; \mathcal{F}_{n+1})\,.$$

We will use the term $-(\omega\,|\,\frac{\partial K}{\partial x})$ to annihilate as many terms of $\pi_{\mathcal{A}_n} V_\lambda(x,y,z)$ as possible. The remaining terms are the normal form terms of V_λ in $\mathcal{A}_n$.

The case $n = 1$ We write: $V_\lambda(x,y,z) = \sum_{j=1}^{p} \lambda_j \cdot U_j(x,z)$ (mod $\mathcal{F}_2$) with $U_j(x,z) = \sum_{k\in\mathbb{Z}^2} u_{jk}(z) e^{i(k\,|\,x)}$. Then the function

$$K(x,y,z,\lambda) \;:=\; \sum_{j=1}^{p} \lambda_j \cdot \sum_{k\neq 0} \frac{u_{jk}(z)}{i(k\,|\,\omega)} e^{i(k\,|\,x)}$$

solves the equation

$$(\omega\,|\,\frac{\partial K}{\partial x}) \;=\; V_\lambda(x,y,z) \,-\, \sum_{j=1}^{p} \lambda_j \cdot \underbrace{\int_{\mathbf{T}} U_j(x,z)\, \mathrm{d}x}_{=u_{j0}(z)} \;\; (\text{mod}\; \mathcal{F}_2)\,.$$

Note that the left hand side already lies in $\mathcal{A}_1$. The convergence of the Fourier series defining K is guaranteed by the diophanticity of ω, compare Arnol'd [1978]. We conclude that

$$H \circ \psi_1 \;=\; H \,-\, (\omega\,|\,\frac{\partial K}{\partial x}) \;=\; H \,-\, V_\lambda \,+\, \sum_{j=1}^{p} \lambda_j \cdot u_{j0}(z)$$

$$= \; t \;+\; (\omega\,|\,y) \;+\; \tau(y) \;+\; \sum_{j=1}^{p} \lambda_j \cdot \int_{\mathbf{T}} U_j(x,z)\,\mathrm{d}x \qquad (\mathrm{mod}\ \mathcal{F}_2)\,.$$

The step $n \to n+1$

$$\pi_{\mathcal{A}_n} V_\lambda(x,y,z) = \sum_{j=1}^{p} \sum_{\substack{|l|+|m|=n \\ |l|\geq 1}} \lambda_j^l\, U_{lmj}(x,z) y^m$$

with

$$U_{lmj}(x,z) = \sum_{k\in\mathbb{Z}^2} u_{lmjk}(z) e^{i(k\,|\,x)},$$

and

$$K(x,y,z,\lambda) \;\;:=\;\; \sum_{j=1}^{p} \sum_{\substack{|l|+|m|=n \\ |l|\geq 1}} \lambda_j^l \cdot \sum_{k\neq 0} \frac{u_{lmjk}(z)}{i(k\,|\,\omega)} e^{i(k\,|\,x)} \cdot y^m$$

again solves

$$(\omega\,|\,\frac{\partial K}{\partial x}) \;\;=\;\; \pi_{\mathcal{A}_n} V_\lambda(x,y,z) \;-\; \int_{\mathbf{T}} \pi_{\mathcal{A}_n} V_\lambda(x,y,z)\,\mathrm{d}x$$

and we conclude

$$(H\circ\psi_1)(x,y,z,\lambda) = T(y) + \int_{\mathbf{T}} V_\lambda(x,y,z)\mathrm{d}x \;(\mathrm{mod}\ \mathcal{F}_{n+1}).$$

□

The manner in which the last equation is written completely hides the fact that, due to the induction procedure, we do not average the original Hamilton function. Instead we average a transformed Hamilton function that was altered by the "higher order terms" on lower levels of the induction procedure. If we really want to calculate the normal form of order n, we have to calculate the first coordinate transformation up to order $n-1$, the second up to order $n-2$ and so on. In our applications we expect n to be quite low. For the case of a linear potential we will even have $n=1$, and in the general case we will stop as soon as

$$H_{y,\lambda}(z) \;:=\; T(y) \;+\; \int_{\mathbf{T}} V_\lambda(x,y,z)\,\mathrm{d}x$$

is structurally stable as a family. The truncated normal form $H_{y,\lambda}(z)$ is independent of x. Consequently the conjugate actions y remain constant and we can regard them as (distinguished) parameters.

Let us see how the presence of symmetries can be incorporated into the Normal Form Theorem. In case the body itself is symmetric, in particular $I_1 = I_2$ *and* the centre of mass lies on the figure axis $\vec{e}_3$, the potential V_λ will not depend on the angle ρ. The conjugate action $\Im = (\mu\,|\,\vec{e}_3)$ is consequently an integral even for the perturbed problem. Thus we only need the Normal Form Theorem to get rid of the precession-angle ξ. We do not need any longer that $(|\mu|_0, \Im_0)$ is diophantine and merely demand $\omega_1 = \dfrac{|\mu|_0}{I_1}$ to be nonzero. Also, we expand our Hamilton function as a Taylor series only in $(|\mu| - |\mu|_0)$ and λ. The normal form theorem can then be applied for $x = x_1 = \xi$, $y = y_1 = |\mu| - |\mu|_0$ and $z = (p,q,\Im,\rho)$. We can even simplify the calculations replacing the integration of $(\omega\,|\,\frac{\partial K}{\partial x})$ by means of a Fourier series by an indefinite integral $K = \int \frac{\partial K}{\partial x} dx$. Since, by construction, K does not depend on ρ, there is no co-ordinate change in the $\Im$-direction. More importantly, the transformed Hamilton function is still independent of ρ. As a consequence the translations in the ρ-direction are of no interest and need not be computed.

Now let the potential be S^1-symmetric, *i.e.* let V_λ be independent of q. Then, by construction, K does not depend on q, so the transformed Hamilton function will still have this S^1-symmetry. For the same reason there is no co-ordinate transformation in the p-direction, and the one in the q-direction is of no interest. The preservation of symmetries in normal forms is a general phenomenon, compare Broer [1981].

3 A linear potential — the Lagrange top

Let m be the total mass of our rigid body. By $\vec{s} = s_1\vec{e}_1 + s_2\vec{e}_2 + s_3\vec{e}_3$ we denote the centre of mass. The constant acceleration field $E(\vec{r}) = -g\vec{e}_z$ gives rise to the potential energy $V_\lambda = \lambda \cdot (\vec{e}_z \,|\, \vec{s})$, where $\lambda := g \cdot m$. From now on we restrict our attention to the dynamically symmetric case $I_1 = I_2$. We obtain the following corollary of the Normal Form Theorem.

Corollary 3.1 *In the above situation the normal form of order one of* $H = T + V_\lambda$ *is*

$$H_{|\mu|,\Im,\lambda}(p,q) \;=\; \frac{|\mu|^2}{2I_1} \;-\; \frac{I_3 - I_1}{I_1 I_3} \cdot \frac{\Im^2}{2} \;+\; \lambda \cdot s_3 \frac{\Im}{|\mu|}\frac{p}{|\mu|} \,.$$

Proof We are not interested in the normal form terms of order 2. Hence, we do not have to calculate the actual co-ordinate transformation that, modulo $\mathcal{F}_2$, gets rid of the terms depending on ξ and ρ. We merely have to average

$$\begin{aligned} H \;=\; & \tfrac{|\mu|^2}{2I_1} \;-\; \tfrac{I_3 - I_1}{I_1 I_3} \cdot \tfrac{\Im^2}{2} \;+\; \lambda \cdot s_3 \tfrac{\Im \cdot p}{|\mu|^2} \\ & - \lambda \cdot s_3 \frac{\sqrt{|\mu|^2 - \Im^2}\sqrt{|\mu|^2 - p^2}}{|\mu|^2} \cdot \cos\xi \\ & + \lambda \cdot \frac{p\sqrt{|\mu|^2 - \Im^2}}{|\mu|^2} \cdot (-s_1 \cos\rho + s_2 \sin\rho) \\ & + \lambda \cdot \frac{\sqrt{|\mu|^2 - p^2}}{|\mu|} \cdot (s_1 \sin\xi \sin\rho + s_2 \sin\xi \cos\rho) \\ & + \lambda \cdot \frac{\Im\sqrt{|\mu|^2 - p^2}}{|\mu|^2} \cdot (-s_1 \cos\xi \cos\rho + s_2 \cos\xi \sin\rho) \end{aligned}$$

along ξ and ρ, compare Arhangel'skiĭ [1977]. Obviously, only the first term in the potential survives this averaging. $\square$

The resulting parametrised Hamilton function $H_{|\mu|,\Im,\lambda}$, up to some constants, is the height function on the (p,q)-sphere. Since this is a Morse-function, the corresponding Hamiltonian vector field is structurally stable in the set of all Hamiltonian vector fields and we do not need to compute the higher order terms of the normal form. They would not alter the qualitative picture.

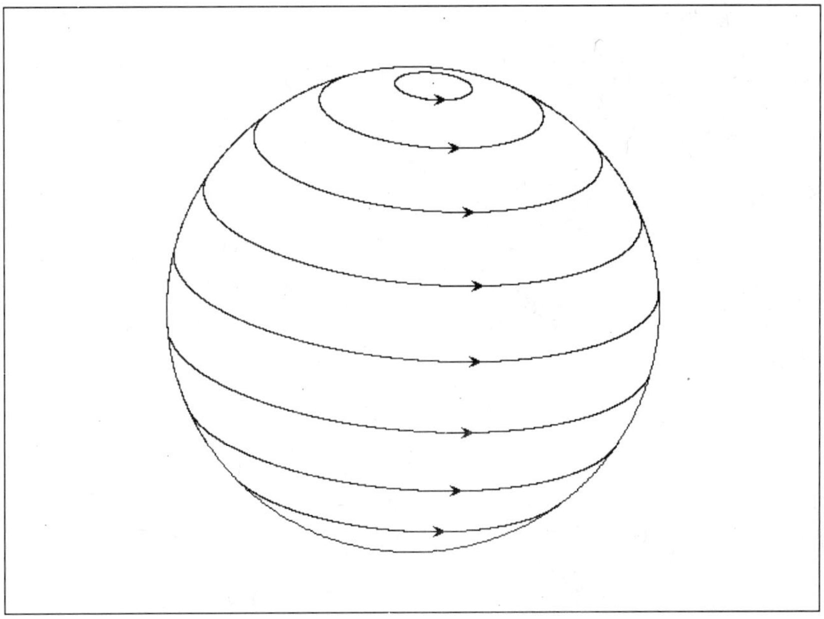

Figure 3 The flow of the one-degree-of-freedom problem on the (p, q)-sphere for a constant force field. $I_1 = I_2$, but the centre of mass $\vec{s}$ may be in general position.

The flow of the Hamiltonian vector field $X_{H_{|\mu|,\Im,\lambda}}$ is very simple,

$$q(t) = q_0 + \lambda s_3 \frac{\Im}{|\mu|} \cdot t,$$

and

$$p(t) = p_0$$

remains constant. North and South Pole are centres, and we have a circular motion along the parallels of latitude, the direction of which is determined by the signs of λ and $\Im$. (See *Figure 3*.)

3.1 The Lagrange case As a special case of our setting $I_1 = I_2$ we first restrict to the case $s_1 = s_2 = 0$. Then the centre of mass lies on the figure axis $\vec{e}_3$, and consequently the Hamilton function does not depend on the angle ρ of rotation about the figure axis. This, together with the S^1-symmetry of the potential (independence of q), shows that this case is integrable. In fact this is the one named after Lagrange. Hence, we do not have to deal with KAM-theory. Equilibria on the (p, q)-sphere correspond to invariant 2-tori in the whole system, and periodic orbits to invariant 3-tori. For fixed parameter values of λ we get a 3-parameter family of invariant 3-tori shrinking down to two 2-parameter families of invariant 2-tori.

Let us recall some facts about the Lagrange top, compare Cushman and Knörrer [1985] or Cushman and Bates [1994]. The range of the energy-momentum mapping $(H, \Im, p) : T^*SO(3) \longrightarrow \mathbb{R}^3$ lies above the set of singularities which looks like a bowl, with edges and with a one-dimensional thread. (See *Figure 4*.) The fibres

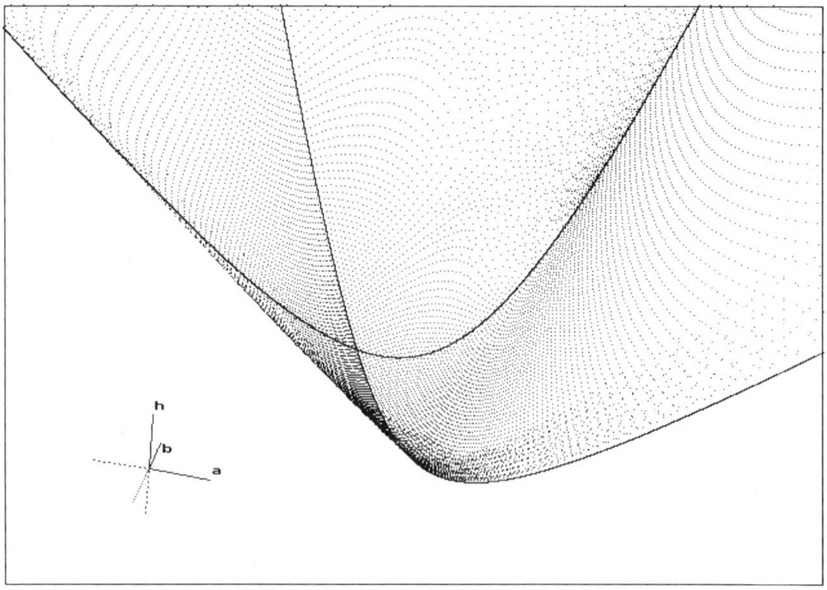

Figure 4 The singular values of the energy-momentum mapping for the Lagrange top. The dark lines are the thread and the edges.

$(H, \Im, p)^{-1}(h, a, b)$ are of the following types, according to the value $(h, a, b) \in I\!R^3$.

1) *regular values:* The fibre is a 3-torus. The motion consists of rotation, precession and nutation.

2) *singular values outside $a = \pm b$:* The fibre is a 2-torus. The corresponding motion is called regular precession.

3) *singular values at $a = -b \neq 0$:* The fibre is a 1-torus, a circle. The top is rotating upside down.

4) *singular values at $a = b = 0$:* For $h = -\lambda$, the top is hanging upside down and not moving at all. The fibre $(H, \Im, p)^{-1}$ (h, a, b) is a degenerate 1-torus. For $h = \lambda$ (on the thread), the fibre consists of a degenerate 1-torus together with its unstable manifold. The points on the 1-torus correspond to a top standing up.

5) *singular values at $a = b \neq 0$:* For high energy (*i.e.* λ small) the fibre is a 1-torus. The top rotates stably while standing up, this is the situation of the 'sleeping top'. When we decrease the energy the top "wakes up": over the thread the fibre is an unstable 1-torus together with its unstable manifold. On the 'bowl' we have again regular precession.

Obviously, we rediscovered the 3-tori and the 2-tori in our approach. We can even find the 1-tori: for $\Im = \pm|\mu|$ the free motion was just steady rotation. Hence, corresponding to North or South Pole of the (p, q)-sphere we get a 1-torus. The poles are given by $p = \pm|\mu|$, so these 1-tori indeed occur for $\Im = \pm p$. However, we cannot rediscover the dynamics related to the one-dimensional thread, let alone the Hamiltonian Hopf bifurcation leading to it. These effects are controlled by the parameter λ, the greatest distance of the thread to the bowl being 2λ, and our analysis is only valid asymptotically as $\lambda \to 0$.

3.2 The general case Let us now turn to the case of a centre of mass in general position $\vec{s} = s_1\vec{e}_1 + s_2\vec{e}_2 + s_3\vec{e}_3$, still keeping $I_1 = I_2$. In view of the results of Sal'nikova [1984] and Kozlov and Treshchev [1985, 1986], we have to work a little bit more to establish the existence of invariant 2- and 3-tori. For the 2-tori we use the theory of Moser [1967] and Broer, Huitema and Takens [1990]. Related to a 2-torus associated with, say, the North Pole, are the internal frequencies

$$\omega_1 = \frac{\partial H_{|\mu|,\Im,\lambda}}{\partial|\mu|}$$

and

$$\omega_2 = \frac{\partial H_{|\mu|,\Im,\lambda}}{\partial\Im}$$

of the motion on the torus and the *normal frequency* Ω. One can define Ω as 2π divided by the limit of the period of the orbits surrounding the North Pole when approaching the North Pole. It is easily shown that this is the eigenvalue of the linearization of $H_{|\mu|,\Im,\lambda}$ at the North Pole.

We have $\omega_1 = \frac{|\mu|}{I_1} - 2\lambda s_3 \frac{\Im p}{|\mu|^3}$ and $\omega_2 = -\frac{I_3 - I_1}{I_1 I_3}\Im + 2\lambda s_3 \frac{p}{|\mu|^2}$. Since the motion on the periodic orbits does not depend on the height p, the normal frequency is just $\Omega = \lambda s_3 \frac{\Im}{|\mu|^2}$. The derivative of the frequency mapping $(\omega_1, \omega_2, \Omega)$ as a function of the parameters $(|\mu|, \Im, \lambda) \in \Gamma = \{\, |\mu| > 0\, ,\, \Im \in]0, |\mu|[\, ,\, \lambda \in]0, \lambda_0[\, \}$ is

$$\begin{pmatrix} \frac{1}{I_1} + 6\lambda s_3 \frac{\Im p}{|\mu|^4} & -2\lambda s_3 \frac{p}{|\mu|^3} & -2 s_3 \frac{\Im p}{|\mu|^3} \\ -2\lambda s_3 \frac{p}{|\mu|^3} & -\frac{I_3 - I_1}{I_1 I_3} & s_3 \frac{p}{|\mu|^2} \\ -2\lambda s_3 \frac{\Im}{|\mu|^3} & \lambda s_3 \frac{1}{|\mu|^2} & s_3 \frac{\Im}{|\mu|^2} \end{pmatrix}.$$

Its determinant

$$-\, s_3 \cdot \left(\frac{1}{I_1} + \frac{2\lambda s_3 \Im p}{|\mu|^4}\right)\left(\frac{I_3 - I_1}{I_1 I_3}\frac{\Im}{|\mu|^2} + \frac{\lambda s_3 p}{|\mu|^4}\right),$$

with $p = \pm|\mu|$ for North and South Pole, vanishes only on a set of codimension one. Corollary 6.2 in Broer, Huitema and Takens [1990] tells us that on a "Cantor set" of large measure we still have invariant 2-tori. This means that we have *strong quasi-periodic stability* of these 2-tori.

For sake of definiteness we restrict our attention to the 2-tori associated with the North Pole. Let $\Gamma_\gamma \subseteq \Gamma$ denote the set of parameters for which the frequency vector $(\omega_1, \omega_2, \Omega)$ satisfies a certain diophantine condition with constant γ. See Broer, Huitema and Takens [1990] for details. There is a Whitney-smooth conjugacy of the foliation

$$\{\, ((\xi, \rho, q, |\mu|, \Im, p), \lambda) \;/\; (|\mu|, \Im, \lambda) \in \Gamma_\gamma \text{ and } (p, q) \text{ is the North Pole} \,\}$$

of 2-tori into the perturbed system. The image of this foliation is the "Cantor set" of large measure; we will colloquially use the expression *"Cantor set"* for any such Whitney-smooth foliation of manifolds (with boundary) throughout this paper. In particular, the conjugacy preserves the internal and normal frequencies. Hence, three unfolding parameters are needed to get strong quasi-periodic stability. By replacing the conjugacy by an equivalence, *i.e.* allowing monotone rescaling of time, one can get rid of one invariant. The corresponding notion of stability is called *weak quasi-periodic stability*.

In physical applications one usually wants to know what happens when λ , *i.e.* the potential is fixed. Since the image of our foliation is only a "Cantor set" we do not know beforehand whether its slice in $\{ ((\xi, \rho, q, |\mu|, \Im, p), \tilde{\lambda}) \; / \; \tilde{\lambda} = \lambda \}$ is still of large measure. The two distinguished parameters $|\mu|$ and $\Im$ do not suffice to unfold the frequency vector $(\omega_1, \omega_2, \Omega)$ transversally. Instead, we consider the ratio $[\omega_1 : \omega_2 : \Omega]$ as a function of $(|\mu|, \Im) \in \{ (|\mu|, \Im) \; / \; |\mu| > 0 \, , \, \Im \in \,]0, |\mu|[\, \}$. This weakens our result, we only have weak quasi-periodic stability.

Proposition 3.2 *Let T be the kinetic energy of a rigid body with mass m, fixed at one point, with centre of mass $\vec{s}$ in general position and with principal moments of inertia $I_1 = I_2 \neq I_3$. Furthermore let V_λ denote its potential energy in the weak constant force field $\begin{pmatrix} 0 \\ 0 \\ -\lambda \end{pmatrix}$. Then the Hamiltonian vector field associated with $H = T + V_\lambda$ has the following properties. In the phase space $T^*SO(3)$ there are nested 3-tori with quasi-periodic motion. In bounded subsets of $T^*SO(3)$ the measure of their complement goes to zero as $\lambda \to 0$. The nested 3-tori shrink down to two "Cantor sets" of 2-tori with quasi-periodic motion. As sets, these 2-tori are stable in the sense of Lyapunov.*

Proof Since the conclusion of the proposition holds for the truncated Hamilton function we only have to show that the stated properties still hold when the higher order terms are added. First we show that $(|\mu|, \Im)$ unfolds the frequency ratio transversally. Since Ω has to be nonzero anyway, we may as well work with

$$\frac{\omega_1}{\Omega} = \frac{|\mu|^3}{\lambda s_3 I_1 \Im} - 2\frac{p}{|\mu|}$$

and

$$\frac{\omega_2}{\Omega} = -\frac{I_3 - I_1}{I_1 I_3} \cdot \frac{|\mu|^2}{\lambda s_3} + \frac{p}{\Im}.$$

The derivative of $(\frac{\omega_1}{\Omega}, \frac{\omega_2}{\Omega})$ with respect to $|\mu|$ and $\Im$ is

$$\begin{pmatrix} \frac{3|\mu|^2}{\lambda s_3 I_1 \Im} + 2\frac{p}{|\mu|^2} & -\frac{|\mu|^3}{\lambda s_3 I_1 \Im^2} \\ -\frac{I_3 - I_1}{I_1 I_3} \frac{2|\mu|}{\lambda s_3} & -\frac{p}{\Im^2} \end{pmatrix}$$

with determinant $-2\frac{p^2}{\Im^2 |\mu|^2} - 3\frac{|\mu|^2 p}{\lambda s_3 I_1 \Im^3} - 2\frac{I_3 - I_1}{I_1 I_3} \frac{|\mu|^4}{\lambda^2 s_3^2 I_1 \Im^2}$. Letting $Y_{|\mu|, \Im, \nu} = \nu \cdot X_{H_{|\mu|, \Im, \lambda}}$ we add a third parameter (recall that we regard λ as fixed). This corresponds to a rescaling of time. Now we can apply Corollary 6.2 of Broer, Huitema and Takens [1990]. Again we have invariant 2-tori on a "Cantor set" of large measure. Since we rescaled time, we only have a (Whitney-smooth) equivalence between 2-tori of the original and the truncated system.

For the 3-tori we use standard KAM-theory as developed in Arnol'd [1963, 1974] or Pöschel [1982] to show that most of the invariant 3-tori survive the perturbation.

Related to such a 3-torus are the frequencies:

$$\omega_1 = \frac{\partial H_{|\mu|,\Im,\lambda}(p,q)}{\partial|\mu|} = \frac{|\mu|}{I_1} - 2\lambda s_3 \frac{\Im p}{|\mu|^3},$$

$$\omega_2 = \frac{\partial H_{|\mu|,\Im,\lambda}(p,q)}{\partial \Im} = -\frac{I_3 - I_1}{I_1 I_3}\Im + 2\lambda s_3 \frac{p}{|\mu|^2},$$

and

$$\omega_3 = \frac{\partial H_{|\mu|,\Im,\lambda}(p,q)}{\partial p} = \lambda s_3 \frac{\Im}{|\mu|^2}.$$

The derivative of these frequencies with respect to the actions $(|\mu|, \Im, p)$ is

$$\begin{pmatrix} \frac{1}{I_1} + 6\lambda s_3 \frac{\Im p}{|\mu|^4} & -2\lambda s_3 \frac{p}{|\mu|^3} & -2\lambda s_3 \frac{\Im}{|\mu|^3} \\ -2\lambda s_3 \frac{p}{|\mu|^3} & -\frac{I_3 - I_1}{I_1 I_3} & \lambda s_3 \frac{1}{|\mu|^2} \\ -2\lambda s_3 \frac{\Im}{|\mu|^3} & \lambda s_3 \frac{1}{|\mu|^2} & 0 \end{pmatrix}.$$

Its determinant

$$\frac{\lambda^2 s_3^2}{|\mu|^4}\left(\frac{I_3 - I_1}{I_1 I_3}\frac{\Im}{|\mu|^2} - \frac{1}{I_1}\right) + 2\lambda^3 s_3^3 \frac{\Im p}{|\mu|^8}$$

vanishes only on a set of codimension one. Hence, 3-tori with quasi-periodic dynamics occupy a large part of the phase space. Since the Hamilton function does not depend on q, the conjugate action p is an integral of H. Consequently, the 3-tori divide the 4-dimensional submanifolds of constant p and H, compare the appendix of Arnol'd [1963]. In particular, a motion starting near a 2-torus will remain close to it. □

We can also give an interpretation for the limiting cases $(|\mu|, \Im, p) = (|\mu|, \pm|\mu|, \pm|\mu|)$ where the truncated Hamiltonian vector field has periodic orbits. These correspond to the Staude rotations of the heavy rigid body, compare Grammel [1950] or Lewis et al. [1992].

4 A quadratic potential

A quadratic potential function $U_\lambda : I\!R^3 \longrightarrow I\!R$ leads to the potential energy $V_\lambda = (DU_\lambda \,|\, M)$, where M is the matrix of second momenta of the mass distribution of the rigid body. Since $\vec{e}_1, \vec{e}_2, \vec{e}_3$ lie along the principal axes of inertia, they automatically diagonalise M, the eigenvalues are $M_1 = \frac{1}{2}(I_2 + I_3 - I_1)$, $M_2 = \frac{1}{2}(I_1 + I_3 - I_2)$ and $M_3 = \frac{1}{2}(I_1 + I_2 - I_3)$. Again we restrict to the case $I_1 = I_2$, so we get $M_1 = M_2 = \frac{1}{2}I_3$ and $M_3 = I_1 - \frac{1}{2}I_3$. We may (and will) assume that $\vec{e}_x, \vec{e}_y, \vec{e}_z$ diagonalise the acceleration field DU_λ and denote the eigenvalues by a, b and c. These are the parameters of our potential energy, and from now on we drop the subscript λ. However, note that a parameter change $(a,b,c) \mapsto (a+d, b+d, c+d)$ does not change the equations of motion (the body is rigid).

Although we do not require the centre of mass to lie on the figure axis, the potential does not depend on the angle ρ of rotation around the figure axis $\vec{e}_3$. The potential can not distinguish a dynamically symmetric body from a really symmetric one. Therefore, we can use the simplifications of the Normal Form Theorem as described at the end of Section 2. We only have to average the Hamilton

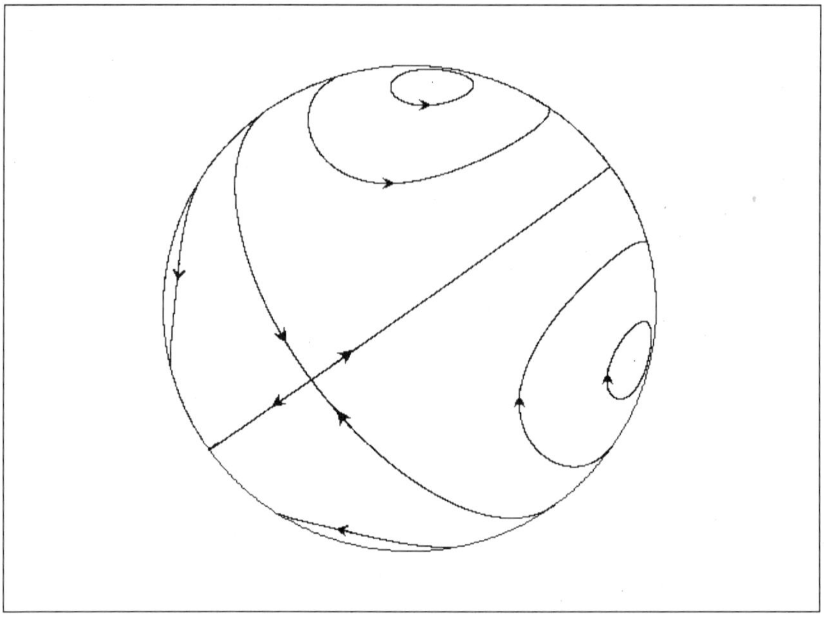

Figure 5 The flow on the (p,q)-sphere to the first order normal form of a quadratic potential, for $3\Im^2 \neq |\mu|^2$.

function $H = T + V$ along the angle ξ. Doing so we get the following corollary to the Normal Form Theorem.

Corollary 4.1 *Let H be the Hamilton function of a rigid body, fixed at one point, with principal moments of inertia $I_1 = I_2 \neq I_3$, and moving in a conservative acceleration field with quadratic potential function. Then the normal form of order one of H is*

$$\begin{aligned} H_{|\mu|,\Im,a,b,c}(p,q) &= \frac{|\mu|^2}{2I_1} - \frac{I_3 - I_1}{I_1 I_3}\frac{\Im^2}{2} + \frac{a+b+c}{2}(I_1 + \frac{\Im^2}{|\mu|^2}(I_3 - I_1)) \\ &- \frac{1}{2}\,(a\,\frac{|\mu|^2 - p^2}{|\mu|^2}\cos^2 q + b\,\frac{|\mu|^2 - p^2}{|\mu|^2}\sin^2 q + c\,\frac{p^2}{|\mu|^2})\;(I_3 - I_1)\;\frac{3\Im^2 - |\mu|^2}{|\mu|^2}. \end{aligned}$$

The flow of the Hamiltonian vector field $X_{H_{|\mu|,\Im,a,b,c}}$ is governed by Euler's equations, restricted to the $|\mu|$-sphere and with principal moments of inertia

$$\frac{-|\mu|^2}{I_3 - I_1}\frac{|\mu|^2}{3\Im^2 - |\mu|^2}\cdot\frac{1}{a},$$

$$\frac{-|\mu|^2}{I_3 - I_1}\frac{|\mu|^2}{3\Im^2 - |\mu|^2}\cdot\frac{1}{b}$$

and

$$\frac{-|\mu|^2}{I_3 - I_1}\frac{|\mu|^2}{3\Im^2 - |\mu|^2}\cdot\frac{1}{c}$$

(See *Figure 5*.) However, for $3\Im^2 = |\mu|^2$ the vector field identically vanishes and we have to calculate more terms in the normal form. We will do so for a special case in the next section. Away from $3\Im^2 = |\mu|^2$ (and for a, b, c sufficiently small) the Hamiltonian vector field is structurally stable in the following sense.

Locally $H_{|\mu|,\Im,a,b,c}$ is a Morse function, *i.e.* all equilibria of $X_{H_{|\mu|,\Im,a,b,c}}$ are non-degenerate. However, globally we have heteroclinic saddle connections. This is due to the reflectional symmetries of the quadratic potential. Hence, within the set of all Hamiltonian vector fields on the (p,q)-sphere with the three symmetries $p \mapsto -p$, $q \mapsto 2\pi - q$ and $q \mapsto \pi - q$, the vector field $X_{H_{|\mu|,\Im,a,b,c}}$, away from $3\Im^2 = |\mu|^2$, is structurally stable. Since the Normal Form Theorem preserves also these symmetries, the higher order terms belong to this set. They cannot lead to qualitative changes of $X_{H_{|\mu|,\Im,a,b,c}}$.

In the truncated system on $T^*SO(3)$ we get four families of 3-tori shrinking down to 2-tori and separated by the unstable manifolds of two normally hyperbolic families of 2-tori. We fix the potential, *i.e.* we fix a, b, and c, and look for persistence of the above scenario in the full system on $T^*SO(3)$.

Proposition 4.2 *Let H be the Hamilton function of a rigid body, fixed at one point, with principal moments of inertia $I_1 = I_2 \neq I_3$, and moving in a weak conservative acceleration field with quadratic potential function. Then the Hamiltonian vector field has, away from the set $\{\ 3\Im^2 = |\mu|^2\ \}$, four 2-parameter families of normally elliptic invariant 2-tori and two 2-parameter families of normally hyperbolic invariant 2-tori. As sets, the normally elliptic 2-tori are stable in the sense of Lyapunov; and they are surrounded by a 3-parameter family of invariant 3-tori.*

Proof As already mentioned, H is independent of ρ. Hence, the conjugate action $\Im$ is an integral of the original problem and we can reduce the number of degrees of freedom by one. In the $(\xi, |\mu|, q, p)$-space the truncated Hamilton function has six 1-parameter families of periodic orbits. For sake of definiteness we restrict to the one associated with the North Pole. As long as the frequency

$$\omega = \frac{\partial H_{|\mu|,\Im,a,b,c}(p,q)}{\partial |\mu|} = \frac{|\mu|}{I_1} - (a+b-2c)\frac{\Im^2}{|\mu|^3}(I_3 - I_1)$$

does not vanish, we can parametrise the periodic orbit by the energy. In other words, on the energy levels of the truncated system we have a periodic orbit. In case the North Pole is a centre the normal frequency of this periodic orbit is

$$\Omega = \sqrt{(c-a)(c-b)}(I_3 - I_1)\frac{3\Im^2 - |\mu|^2}{|\mu|^3};$$

in case it is a saddle it is hyperbolic with eigenvalues

$$\pm\sqrt{-(c-a)(c-b)}(I_3 - I_1)\frac{3\Im^2 - |\mu|^2}{|\mu|^3}.$$

In both cases the periodic orbit will survive a small perturbation, *i.e.* in the original system we have respectively a normally elliptic or hyperbolic periodic orbit on the level sets of H.

The full (untruncated) system itself is Liouville-integrable, see Bogoyavlenskiĭ [1984,1992]. This means that the phase space $T^*SO(3)$, with the exception of a set of measure zero, is foliated by invariant 3-tori. In particular, a motion starting near one of the normally elliptic 2-tori will remain close to it. □

Let us close this section considering the case that the symmetries inherent to H are perturbed away in a generic way. The 2-parameter family of normally elliptic invariant 2-tori is then weakly quasi-periodically stable, the continuum of 2-tori breaks up into a "Cantor set" of 2-tori with quasi-periodic motion. For the normally hyperbolic 2-tori we restrict to a (4-dimensional) centre manifold and use standard KAM-theory to show the persistence of a "Cantor set" of 2-tori therein. Finally, we can use standard KAM-theory to conclude that most of the phase space remains filled by "Cantor sets" of invariant 3-tori.

5 The Kirchhoff top

We now consider the special case of a quadratic potential with S^1-symmetry, *i.e.* with $a = b$. By using the translation $(a, b, c) \mapsto (a - b, b - b, c - b)$ we can ensure that $a = b = 0$.

Corollary 5.1 *Let H be the Hamilton function of a rigid body, fixed at one point, with principal moments of inertia $I_1 = I_2 \neq I_3$, and moving in a conservative acceleration field with S^1-symmetric quadratic potential function. Then the normal form of order two of H is*

$$H_{|\mu|,\Im,c}(p,q) = \frac{|\mu|^2}{2I_1} - \frac{I_3 - I_1}{I_1 I_3}\frac{\Im^2}{2} + \frac{c}{2}(I_1 + \frac{\Im^2}{|\mu|^2}(I_3 - I_1) - \frac{3\Im^2 - |\mu|^2}{|\mu|^2}(I_3 - I_1)\frac{p^2}{|\mu|^2}) + \frac{c^2 I_1 (I_3 - I_1)^2}{16|\mu|^6}(|\mu|^4 - 6|\mu|^2\Im^2 + 5\Im^4 - (6|\mu|^4 - 100|\mu|^2\Im^2 + 126\Im^4)\frac{p^2}{|\mu|^2} + (5|\mu|^4 - 126|\mu|^2\Im^2 + 153\Im^4)\frac{p^4}{|\mu|^4}) .$$

Proof The co-ordinate transformation:

$$\xi \mapsto \xi \, , \; y \mapsto y - \frac{cI_1}{|\mu|_0}(I_3 - I_1)\frac{|\mu|_0^2 - \Im^2}{|\mu|_0^2}\frac{|\mu|_0^2 - p^2}{|\mu|_0^2} \cdot (\sin^2\xi - \frac{1}{2}) - 2\frac{cI_1}{|\mu|_0}(I_3 - I_1)\frac{\Im\sqrt{|\mu|_0^2 - \Im^2}}{|\mu|_0^2}\frac{p\sqrt{|\mu|_0^2 - p^2}}{|\mu|_0^2} \cdot \cos\xi \, , \; \Im \mapsto \Im \, , \; p \mapsto p$$

(the transformations in ρ and q are of no interest) changes the Hamilton function, modulo the higher order terms in $\mathcal{F}_2$, to

$$H \equiv \frac{|\mu|^2}{2I_1} - \frac{I_3 - I_1}{I_1 I_3}\frac{\Im^2}{2} + \frac{c}{2}(I_1 + \frac{\Im^2}{|\mu|^2}(I_3 - I_1) - \frac{3\Im^2 - |\mu|^2}{|\mu|^2}(I_3 - I_1)\frac{p^2}{|\mu|^2}) .$$

Averaging the quadratic part of the higher order terms gives the second term of the normal form. We now truncate and get the desired result. □

The Hamiltonian vector field $X_{H_{|\mu|,\Im,c}}$ is given by the equations

$$\dot{p} = 0$$

$$\dot{q} = -c\frac{I_3 - I_1}{|\mu|}\frac{p}{|\mu|}[\frac{3\Im^2 - |\mu|^2}{|\mu|^2} + c\frac{I_1(I_3 - I_1)}{|\mu|^2}\frac{3|\mu|^4 - 50|\mu|^2\Im^2 + 63\Im^4}{4|\mu|^4} - c\frac{I_1(I_3 - I_1)}{|\mu|^2}\frac{5|\mu|^4 - 126|\mu|^2\Im^2 + 153\Im^4}{4|\mu|^4}\frac{p^2}{|\mu|^2}] .$$

We conclude that the equator consists of (degenerate) equilibria, and we have two equilibria at North and South Pole. For $3\Im^2 = |\mu|^2$ the second equation simplifies

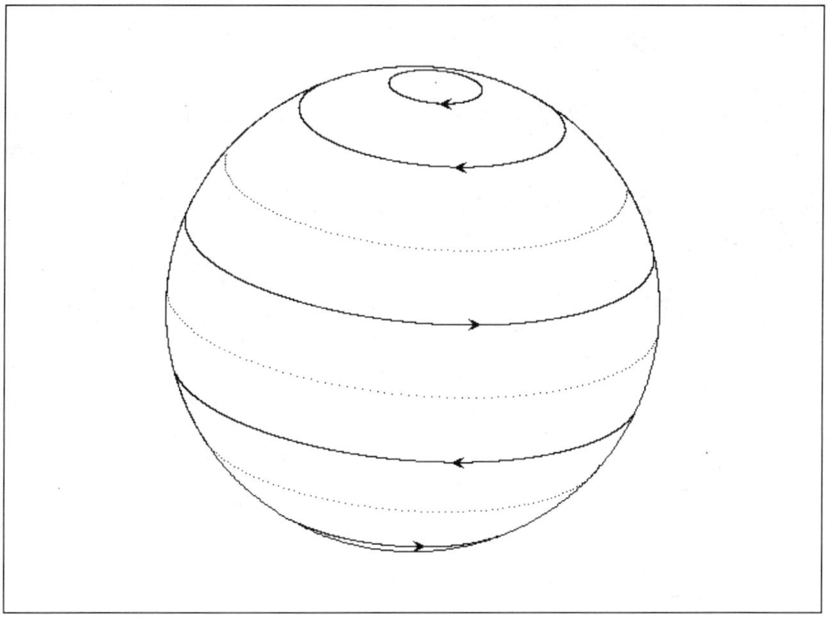

Figure 6 The flow on the (p,q)-sphere to the second order normal form in the Kirchhoff case $a = b$.

to $\dot{q} = c^2 \frac{I_1(I_3 - I_1)^2}{|\mu|^3} \frac{p}{|\mu|}(\frac{5}{3} - 5\frac{p^2}{|\mu|^2})$ and we have in addition two parallels of latitude consisting of equilibria at $p = \pm\sqrt{\frac{1}{3}} \cdot |\mu|$, compare *Figure 6*. Since these are not present for values of the distinguished parameters away from $3\Im^2 = |\mu|^2$, we expect them to vanish in the equator or the poles under variation of these parameters. The equilibria at the North and South Pole are due to the symmetry. When the parallels of latitude consisting of equilibria shrink down to the poles, we will get $\dot{q} = 0$ at $p = \pm|\mu|$. The zeros of $\dot{q}$ at $|p| > |\mu|$ are no longer on the sphere. Due to the symmetry $p \mapsto -p$ the two parallels of latitude consisting of equilibria will reach the equator and vanish both at the same time. $\dot{q}$ will have a double zero at $p = 0$.

Let us first look at the bifurcation at the equator. The expression in square brackets [ ] in $\dot{q}$ vanishes for

$$-\frac{4|\mu|^4(3\Im^2 - |\mu|^2)}{cI_1(I_3 - I_1)} = 3|\mu|^4 - 50|\mu|^2\Im^2 + 63\Im^4 \ .$$

This leads to $\Im^2 = \frac{1}{63}(6\alpha + 25 - \sqrt{36\alpha^2 + 48\alpha + 436}) \cdot |\mu|^2$ with $\alpha = \frac{-|\mu|^2}{cI_1(I_3-I_1)}$. For sake of definiteness we take $c < 0$ and $I_3 > I_1$. Since $\Im^2$ increases with α (with limit $\frac{|\mu|^2}{3}$), the bifurcation from the equator takes place for $3\Im^2 < |\mu|^2$. The smaller we take c, the closer the bifurcation parameters come to $3\Im^2 = |\mu|^2$.

For North and South Pole we have to consider

$$\frac{4|\mu|^4(3\Im^2 - |\mu|^2)}{cI_1(I_3 - I_1)} = 2|\mu|^4 - 76|\mu|^2\Im^2 + 90\Im^4 \ .$$

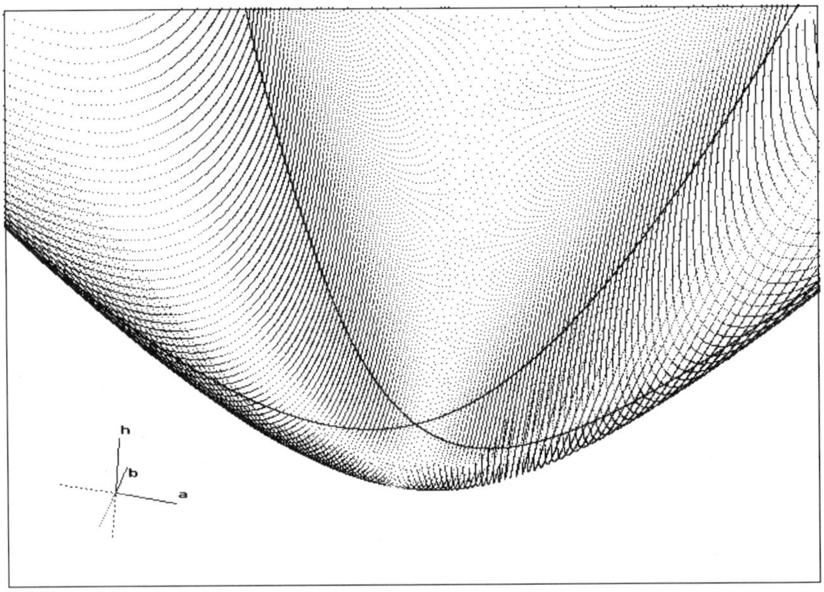

Figure 7 The singular values of the energy-momentum mapping for the Kirchhoff top. The dark lines are the threads. In the upper part of the picture these threads meet the surface and form edges.

This leads to $\Im^2 = \frac{1}{90}(38 - 6\alpha + \sqrt{36\alpha^2 - 816\alpha + 1264}) \cdot |\mu|^2$, again with $\alpha = \frac{-|\mu|^2}{cI_1(I_3-I_1)}$. As was to be expected, $\Im^2$ decreases now with α to $\frac{|\mu|^2}{3}$, the parallels of latitude consisting of equilibria reach the poles for $3\Im^2 > |\mu|^2$.

The above scenario also holds for $c > 0$, $I_3 < I_1$. In the other two cases $c > 0$, $I_3 > I_1$ and $c < 0$, $I_3 < I_1$ the scenario reverses: increasing $\Im^2$ we have parallels of latitude consisting of equilibria emerging from the poles before the line $3\Im^2 = |\mu|^2$ and vanishing in the equator after crossing this line.

As a particular case of a rigid body in a quadratic potential this case is integrable, it is the case named after Kirchhoff. The integrals p and $\Im$ are induced by the S^1-symmetry of the potential and the dynamical symmetry of the rigid body. This facilitates the interpretation of our results. Equilibria correspond to 2-tori and periodic orbits to 3-tori. Together the 2-tori corresponding to the equilibria on a parallel of latitude form a 3-torus with degenerate frequency ratio $[\,\omega_1 : \omega_2 : 0\,]$. As in the Lagrange case we have a 3-parameter family of 3-tori shrinking down to two 2-parameter families of 2-tori. Again we can interpret the limiting cases $\Im = \pm|\mu|$ where the 2-tori of the Euler case degenerate to 1-tori; we expect four 1-parameter families of periodic orbits to occur.

Let us check this using some known facts about the Kirchhoff top, compare Bates and Zou [1993]. The range of the energy-momentum mapping $(H, \Im, p) : T^*SO(3) \longrightarrow I\!R^3$ lies above the set of singularities which looks like a bowl, with edges and with two one-dimensional threads that meet at the point $(H, \Im, p) = (\frac{1}{2}c, 0, 0)$. (See *Figure 7*.) The fibres $(H, \Im, p)^{-1}(h, a, b)$ are of the following types, according to the value (h, a, b).

1) regular values: The fibre is a 3-torus. The motion consists of rotation, precession and nutation.

2) singular values outside $a = \pm b$: The fibre is a 2-torus. The corresponding motion is a regular precession.

3) singular values at $a = \pm b$: For high energy the fibre is a 1-torus, *i.e.* the top rotates stably. When we decrease the energy, we pass a Hamiltonian Hopf bifurcation. Along the 'bowl' we again have regular precession, and the fibre associated to points on the thread is an unstable 1-torus together with its unstable manifold.

Like in the Lagrange case, we could not expect to rediscover the subtleties of the dynamics corresponding to the one-dimensional threads. These effects are controlled by the parameter c, and our analysis is valid only asymptotically as $c \to 0$.

On the other hand we discovered resonant 3-tori. I do not know whether these have already been detected by other methods.

6 Conclusion

Sections 3–5 show that the proposed program can lead to interesting results. A first glance at the situation around $3\Im^2 = |\mu|^2$ of the (nonsymmetric) quadratic potential shows that a series of bifurcations of the invariant 2-tori takes place. It would be interesting to compare such a scenario with the solutions given by direct integration in Bogoyavlenskiĭ [1984,1992]. In the (still integrable) case of an asymmetric rigid body in a quadratic potential the proposed approach can only detect "Cantor sets" of invariant 2-tori and one would expect that the integrability forces also the resonant 2-tori in the "holes" of the "Cantor set" to persist. However, it should be noted that the presented program is mainly designed to detect invariant 2-tori as singularities in the (Cantor) foliation of invariant 3-tori in nonintegrable perturbations of the Euler top. Here the resonant tori will generally break up, so a "Cantor set" of invariant 2-tori will be the result of weak quasi-periodic stability. It is also very interesting to study situations which fail to be stable. I expect local bifurcations like the Hamiltonian saddle node to occur, as well as global bifurcations like hetero- and homoclinic bifurcations. Another interesting problem is to study small perturbations of the cases presented here, destroying the inherent symmetries. For example, how will the resonant 3-tori of the Kirchhoff case break up into finitely many normally elliptic and normally hyperbolic invariant 2-tori? One possibility is the situation related to *Figure 5*, an investigation à la Broer and Vegter [1992] or Cushman and Sanders [1986] should reveal alternative scenarios.

Acknowledgements

I want to thank Henk Broer, who introduced me to the subject and is always there when I have questions. I also thank Richard Cushman for helpful discussions about the Lagrange top, Hinke Osinga for helping me to check the formulas with `MATHEMATICA`, Bernd Krauskopf who carefully read several versions of this text and the referee for stimulating discussions. The pictures were made using the software package `DYNPAO` of Rense Posthumus, for a description see Posthumus and Scholtmeijer [1990]. Finally I would like to acknowledge the support of The Fields Institute where part of this paper was written.

References

Abraham, R. and Marsden, J.E. [1978] *Foundations of Mechanics*, 2nd ed. Benjamin.

Arhangel'skiĭ, Y.A. [1977] *Analytic dynamics of solids*, (in Russian) Izdat. Nauka.

Arnol'd, V.I. [1963] *Proof of a theorem of A. N. Kolmogorov on the invariance of quasi-periodic motions under small perturbations of the Hamiltonian*, Russ. Math. Surv., **18**(5), 9–36.

Arnol'd, V.I. [1974] *Mathematical Methods of Classical Mechanics*, Springer Verlag.

Arnol'd, V.I. [1978] *Geometrical Methods in the Theory of Ordinary Differential Equations*, Springer Verlag.

Arnol'd, V.I., Kozlov, V.V. and Neishtadt, A.I. [1985] *Mathematical Aspects of Classical and Celestial Mechanics*; in Dynamical Systems III (V.I. Arnol'd, ed.), Springer Verlag.

Ashbaugh, M., Chicone, C. and Cushman, R. [1991] *The twisting tennis racket*, Journal of Dynamics and Differential Equations, **3**, 67–86.

Bates, L. and Zou, M. [1993] *Degeneration of Hamiltonian monodromy cycles*, Nonlinearity, **6**,(2), 313–335.

Bogoyavlenskiĭ, O.I. [1984] *New integrable problem of classical mechanics*, Commun. Math. Phys., **94**, 255–269.

Bogoyavlenskiĭ, O.I. [1992] *Euler equations on finite-dimensional Lie coalgebras arising in problems of mathematical physics*, Russ. Math. Surv., **47**(1), 117–189.

Braaksma, B.J.L., Broer, H.W. and Huitema, G.B. [1990] *Toward a quasi-periodic bifurcation theory*, Mem. AMS, **83**, #421, 83–167.

Broer, H.W. [1981] *Formal normal forms of vector fields and some consequences for bifurcations in the volume preserving case*; in Dynamical Systems and Turbulence, Warwick 1980, pp. 54–74. (D. Rand and L-S. Young, eds.), SLN, **898**, Springer Verlag.

Broer, H.W., Chow, S.-N., Kim, Y. and Vegter, G. [1993] *A normally elliptic Hamiltonian bifurcation*, Z. Angew. Math. Phys., **44**, 389–432.

Broer, H.W., Huitema, G.B. and Takens, F. [1990] *Unfoldings of quasi-periodic tori*, Mem. AMS, **83**, #421, 1–82.

Broer, H.W. and Vegter, G. [1992] *Bifurcational aspects of parametric resonance*, Dynamics Reported, New Series, **1**, 1-53.

Cushman, R. and Bates, L. [1994] *Classical Integrable Systems*, to appear.

Cushman, R. and Knörrer, H. [1985] *The momentum mapping of the Lagrange top*; in Differential Geometric Methods in Physics (H. Doebner et. al., eds.), pp. 12–24, SLN **1139**, Springer Verlag.

Cushman, R. and Sanders, J.A. [1986] *Limit cycles in the Josephson equation*, SIAM J. Math. Anal., **17**(3), 495–511.

Deprit, A. [1967] *Free rotation of a rigid body studied in the phase plane*, Am. J. Phys., **35**, 424–428.

Goldstein, H. [1950] *Classical Mechanics*, Addison-Wesley.

Grammel, R. [1950] *Der Kreisel*, 2. Aufl., Springer Verlag.

Kozlov, V.V. and Treshchev, D.V. [1985] *Nonintegrability of the general problem of rotation of a dynamically symmetrically heavy rigid body with a fixed point 1*, Mosc. Univ. Mech. Bull., **40**(6), 27–36.

Kozlov, V.V. and Treshchev, D.V. [1986] *Nonintegrability of general problem of rotation of a dynamically symmetrical heavy rigid body with a fixed point 2*, Mosc. Univ. Mech. Bull., **41**(1), 11–16.

Lewis, D., Ratiu, T., Simo, J.C. and Marsden, J.E. [1992] *The heavy top: a geometric treatment*, Nonlinearity, **5**, 1–48.

Mishchenko, A.S. and Fomenko, A.T. [1978] *Generalized Liouville method of integration of Hamiltonian systems*, Funct. Anal. Appl., **12**(2), 113–121.

Moser, J. [1967] *Convergent series expansion for quasi-periodic motions*, Math. Ann., **169**, 136–176.

Nehorošev, N.N. [1972] *Action-angle variables and their generalizations*, Trans. Moscow Math. Soc., **26**, 180–198.

Pöschel, J. [1982] *Integrability of Hamiltonian systems on Cantor sets*, Comm. Pure Appl. Math., **35**, 653–696.

Posthumus, R.A. and Scholtmeijer, J. [1990] *Chaos and Fractals*, Practicumhandleiding bij DYNPAO, Rijksuniversiteit Groningen.

Sadov, Y.A. [1970] *The action-angle variables in the Euler-Poinsot problem*, J. Appl. Math. Mech., **34**, 922–925.

Sal'nikova, T.V. [1984] *Nonintegrability of perturbed Lagrange problem*, Mosc. Univ. Mech. Bull., **39**(4), 32–39.

Tantalo, P. [1993] *Geometric Phases for the Free Rigid Body with Variable Inertia Tensor*, Ph.D. dissertation, University of California, Santa Cruz.

Added in proof: An application of the method presented in this paper to a non-integrable problem will appear in my Ph.D. thesis *Quasi-periodic motions of a rigid body – a case study in perturbations of superintegrable systems* at the Rijksuniversiteit Groningen.

Fields Institute Communications
Volume 4, 1995

A Homoclinic Orbit of the Lorenz System by Precise Shooting

Brian Hassard
Department of Mathematics
State University of New York, Buffalo
106 Diefendorf Hall
Buffalo, NY, USA 14214

Jianhe Zhang
Department of Mathematics
State University of New York, Buffalo
106 Diefendorf Hall
Buffalo, NY, USA 14214

Abstract. A proof of the existence of a homoclinic orbit of the Lorenz system is outlined, based on shooting using precise numerical integrations.

1 Introduction

The Lorenz system (Lorenz [1963])

$$x' = s(y - x) \tag{1.1}$$
$$y' = (R - z)x - y \tag{1.2}$$
$$z' = xy - qz \tag{1.3}$$

is often cited as a simple ordinary differential system which exhibits complex dynamical behavior, including chaos. However, remarkably little of this behavior has been actually proven to exist. Our knowledge of the more interesting phenomena in the system, derives from traditional numerical integrations lacking rigorous control of truncation and round off error.

Only recently has even existence of a homoclinic orbit for the system been proven. Hastings and Troy [1992] show for each (s, q) in some neighborhood of (10,1), there is an R in the range (1,1000) for which (1.1)–(1.3) has a homoclinic orbit.

1991 *Mathematics Subject Classification*. Primary 34A50; Secondary 34C37, 65L70.

First published in the SIAM Journal on Mathematical Analysis in Vol. 25, Number 1, pp. 179-196, January 1994.

Here we outline a proof that a homoclinic orbit exists for $s = 10$, $q = \frac{8}{3}$ which is essentially a rigorous version of a numerical shooting argument due to Kaplan and Yorke [1979] which appears in Sparrow's monograph [1982]. The details of the analysis will be given in Hassard and Zhang (to appear). Here we concentrate on the numerical part of the proof which uses a precise numerical integrator with true error bounds. As in Hastings and Troy (to appear), the existence of a homoclinic orbit is inferred as a consequence of properties of just two computed orbits (shots). As in Sparrow [1982], the estimate of a value of R for which a homoclinic orbit exists is sharpened by additional shooting.

The numerical argument (Kaplan and Yorke [1979] and Sparrow [1982]) is illustrated in the following figure for $s = 10$, $q = \frac{8}{3}$. Integrations are performed for initial conditions displaced from the origin by a small distance along the eigenvector for the positive eigenvalue. The computed trajectories are observed to return close to the origin but are then swept away in one of two different directions, depending on the value of R. For R too large, trajectories are swept away to negative x and y values, and for R too small, they are swept away to positive x and y values. Bisection of R values is performed to locate more precisely the (assumed) R^* for which a homoclinic orbit exists.

To make precise the different qualitative behaviors, we define the following properties.

Property P As $t \to -\infty, (x, y, z)^T \to (0,0,0)^T$. Also, there exist numbers $\tau_1 < s_1 < t_1$ such that:

$$\begin{aligned} x' &> 0 \text{ on } (-\infty, \tau_1); \\ x' &< 0 \text{ on } (\tau_1, t_1]; \\ x(t_1) &= 0; \\ y' &< 0 \text{ on } [\tau_1, s_1]; \\ y(s_1) &= 0; \\ y &< 0 \text{ on } (s_1, t_1]; \\ x &> 0 \text{ on } (-\infty, t_1). \end{aligned}$$

The above trajectory for $R = 15$ illustrates Property P. It is also useful to define:

Property Q As $t \to -\infty, (x, y, z)^T \to (0,0,0)^T$. There exists a τ_1 such that $x' > 0$ on $(-\infty, \tau_1)$. Also, the orbit $(x, y, z)^T$ does not have property P.

In Section 2 we shall show how the existence of orbits having properties P and Q leads to the existence of a homoclinic orbit. We incorporate more information about orbits into property P than in the hypotheses of (Hastings and Troy (to appear) Lemma 2.2) so that we can separate a central argument (Lemma 2.1) from the body (Lemma 2.2) of the proof, and organize the proofs around the concept of sequences of orbits having property P.

In Section 3 and Section 4 we construct orbits with properties P and Q. On $(-\infty, 0]$, orbits are approximated by polynomials in $e^{\lambda_3 t}$, where λ_3 is the positive eigenvalue of the Jacobian at $(0,0,0)^T$. Initial conditions for the numerical integrations derive from these approximations at $t = 0$. The origin of t is defined such that on the time interval $(-\infty, 0]$, the orbits are in a neighborhood of $(0,0,0)^T$ sufficiently small that local theory (Lemma 3.7) applies to show existence and to

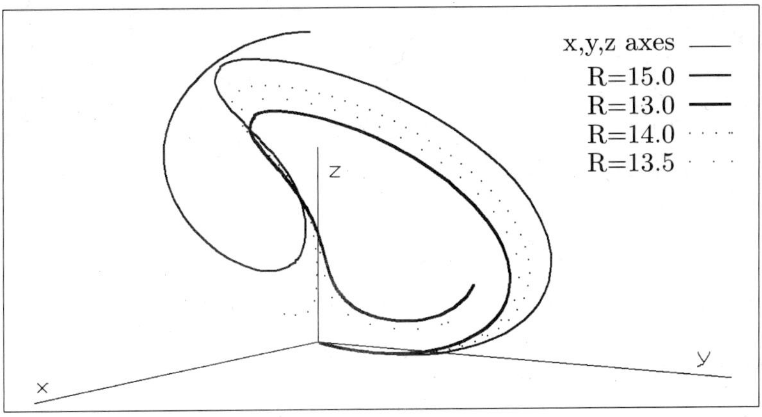

Figure 1

construct highly accurate approximations. High accuracy at $t = 0$ is necessary to compensate for the large loss of precision during the integrations for $t \geq 0$.

In Section 4 we describe precise numerical integrations of using an algorithm due to Aberth [1988]. The actual code we used is based on Aberth's C++ program called "difsys.cc", which generates precise solutions of initial value problems of ordinary differential systems for which the component functions are elementary. In Section 4 we first describe the modifications we had to make to Aberth's algorithm to achieve our goals, and then present the results in the form of Lemma 4.1. Finally we state the Theorem.

2 Conditions for existence of a homoclinic orbit

We fix $q > 0$, $s > 0$ and restrict our attention to some interval $[R_a, R_b]$ of R values, $R_a > 1$. The Jacobian matrix for (1.1)–(1.3) at the origin then has two negative and one positive eigenvalues. The unstable manifold at the origin has components $\gamma^+(R)$, $\gamma^-(R)$ where locally $x > 0$ on γ^+, $x < 0$ on γ^-. We will be concerned only with γ^+.

We let $p^+(R) = (\sqrt{q(R-1)}, \sqrt{q(R-1)}, R-1)^T$ denote the stationary point in the positive octant. Let $p(t; R) = (x(t; R), y(t; R), z(t; R))^T$ denote a family of orbits of the Lorenz system, one for each $R \in [R_a, R_b]$, jointly smooth in t and R, known to exist on the interval $(-\infty, 0]$, and such that $\lim_{t \to -\infty} p(t; R) = (0, 0, 0)^T$ and $x'(t, R) > 0$ on $(-\infty, 0]$. Such orbits will be constructed in Section 3. Then $p(.; R) \in \gamma^+(R)$. It is well known (Lorenz [1963] or Sparrow [1982], Appendix C) that for fixed parameters q,R,s, there is a region V, the interior of an ellipse, which contains the origin and is a (forward) invariant set for the flow. Because of the invariant region, each orbit $p(t; R)$ on $(-\infty, 0]$ may be continued to the entire line $(-\infty, \infty)$. Furthermore the partial derivatives of x,y and z with respect to t of

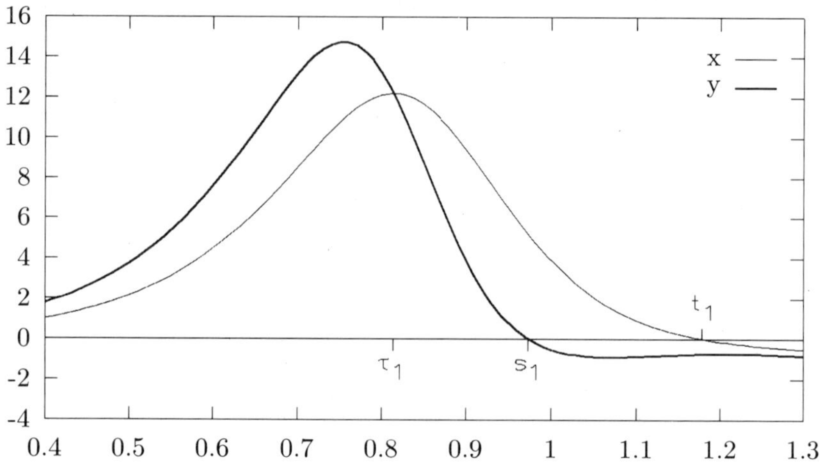

Figure 2

any fixed order, are bounded for all t, and bounds which are uniform in R may be constructed.

The main task in establishing the existence of a homoclinic orbit is to show that for the sequence $\{p(.;R_j)\}$ of orbits with property P constructed in Lemma 2.2, the sequence $\{t_{1,j}\}$ is unbounded. This will be done using:

Lemma 2.1 *Suppose* $0 < q < 2s$ *and* $1 < R_a < R < R_b$. *Let* $R_j \in (R_a, R_b)$, $R_j \to R$ *and suppose each orbit* $p(.;R_j) = (x(.;R_j), y(.;R_j), z(.;R_j))^T$ *has property P. Let* $\tau_{1,j} < s_{1,j} < t_{1,j}$ *be from property P such that* $x'(\tau_{1,j}; R_j) = 0$, $y(s_{1,j}; R_j) = 0$ *and* $x(t_{1,j}; R_j) = 0$. *If the sequences* $\{\tau_{1,j}\}, \{s_{1,j}\}, \{t_{1,j}\}$ *have finite limits* τ_1, s_1, t_1, *then the orbit* $(x(.;R), y(.;R), z(.;R))$ *also has property P.*

The basic existence result is then:

Lemma 2.2 *Suppose* $0 < q < 2s$ *and* $1 < R_a < R_b$. *Suppose that for* $R = R_b$, *there is an orbit with property P and for* $R = R_a$ *there is an orbit with property Q. Suppose further that for all* $R \in [R_a, R_b]$, $p^+(R)$ *is linearly stable. Then there is a value* R^*, $R_a < R^* < R_b$ *for which there is an orbit* $p(.;R^*) = (x, y, z)^T$ *with the following properties:*

As $t \to \pm\infty$, $(x,y,z)^T \to (0,0,0)^T$. *Also, there exist* τ_1, s_1,$-\infty < \tau_1 < s_1 < \infty$ *such that*

$$\begin{aligned} x' &> 0 \text{ on } (-\infty,\tau_1) \\ x' &< 0 \text{ on } (\tau_1,\infty) \\ y' &< 0 \text{ on } [\tau_1,s_1] \\ y(s_1) &= 0 \\ y &< 0 \text{ on } (s_1,\infty) \\ x &> 0 \text{ on } (-\infty,\infty). \end{aligned}$$

3 Local approximation of the unstable manifold

Our objective in this section is to construct highly accurate approximations to solutions $p(t)$ of (1.1)–(1.3) with the property $\lim_{t\to-\infty} p(t) = (0,0,0)^T$. The construction also shows existence. The first step is a coordinate transformation.

Let $s' = s-1$ and $d = ((s')^2 + 4sR)^{\frac{1}{2}}$. The eigenvalues of the Jacobian matrix for (1.1)–(1.3) at the origin are then $\lambda_1 = -q < 0$, $\lambda_2 = \frac{(-(s+1)-d)}{2} < 0$, and $\lambda_3 = \frac{(-(s+1)+d)}{2} > 0$. Let P denote the matrix whose columns are the corresponding eigenvectors $(0,0,1)^T$, $(1,c_2,0)^T$ and $(1,c_3,0)^T$. Here $c_2 = \frac{(s'-d)}{2s} < 0$ and $c_3 = \frac{(s'+d)}{2s} > 0$. Let $c_1 = \frac{s}{d} > 0$. Setting $(x,y,z)^T = Pw$, (1.1)–(1.3) becomes:

$$\frac{dw_1}{dt} = \lambda_1 w_1 + f_1 = \lambda_1 w_1 + (c_2 w_2 + c_3 w_3)(w_2 + w_3); \tag{3.1}$$

$$\frac{dw_2}{dt} = \lambda_2 w_2 + f_2 = \lambda_2 w_2 + c_1 w_1 (w_2 + w_3); \tag{3.2}$$

$$\frac{dw_3}{dt} = \lambda_3 w_3 + f_3 = \lambda_3 w_3 - c_1 w_1 (w_2 + w_3). \tag{3.3}$$

We shall find solutions of (3.1)–(3.3) with the property $\lim_{t\to-\infty} w = (0,0,0)^T$, by solving the system of integral equations

$$w_1(t) = \int_{\tau=-\infty}^{t} e^{\lambda_1(t-\tau)}(c_2 w_2(\tau) + c_3 w_3(\tau))(w_2(\tau) + w_3(\tau))d\tau, \tag{3.4}$$

$$w_2(t) = \int_{\tau=-\infty}^{t} e^{\lambda_2(t-\tau)} c_1 w_1(\tau)(w_2(\tau) + w_3(\tau))d\tau, \tag{3.5}$$

$$w_3(t) = \eta e^{\lambda_3 t} - \int_{\tau=t}^{0} e^{\lambda_3(t-\tau)} c_1 w_1(\tau)(w_2(\tau) + w_3(\tau))d\tau. \tag{3.6}$$

The choice range of integration in (3.6) is partly motivated by the requirement that $e^{\lambda_3(t-\tau)} < 1$ for τ in the range. We next define some functionals:

Definition 3.1 For bounded continuous R^3-valued functions $u(t), v(t)$ on $(-\infty, 0]$ and for real η, let $\beta(u,v) = (\beta_1,\beta_2,\beta_3)^T$, $\varphi(u)$ and $\psi(u;\eta)$ denote the functionals

$$\beta_1(u,v)(t) = \int_{-\infty}^{t} e^{\lambda_1(t-\tau)}(c_2u_2(\tau)+c_3u_3(\tau))(v_2(\tau)+v_3(\tau))d\tau, \quad (3.7)$$

$$\beta_2(u,v)(t) = \int_{-\infty}^{t} e^{\lambda_2(t-\tau)}c_1u_1(\tau)(v_2(\tau)+v_3(\tau))d\tau, \quad (3.8)$$

$$\beta_3(u,v)(t) = -\int_{t}^{0} e^{\lambda_3(t-\tau)}c_1u_1(\tau)(v_2(\tau)+v_3(\tau))d\tau, \quad (3.9)$$

$$\varphi(u) = \beta(u,u), \quad (3.10)$$

$$\psi(u;\eta) = (0,0,\eta e^{\lambda_3 t})^T + \varphi(u). \quad (3.11)$$

The system (3.4)–(3.6) then becomes simply

$$w = \psi(w;\eta). \quad (3.12)$$

Definition 3.2 Let $W^1 = (0,0,\eta e^{\lambda_3 t})^T$, and for each $J \geq 2$, let

$$W^J(t) = W^{J-1}(t) + \frac{\eta^J}{J!}\frac{\partial^J \varphi(W^{J-1})(t)}{\partial \eta^J}\Big|_{\eta=0}. \quad (3.13)$$

We use W^J, for some fixed J as the approximation to the unstable manifold for (3.1)–(3.3) at the origin. Our objective in the remainder of this section is to establish the error bound in Lemma 3.7. (Remark: U^J defined by $U^1 = W^1$, $U^j = \psi(U^{j-1};\eta)$ for $2 \leq j < J$, has similar approximation properties to W^J but is impractical here since U^J is a polynomial of degree 2^{J-1} in η.)

Lemma 3.3 *For each $J \geq 1$, the component functions $W_i^J, i = 1,2,3$ of W^J have the form*

$$W_i^J = \sum_{j=1}^{J} \eta^j \sum_{k=1}^{j} W_{ijk}\xi^k, \quad (3.14)$$

where $\xi = e^{\lambda_3 t}$ and the coefficients W_{ijk} are constants given recursively by (3.15) and (3.16)–(3.18) below.

$$W_{111} = W_{211} = 0 \text{ and } W_{311} = 1, \quad (3.15)$$

$$W_{iJk} = \frac{f_{iJk}^{J-1}}{k\lambda_3 - \lambda_i}, \quad 1 \leq i \leq 2, \ 1 \leq k \leq J, \quad (3.16)$$

$$W_{3J1} = -\sum_{k=2}^{J} \frac{f_{3Jk}^{J-1}}{k\lambda_3 - \lambda_3}, \quad (3.17)$$

$$W_{3Jk} = \frac{f_{3Jk}^{J-1}}{k\lambda_3 - \lambda_3}, \quad 2 \leq k \leq J. \quad (3.18)$$

It is easy to show that the components of $\psi(W^J;\eta)$ are of the general form

$$\psi_i(W^J;\eta) = \sum_{j=1}^{2J} \eta^j \sum_{k=1}^{j} \xi^k \psi^J_{ijk}.$$

The purpose of the following two Lemmas is to show that in fact $\psi^J_{ijk} = W_{ijk}$ for $1 \le j \le J$ and $1 \le k \le j$. The "$O(\eta^{J+1})$" reasoning is simply a quick way to achieve this result. The actual bounding argument does not appear until Lemma 3.7.

Lemma 3.4 *For each $J \ge 2$, $\psi(W^{J-1};\eta) = W^J + O(\eta^{J+1})$ for any fixed $t \le 0$.*

Lemma 3.5 *For each $J \ge 2$, $\psi(W^J;\eta) = W^J + O(\eta^{J+1})$ for any fixed $t \le 0$, and the components of $\psi(W^J;\eta)$ are given by (3.19)-(3.21) below.*

$$\psi^J_{ijk} = \frac{f^J_{ijk}}{k\lambda_3 - \lambda_i}, \quad 1 \le i \le 2, 1 \le k \le j, \tag{3.19}$$

$$\psi^J_{3j1} = -\sum_{k=2}^{J} \frac{f^J_{3jk}}{k\lambda_3 - \lambda_3}, \tag{3.20}$$

$$\psi^J_{3jk} = \frac{f^J_{3jk}}{k\lambda_3 - \lambda_3}, \quad 2 \le k \le J. \tag{3.21}$$

Definition 3.6 of norms For $u \in R^3$, we write $|u| = |u_1| + |u_2| + |u_3|$, and for continuous, R^3-valued functions $u(t)$ on $(-\infty, 0]$ we write $\|u\| = \sup_{t \le 0} |u(t)|$.

Lemma 3.7 *Fix $J \ge 1$. Let $\rho > 0$ be such that $0 < K(\rho) < \frac{1}{2}$, where $K(\rho) = 2\rho(\frac{c_3}{|\lambda_1|} + \frac{c_1}{|\lambda_2|} + \frac{c_1}{\lambda_3})$. Choose $\eta_0 > 0$ sufficiently small so that for all $|\eta| \le \eta_0$, the conditions $\|w^0\| \le \frac{\rho}{2}$ and $\|\psi(w^0;\eta) - w^0\| \le \frac{\rho}{4}$ are satisfied, where $w^0 = W^J(t,\eta)$. Then the sequence $w^n = \psi(w^{n-1};\eta)$, $n = 1, 2, \ldots$ converges uniformly in t and η to a continuous, bounded solution $w(t)$ of the integral equation $w = \psi(w;\eta)$. Moreover, for each $t \le 0$ and $|\eta| \le \eta_0$,*

$$|w(t) - w^0(t)| \le (1 - K(\rho))^{-1} \sum_{i=1}^{3} \sum_{j=J+1}^{2J} |\eta|^j \sum_{k=1}^{j} \xi^k |\psi^J_{ijk}|, \tag{3.22}$$

where $\xi = e^{\lambda_3 t}$ and the coefficients ψ^J_{ijk} are given by (3.19)-(3.21).

To apply Lemma 3.7, J is chosen and the coefficients

$$W_{ijk}, \quad 1 \le i \le 3, \quad 1 \le j \le J, \quad 1 \le k \le j$$

and

$$\psi^J_{ijk}, \quad 1 \le i \le 3, \quad J+1 \le j \le 2J, \quad 1 \le k \le j$$

are computed. Then η is chosen, and the expressions

$$B_0 = \sum_{i=1}^{3} \sum_{j=1}^{J} |\eta|^j \sum_{k=1}^{j} |W_{ijk}|$$

$$B_1 = \sum_{i=1}^{3} \sum_{j=J+1}^{2J} |\eta|^j \sum_{k=1}^{j} |\psi_{ijk}^J|$$

are computed. These bound $\|w^0\|$ and $\|w^1 - w^0\|$ respectively. Then for $\rho = \max(2B_0, 4B_1)$, $K(\rho)$ is evaluated. If $K(\rho) \geq \frac{1}{2}$, it is necessary to choose a smaller η. Assuming $K(\rho) < \frac{1}{2}$, the expressions

$$W_i^J(0) = \sum_{j=1}^{J} \eta^j \sum_{k=1}^{j} W_{ijk},\ 1 \leq i \leq 3$$

are evaluated, and then

$$w_i(0) \in \bar{w}_i = [W_i^J(0) - \frac{B_1}{1-K(\rho)}, W_i^J(0) + \frac{B_1}{1-K(\rho)}],\ 1 \leq i \leq 3.$$

For example, if $s = 10$, $R = 15$, and $q = \frac{8}{3}$, one finds that $\lambda_1 = -2.666667$, $\lambda_2 = -18.548$, $\lambda_3 = 7.548$, $c_1 = 0.3832$, $c_2 = -0.8548$, and $c_3 = 1.7548$, the numeric values in this example being precise in the sense that the absolute errors do not exceed $\frac{1}{2}$ unit in the last decimal place shown. Then for $J = 2$,

$$w^0 = (0.0988\eta^2\xi^2, 0, \eta\xi)^T$$

$$\psi(w^0; \eta) = w^0 + (0, 0.000919\eta^3\xi^3, 0.00251\eta^3(1-\xi^3))^T.$$

Choosing $\eta = \frac{1}{10}$, we find that:

$$\|w^0\| \leq B_0 = 0.10098815$$

and

$$\|w^1 - w^0\| \leq B_1 = 0.0000059367.$$

Then for

$$\rho = \max(2B_0, 4B_1) = 0.2019763, \quad K(\rho) = 0.2947 < \frac{1}{2},$$

so Lemma 3.7 applies. Now

$$w^0(0) = (0.000988, 0, 0.1)^T,$$

and $w(0)$ obeys

$$|w(0) - w^0(0)| \leq \frac{B_1}{1-K(\rho)} \leq 0.0000084171,$$

from which

$$x(0) = w_2(0) + w_3(0) \in \bar{x} = [0.0999, 0.1001],$$
$$y(0) = c_2w_2(0) + c_3w_3(0) \in \bar{y} = [0.1754, 0.1756]$$

and

$$z(0) = w_1(0) \in \bar{z} = [0.0009, 0.0011].$$

Lemma 3.7 shows existence of solutions $p(t) = (x, y, z)^T = Pw$ of (1.1)–(1.3), defined on $(-\infty, 0]$ and such that $\lim_{t\to-\infty} p(t) = 0$. Then, thanks to the invariant set V, the solutions $p(t)$ may be continued to $(-\infty, \infty)$. To have property P or Q, the solutions must also satisfy $x' > 0$ for $t < \tau_1$, where $x'(\tau_1) = 0$. By choosing η sufficiently small positive, we can arrange that $x' > 0$ on $(-\infty, 0]$ as in the following lemma, which implies that any τ_1 such that $x'(\tau_1) = 0$ must be positive.

Lemma 3.8 *Let $p(t) = (x(t), y(t), z(t))^T$, $t \in (-\infty, 0]$ be a solution of (1.1)–(1.3) such that $0 < x(0) < y(0)$, $z(t) < R-1$ on $(-\infty, 0]$ and $x(t) \to 0$ as $t \to -\infty$. Then $x' > 0$ on $(-\infty, 0]$.*

4 The solutions for $t > 0$

Aberth [1988] describes an algorithm for computing the solutions of initial value problems "precisely". The algorithm is a Taylor series method using interval arithmetic, and applies to ordinary differential systems in which the functions defining the derivatives are elementary. In Aberth [1988], algorithms and programs are given for first and second order systems. The generalization to systems of n first order equations is more recent and is coded as program difsys.cc. (See the discussion.) Modifications of difsys were required to obtain the information needed for present purposes. The basic task in difsys is to advance the solution to $t_{k'} = t_k + h$ and determine bounds for $x(t_{k'}), y(t_{k'}), z(t_{k'})$. Before a trial interval $[t_k, t_k + h]$ is accepted as a "bounding interval", an associated containment region is constructed. The first two modifications we made to the algorithm, were

M_1 reject a trial interval unless the containment region guarantees

- at least one of x, x' is of one sign on the interval, and
- at least one of y, y' is of one sign on the interval, and
- at least one of z, z' is of one sign on the interval.

M_2 reject a trial interval unless the containment region guarantees

- at least one of x', x'' is of one sign on the interval, and
- at least one of y', y'' is of one sign on the interval, and
- at least one of z', z'' is of one sign on the interval.

We also added code to compute intervals $I_k^0 \supset x(t_k)x(t_{k'})$, $J_k^0 \supset y(t_k)y(t_{k'})$, $K_k^0 \supset z(t_k)z(t_{k'})$, $I_k^1 \supset x'(t_k)x'(t_{k'})$, $J_k^1 \supset y'(t_k)y'(t_{k'})$, and $K_k^1 \supset z'(t_k)z'(t_{k'})$. The last two modifications are

M_3 if $I_k^0 < 0$, print a message "x changes sign"

- if $J_k^0 < 0$, print a message "y changes sign"
- if $K_k^0 < 0$, print a message "z changes sign"
- if I_k^0, J_k^0 or K_k^0 overlaps 0, print a message that this happens.

M_4 if $I_k^1 < 0$, print a message "x' changes sign"

- if $J_k^1 < 0$, print a message "y' changes sign"
- if $K_k^1 < 0$, print a message "z' changes sign"
- if I_k^1, J_k^1 or K_k^1 overlaps 0, print a message that this happens.

Here $I_k^0 > 0$ (resp. < 0) means every point in I_k^0 is positive (resp. negative), and "I_k^0 overlaps zero" means $0 \in I_k^0$.

Suppose that the modified code has accepted a bounding interval $[t_k, t_{k'}]$ and has successfully computed an approximate solution on this interval, and no messages from M_3 appear. Then we claim that $x(t)$ is either strictly positive or strictly negative on $[t_k, t_{k'}]$. For, as M_3 issued no message, neither $I_k^0 < 0$ nor I_k^0 overlaps zero, so $I_k^0 > 0$. Therefore $x(t_k)x(t_{k'}) > 0$ and x does not vanish at either endpoint. Since the interval $[t_k, t_{k'}]$ passed the test M_1, the containment region guarantees that at least one of x, x' is of one sign on the interval. If x is of one sign, then $\operatorname{sgn}(x(t)) = \operatorname{sgn}(x(t_k)) \neq 0$ for all $t \in [t_k, t_{k'}]$ and we are done. If x' is of one sign, so then $x(t)$ is monotonic (increasing or decreasing) from $x(t_k)$ to $x(t_{k'})$, where $x(t_k)$ and $x(t_{k'})$ are of the same sign, and so again $\operatorname{sgn}(x(t)) = \operatorname{sgn}(x(t_k))$ for all $t \in [t_k, t_{k'}]$. Similarly if no messages from M_3 appear, $y(t)$ and $z(t)$ are strictly positive or negative on $[t_k, t_{k'}]$.

If the message "x changes sign" appears, then $x(t_k)x(t_{k'}) < 0$ so $x(t_k)$ and $x(t_{k'})$ are of opposite signs and $x(t)$ has at least one zero in the interval. In this case, since the interval passed the test M_1, necessarily x' is of one sign on the interval and it follows that $x(t)$ changes sign at exactly one point, an interior point, of the interval. Messages "y changes sign" and "z changes sign" are similarly interpreted. If no messages from M_4 appear for a bounding interval $[t_k, t_{k'}]$, by a similar argument $x'(t), y'(t)$ and $z'(t)$ are each of one sign on the interval. If a message "x' changes sign", appears, then x' is strictly monotonic on the interval and has a zero at an interior point. Messages "y' changes sign" and "z' changes sign" are interpreted similarly. If any messages that I_k^0, J_k^0, K_k^0, I_k^1, J_k^1 or K_k^1 overlap 0 should appear, it is necessary to repeat the computation at a higher precision.

When the precision of x, y, z values becomes inadequate for difsys to write n correct digits, the original version of difsys attempts to repeat the entire computation over again at higher precision. We disabled this feature and instead allowed the code to continue the current computation while printing fewer correct digits. We also 'tuned' the code for better efficiency in the current problem by increasing the degree of the Taylor polynomial approximations from 12 to 22.

We now state the numerical results as:

Lemma 4.1 *For each set of values s, q, R in Table 1, the Lorenz system has an orbit p(t) with the property Q, and for each set of values s, q, R in Table 2, the Lorenz system has an orbit p(t) with property P.*

Proof For each triple (s, q, R) shown in Tables 1 and 2, we chose J and computed intervals $\bar{x}(0), \bar{y}(0), \bar{z}(0)$ guaranteed by Lemma 3.7 to contain a point $x(0), y(0), z(0)$ of a trajectory of the Lorenz system such that $x(t) \to 0$ as $t \to -\infty$. This was done with a short program in C++ using the Aberth-Schaefer interval arithmetic package. Input data consists of the parameters q, R, s, the integer J, $\eta = w_3(0) > 0$, and an integer determining the precision of the arithmetic. Output consists of the information whether or not η was chosen sufficiently small for the hypotheses of Lemma 3.7 to be satisfied, and if so, the initial condition intervals $\bar{x}(0), \bar{y}(0), \bar{z}(0)$ and bounds B_0, B_1 on $\|w^0\|$ and $\|w^1 - w^0\|$. All this data is shown in the tables. Examining $\bar{x}(0)$ and $\bar{y}(0)$, since $x(0) \in \bar{x}(0)$ and $y(0) \in \bar{y}(0)$ it follows that $0 < x(0) < y(0)$. Since $|z(t)| = |w_1(t)|$, it follows that:

$$|z(t)| < \|p(t)\| \leq \rho = \max(2B_0, 4B_1) = 2B_0 < R - 1$$

for all $t < 0$. From Lemma 3.8 we conclude that $x'(t) > 0$ on $(-\infty, 0]$, which implies $x(t) > 0$ on $(-\infty, 0]$.

Using Aberth's code difsys with the modifications described above, we computed trajectories of the Lorenz system for $t > 0$, taking $\bar{x}(0), \bar{y}(0), \bar{z}(0)$ as interval-valued initial conditions. The integrations were carried out until either:

(1) the sign change messages showed that the orbit had property P

(2) the sign change messages showed that the orbit had property Q

(3) the computation failed to determine whether the orbit had property P or Q.

Data corresponding to successful computations was then entered in Tables 1 and 2. □

We now state our result concerning a homoclinic orbit for (1.1)–(1.3).

Theorem 4.2 *For* $s = 10$, $q = \frac{8}{3}$ *and some* R^*, $R_a < R^* < R_b$ *where* $R_a = 13.9265$ *is the maximum of the R values in Table 1 and* $R_b = 13.927$ *is the minimum of the values in Table 2, the Lorenz system has an orbit* $p(t)$ *such that* $\lim_{t\to-\infty} p(t) = \lim_{t\to\infty} p(t) = (0,0,0)^T$, *and there are numbers* $\tau_1 < s_1$ *such that*

$$\begin{aligned} x' &> 0 \text{ on } (-\infty, \tau_1) \\ y(\tau_1) &< 0 \\ x' &< 0 \text{ on } (\tau_1, \infty) \\ y' &< 0 \text{ on } [\tau_1, s_1] \\ y(s_1) &= 0 \\ y &< 0 \text{ on } (s_1, \infty) \\ x &> 0 \text{ on } (-\infty, \infty). \end{aligned}$$

Proof It is easily verified that $p^+(R)$ is linearly stable for all $R \in [R_a, R_b]$. The proof then follows immediately from Lemmas 2.2 and 4.1. □

5 Discussion

The condition in Lemma 2.2 that $p^+(R)$ is linearly stable could be dropped if we adopted part of the argument in Hastings and Troy (to appear). We retain the condition because it simplifies the argument and holds for the present parameter values.

We tried a scheme using expansion in a single variable for approximation of the unstable manifold. We adopted the two variable expansion scheme of Section 3 because it appeared that we could better exploit the special nature of the bilinear nonlinearity with this scheme.

To obtain an approximation to the homoclinic orbit uniformly valid on $(-\infty, \infty)$ would require approximation of both the stable and unstable manifolds and the solution of a two-point boundary value problem. Such a uniform approximation would be a step towards proving the existence of chaotic solutions of the Lorenz system, by showing that there exist horseshoes in a Poincaré map associated with the homoclinic orbit.

The existence of chaotic solutions of the Lorenz system is established by a different argument in our joint work with Hastings and Troy (Hassard et al. (to appear)). In Hassard et al. (to appear), the techniques described here are used to verify conditions outlined in Hastings and Troy (preprint), which imply that for the parameter values $q = 9$, $s = 10$, $R = 76$, the system (1.1)–(1.3) possesses a infinite set of solutions. Each solution corresponds to a symbol sequence, where the symbols can be any odd positive integers. The symbols specify the number of sign changes of x' between consecutive zeros of x.

The result in Hassard et al. (to appear) most analogous to the existence of the homoclinic orbit here, is the existence of an orbit p such that $\lim_{t\to-\infty} p(t) =$

$(0,0,0)^T$ and at some $T > 0$, p enters a (forward) invariant region K^-. This region lies in $x < 0, y < 0, z > 0$ and contains the stationary point $p^-(R) = (-\sqrt{q(R-1)}, -\sqrt{q(R-1)}, R-1)^T$.

The present work is ostensibly numerical, while Hastings and Troy (to appear) is analytical. However, this distinction is somewhat artificial. Hastings and Troy use calculator computations, and include margins in their bounds to account for possible roundoff errors. They approximate the unstable manifold and shoot, generating bounds on the trajectories. Here we do the same, except that we use an interval-arithmetic package and methods of precise numerical analysis to obtain the bounds automatically.

The initial uncertainty box for the first integration in Table 1, has sides of lengths 2×10^{-44},2×10^{-43} and 2×10^{-44} in x,y and z respectively. Internally, the programs represent interval variables in the form

$$\sigma(0.d_1d_2..d_{k-4}d_{k-3}d_{k-2}d_{k-1}d_k \pm 0.00..0r_1r_2r_3r_4) \times 10^p$$

where σ indicates the sign, k and p are multiples of 4, and $d_1, .., d_k$,$r_1, .., r_4$ are decimal digits. Each variable has its own precision k. Setting $n = 40$ to control the precision as in the tables below limits the precisions k of results to $k \leq n+12 = 52$. The first entries $x(0)$,$y(0)$,$z(0)$ in Table 1 are each represented internally using $k = 44$. The precision of the computation is effectively determined by the uncertainty in the initial conditions rather than the constraint $k \leq 52$. As the integration proceeds, the uncertainty box associated with the computed x,y,z grows. The precision k associated with each variable is decreased automatically, roughly each time the range of the interval has grown by a factor of 10^4.

Before performing the precise integrations, we greatly underestimated the rate of growth of the uncertainty box. We found that this box grows by a factor of 10^{10} per unit of time. The rate depends on the integrator as well as the system and initial conditions. Since we had chosen an arbitrary precision arithmetic package we were able to compensate for this rapid growth by using a tiny initial box and initially high precision interval arithmetic. An integration fails for our purposes when there is inadequate precision to distinguish between properties P and Q. As successive bisections are performed, increasingly greater precision is needed to make this distinction.

The Aberth-Schaefer interval arithmetic package, including the code difsys.cc, is available by anonymous ftp from math.tamu.edu, as file pub/range/range.tar.Z. This package extends the language C++, but still uses standard C++ compilers. Similar packages extending FORTRAN and PASCAL, but which use special compilers are also available; see the bibliography of literature on enclosure methods and related topics by G. Boblender (June 1993), available by anonymous ftp from garbo.uwasa.fi, as file pc/doc-soft/enclbib1.zip.

We chose the Aberth-Schaefer package because it offered arbitrary precision interval arithmetic and because we understood the algorithm and implementation of Aberth's Taylor series-based precise integrator well enough to perform the required modifications.

Shadowing (Sauer and Yorke [1991] and Van Vleck [1992]), if applied to our initial value problem by itself, would show the existence of a true solution near an approximate solution but for unknown initial conditions. Since this true solution cannot be expected to join up with the unstable manifold, nothing is gained. To obtain meaningful results in this problem, both the initial value problem and a system of integral equations for the unstable manifold for $t < 0$ must be considered simultaneously, a harder task.

6 Tables

The notation '$\sim$' in the following tables indicates intervals having midpoint the given decimal value and half-width one unit in the last decimal place. Other decimal values have an absolute error not exceeding $\frac{1}{2}$ unit in the last decimal place.

Table 1. Orbits with property Q

$R = 13.926$ (exact)
$s = 10$ (exact)
$q = \frac{8}{3}$
$J = 25$
$\eta = 0.1$ (exact)
$B_0 = 0.10101874785080090852561365779491980762459271357300$
$B_1 = 2.408204904263048501757000E - 51$
$x(0)$ in $0.10000101387916306034307402007221470577302262 \sim$
$y(0)$ in $0.1712964441229470378484705897787410542299221 \sim$
$z(0)$ in $0.00101205806115328023910366159554912883232786 \sim$
$n = 40$
y' changes sign on the interval 0.6938476563 to 0.6943359375
x' changes sign on the interval 0.7558593750 to 0.7578125000
z' changes sign on the interval 0.8242187500 to 0.8251953125
y changes sign on the interval 0.9521484375 to 0.9560546875
y' changes sign on the interval 1.0224609375 to 1.0263671875
y changes sign on the interval 1.4863281250 to 1.4873046875
x' changes sign on the interval 1.5744628906 to 1.5747070313
$t_f = 1.6$ (exact)

$R = 13.9265$ (exact)
$s = 10$ (exact)
$q = \frac{8}{3}$
$J = 25$
$\eta = 0.1$ (exact)
$B_0 = 0.10101873551056010840559137742027929504345305451900$
$B_1 = 2.4065991909594552011740119000E - 51$
$x(0)$ in $0.100001013830960354945278654625074601199213577394 \sim$
$y(0)$ in $0.17129842358345975564593357177092041637153684 \sim$
$z(0)$ in $0.001012046083589028516630187386761869868147368 \sim$
$n = 25$
y' changes sign on the interval 0.6938476563 to 0.6943359375
x' changes sign on the interval 0.7558593750 to 0.7578125000
z' changes sign on the interval 0.8242187500 to 0.8251953125
y changes sign on the interval 0.9521484375 to 0.9560546875
y' changes sign on the interval 1.0224609375 to 1.0263671875
y changes sign on the interval 1.5902099609 to 1.5903320313
x' changes sign on the interval 1.6719360352 to 1.6719970703
$t_f = 1.72$ (exact)

Table 2. Orbits with property P

$R = 13.93$ (exact)
$s = 10$ (exact)
$q = \frac{8}{3}$
$J = 25$
$\eta = 0.1$ (exact)
$B_0 = 0.10101864915078526087165201466934523185017094722900$
$B_1 = 2.3953909886544034317548524000E - 51$
$x(0)$ in $0.100001013493656816759885982974650444100994598921 \sim$
$y(0)$ in $0.17131227893846698671597596429068023867274433 \sim$
$z(0)$ in $0.00101196226158061216784055167835988349168981 \sim$
$n = 25$
y' changes sign on the interval 0.6938476563 to 0.6943359375
x' changes sign on the interval 0.7558593750 to 0.7578125000
z' changes sign on the interval 0.8242187500 to 0.8251953125
y changes sign on the interval 0.9521484375 to 0.9560546875
y' changes sign on the interval 1.0224609375 to 1.0263671875
x changes sign on the interval 1.4384765625 to 1.4404296875
y' changes sign on the interval 1.4511718750 to 1.4521484375
$t_f = 1.46$ (exact)

$R = 13.927$ (exact)
$s = 10$ (exact)
$q = \frac{8}{3}$
$J = 25$
$\eta = 0.1$ (exact)
$B_0 = 0.10101872317110199349682497404003229580910620678400$
$B_1 = 2.4049946147446155737298037000E - 51$
$x(0)$ in $0.10000101378276177196766847317506571067572971416 2 \sim$
$y(0)$ in $0.17130040301294877808616205243979664937742171 \sim$
$z(0)$ in $0.00101203410677283095541051420081399240699564 1 \sim$
$n = 25$
y' changes sign on the interval 0.6938476563 to 0.6943359375
x' changes sign on the interval 0.7558593750 to 0.7578125000
z' changes sign on the interval 0.8242187500 to 0.8251953125
y changes sign on the interval 0.9521484375 to 0.9560546875
y' changes sign on the interval 1.0224609375 to 1.0263671875
x changes sign on the interval 1.5346679688 to 1.5351562500
$t_f = 1.54$ (exact)

Acknowledgements We wish to thank S. P. Hastings for suggesting the problem, O. Aberth for his most useful interval analysis codes, and both S. P. Hastings and O. Aberth for helpful discussions. We also wish to thank the referee for reading this work so carefully and for suggestions which greatly improved the presentation.

References

Aberth, O. [1988] *Precise Numerical Analysis*, William C. Brown Publishers, Dubuque, Iowa.

Hassard, B. Hastings, S. Troy, W.and Zhang, J. *A Computer Proof of Chaos in the Lorenz System*, Applied Math. Letters (to appear).

Hassard, B. and Zhang, J. [1994], *Existence of a Homoclinic Orbit of the Lorenz System by Precise Shooting*, SIAM J. Math. Anal, Vol. 25, **1**, 179-196.

Hastings, S. and Troy, W. [1994], *A Proof that the Lorenz Equations Have a Homoclinic Orbit*, J. Diff. Eqns., Vol. 113, **1**, 166-188.

Hastings, S. and Troy, W. *On the Chaotic Solutions of the Lorenz Equations*, (preprint).

Kaplan, J.L. and Yorke, J.A. [1979] *The onset of chaos in a fluid flow model of Lorenz*, In Bifurcation Theory and Applications in Scientific Disciplines, (O. Gurel and O. E. Röessler, eds.), Annal. N. Y. Acad. Sci., **316**, 400–407.

Lorenz, E. N. [1963] *Deterministic Non-periodic Flow.* J. Atmospheric Sci., **20**, 130–141.

Sauer, T. and Yorke, J.A. [1991] *Rigorous Verification of Trajectories for the Computer Simulation of Dynamical Systems*, Nonlinearity, **4**, 961–969.

Sparrow, C. [1982] *The Lorenz Equations: Bifurcations, Chaos, and Strange Attractors*, Springer-Verlag, Applied Mathematical Sciences, **41**.

Van Vleck, E. *A Shadowing Lemma Approach to Global Error Analysis for Initial Value ODE's* (to appear).

Fields Institute Communications
Volume 4, 1995

Homoclinic Intermittency

Ale Jan Homburg
Department of Mathematics
University of Groningen
P.O. Box 800
9700 AV Groningen
The Netherlands

Abstract. We study a one parameter family of vector fields in $\mathbb{R}^3$, with a homoclinic bifurcation. Under some conditions on the global dynamics, we show that breaking the homoclinic loop leads to intermittent dynamics.

1 Introduction

We consider a class $\mathcal{C}$ of smooth vector fields on $\mathbb{R}^3$ consisting of vector fields with a hyperbolic singularity of a kind we describe below and a homoclinic orbit, i.e. an orbit, different from the singularity, converging to the singularity both for positive and negative time. Let X be a vector field in $\mathcal{C}$. We denote by $\Gamma = \{\Gamma(t)|t \in \mathbb{R}\}$ the homoclinic orbit of X, parametrized as a solution. We may and do assume that the distinguished singularity to which Γ converges, is the origin 0. We assume $DX(0)$ has eigenvalues $\lambda^{ss}, \lambda^s, \lambda^u$ with $\lambda^{ss} < \lambda^s < 0 < \lambda^u$. We use corresponding coordinates (x_{ss}, x_s, x_u) in $\mathbb{R}^3$, where, for example, the x_s-axis equals the eigenspace belonging to the eigenvalue λ^s. We consider generic one parameter families X_μ of smooth vector fields with $X_0 = X \in \mathcal{C}$. We may assume that the origin 0 is a singularity of X_μ for all small values of μ. Shil'nikov [1968] proved that generically the following bifurcation occurs. On one side of $\mu = 0$, say for $\mu < 0$, X_μ has a periodic orbit γ whose period goes to infinity as $\mu \to 0$. This periodic orbit disappears in the homoclinic orbit at $\mu = 0$.

We assume in this paper the eigenvalue condition $\lambda^s + \lambda^u < 0$. This implies that γ is an attracting periodic orbit. We further assume the existence for $\mu = 0$, of a heteroclinic connection from 0 to $\bar{\Gamma}$. The existence of such a heteroclinic connection is a persistent property for generic families X_μ intersecting $\mathcal{C}$. We shall see that a consequence of this assumption is the occurence of attractors for $\mu > 0$. Each branch of the unstable manifold of 0 is contained in the stable set of such an

1991 *Mathematics Subject Classification.* Primary 34C37; Secondary 58F13.

attractor (for $\mu > 0$, there can be either one or two attractors). If $\mu < 0$, both branches of the unstable manifold of 0 are contained in the stable set of γ.

We mention that the study of bifurcations from a vector field with two homoclinic orbits in the shape of a 'butterfly' (see Gambaudo [1987]) is related to the bifurcation studied in this paper. Near such a codimension two bifurcation point, there are families of vector fields as studied here.

In Section 2 we describe the situation more precisely. The third and fourth sections are then devoted to a study of the bifurcations. In the last section we discuss a specific feature of the bifurcations, namely that it gives rise to intermittency. We shall explain this phenomena and discuss different quantitative aspects (scaling properties) of it.

We use the following notation for invariant manifolds: $W^{ss}(0)$ denotes the strong stable manifold of 0, tangent to $\{(x_s, x_u) = 0\}$ at 0, $W^u(0)$ denotes the unstable manifold of 0, tangent to $\{(x_{ss}, x_s) = 0\}$ at 0, etc. The local versions of these manifolds are denoted by $W^{ss}_{loc}(0), \ldots$ Observe that we suppress the dependence on the parameter from the notation.

This work has substantially benefitted from discussions with Björn Sandstede, Bernold Fiedler, Valya Afraimovich, Pavol Brunovsky and especially Floris Takens.

2 Generalized homoclinic orbits

Let X be a vector field in $\mathcal{C}$, as in the introduction. After a coordinate change, we may write for $x = (x_{ss}, x_s, x_u)$,

$$DX(0)x = \lambda^{ss} x_{ss} \frac{\partial}{\partial x_{ss}} + \lambda^s x_s \frac{\partial}{\partial x_s} + \lambda^u x_u \frac{\partial}{\partial x_u},$$

with $\lambda^{ss} < \lambda^s < 0 < \lambda^u$. We assume the following eigenvalue condition:

$$\lambda^s + \lambda^u < 0. \tag{2.1}$$

There is a two-dimensional stable manifold $W^{ss,s}(0)$ and a one-dimensional unstable manifold $W^u(0)$ of the singularity. We assume that one branch of $W^u(0)$ is contained in $W^{ss,s}(0)$ so X has a homoclinic orbit. Denote this homoclinic orbit by $\Gamma = \{\Gamma(t) | t \in \mathbb{R}\}$. We will consider one parameter families X_μ with $X_0 = X \in \mathcal{C}$.

In order to get a codimension one unfolding, we must assume some generic conditions on X_μ. We now introduce these conditions. First we assume

$$\Gamma \not\subset W^{ss}(0), \tag{2.2}$$

which implies that $\lim_{t \to \infty} \dot{\Gamma}(t)/\|\dot{\Gamma}(t)\| = \pm(0,1,0)$. Near the origin, there are local centre unstable manifolds $W^{s,u}_{loc}(0)$ tangent to $\{x \in \mathbb{R}^3 | x_{ss} = 0\}$ at the origin. These local centre unstable manifolds are not unique, but their tangent spaces in points of $W^u_{loc}(0)$ are unique. Denote by $W^{s,u}(0)$ an invariant manifold, equal to a local centre unstable manifold near 0. We assume that

$$W^{s,u}_{loc}(0) \text{ intersects } W^{ss,s}(0) \text{ transversally along } \Gamma \cap W^u_{loc}(0). \tag{2.3}$$

Since the tangent spaces of $W^{s,u}_{loc}(0)$ along $\Gamma \cap W^u_{loc}(0)$ are unique, this condition is well defined. Conditions (2.2) and (2.3) imply that the bundle of planes $\{T_x W^{s,u}(0), x \in \bar{\Gamma}\}$, obtained by moving the tangent planes of $W^{s,u}_{loc}(0)$ along $W^u_{loc}(0)$ by the derivative of the flow, is a continuous bundle along $\bar{\Gamma}$.

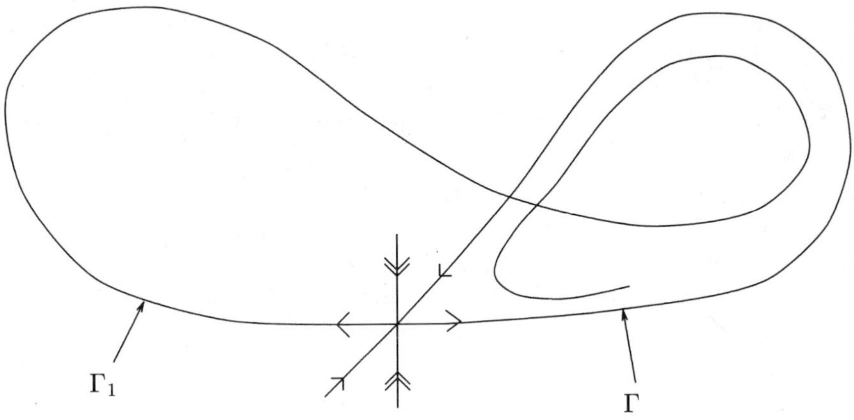

Figure 1 The homoclinic orbit Γ and the generalized homoclinic orbit Γ_1.

We have now introduced the generic conditions we assume to hold for X. To get a generic unfolding, one more condition on the dependence of X_μ on μ must hold. Denote by $\tilde{W}^s, \tilde{W}^u$ the union of the stable resp. unstable manifolds of 0 in the product $\mathbb{R}^3 \times \mathbb{R}$ of state space and parameter space. We assume X_μ is such that

$$\tilde{W}^s \text{ intersects } \tilde{W}^u \text{ transversally along } \Gamma \times 0. \tag{2.4}$$

Bifurcations of one parameter families X_μ for which properties (2.1),...,(2.4) hold, were studied by Shil'nikov [1968], who proved that a periodic orbit disappears in the homoclinic orbit. By (2.1), this periodic orbit is attracting.

Consider again the vector field X_0 for $\mu = 0$. Suppose that

$$\text{the bundle of planes } \{T_x W^{s,u}(0), x \in \bar{\Gamma}\} \text{ is orientable.} \tag{2.5}$$

By this condition, for $\mu = 0$, $W^{ss,s}(0)$ divides a small tubular neighbourhood $\mathcal{V}$ of $\bar{\Gamma}$ into two parts. By the eigenvalue condition (2.1), one of these parts consists of orbits converging to $\bar{\Gamma}$ for positive time. So the set of orbits whose ω limit set equals $\bar{\Gamma}$, is a three dimensional manifold bounded by $W^{ss,s}(0)$. We denote this manifold by $V^s(\bar{\Gamma})$. If either $\lambda^s + \lambda^u > 0$ or if the bundle of planes $\{T_x W^{s,u}(0), x \in \bar{\Gamma}\}$ is nonorientable, there are no orbits whose ω limit set equals $\bar{\Gamma}$.

Suppose that the branch of $W^u(0)$ which does not form Γ, is contained in $V^s(\bar{\Gamma})$. In other words, suppose there is a heteroclinic connection from 0 to $\bar{\Gamma}$. Let $\Gamma_1 = \{\Gamma_1(t) | t \in \mathbb{R}\}$ denote this orbit, see Figure 1. The orbit Γ_1 is an example of what we call a *generalized homoclinic orbit.* This terminology is inspired by the observation that the singularity 0 is contained in both the α and the ω limit set of such an orbit. We refer to Homburg [1993] for a general study of generalized homoclinic orbits and their role in the dynamics and bifurcations of one parameter families X_μ.

Of fundamental importance for the bifurcation study is the following result, which enables a reduction to a two-dimensional vector field. The proof of this theorem can be found in Homburg [1993], below we merely sketch the general idea of the proof.

Theorem 2.1 *Let X_μ be a one parameter family of vector fields with $X_0 = X \in \mathcal{C}$, satisfying properties (2.1),...,(2.5). Suppose X has a generalized homoclinic orbit Γ_1, as described above. Then, in a small neighbourhood $\mathcal{U}$ of $\bar{\Gamma} \cup \bar{\Gamma}_1$, there is a C^1 locally invariant and normally hyperbolic centre manifold $W^{s,u}_{loc}(\bar{\Gamma} \cup \bar{\Gamma}_1)$, depending on μ and tangent to $\{x_{ss} = 0\}$ at 0.* □

Remark 2.2 Local invariance of $W^{s,u}_{loc}(\bar{\Gamma} \cup \bar{\Gamma}_1)$ means that X_μ is tangent to it. The consequence of normal hyperbolicity (Hirsch, Pugh and Shub [1977]) is here that any orbit starting in $\mathcal{U}$ and not on $W^{s,u}_{loc}(\bar{\Gamma} \cup \bar{\Gamma}_1)$, will converge to $W^{s,u}_{loc}(\bar{\Gamma} \cup \bar{\Gamma}_1)$. To understand the dynamics of X_μ in $\mathcal{U}$, we may thus indeed restrict out attention to $W^{s,u}_{loc}(\bar{\Gamma} \cup \bar{\Gamma}_1)$, having reduced the problem to a bifurcation problem on a two-dimensional manifold.

The centre manifold $W^{s,u}_{loc}(\bar{\Gamma} \cup \bar{\Gamma}_1)$ can be orientable or nonorientable, although its restriction to a small neighbourhood $\mathcal{V}$ of $\bar{\Gamma}$, by (2.5), is orientable. We remark that an orientable centre manifold near $\bar{\Gamma} \cup \bar{\Gamma}_1$ is a torus with a hole. So the bifurcation phenomena described below can be encountered in a family of vector fields on a torus. Actually, there are vector fields on the torus possessing two singularities, one source and one saddle, such that removing a small disk around the source yields vector fields of the type considered here. These vector fields are reminiscent of Cherry vector fields (Palis and de Melo [1982]). A nonorientable centre manifold near $\bar{\Gamma} \cup \bar{\Gamma}_1$ is a Klein bottle with a hole. The bifurcation phenomena described in Section 4 below can thus be found in a family of vector fields on a Klein bottle. Also here, there are vector fields on the Klein bottle with two singularities, one source and one saddle, such that removing a small disk around the source yields vector fields as studied in this paper.

We end this section with a sketch of the construction of $W^{s,u}_{loc}(\bar{\Gamma} \cup \bar{\Gamma}_1)$, referring the reader to Homburg [1993] for justifications. By the conditions imposed on X_0, the bundle of tangent planes $T_x W^{s,u}(0)$ along $\bar{\Gamma} \cup \bar{\Gamma}_1$ is continuous. We call this the bundle of centre directions along $\bar{\Gamma} \cup \bar{\Gamma}_1$. Further, there is a continuous bundle of strong stable directions along $\bar{\Gamma} \cup \bar{\Gamma}_1$, where the two bundles together form a splitting of $T\mathbb{R}^3$ along $\bar{\Gamma} \cup \bar{\Gamma}_1$. By strong stable we mean that the derivative of the time t map of the flow, for t large enough, contracts vectors in the space of strong stable directions stronger then it can possibly contract any vector in the space of centre directions. These two bundles provide a system of coordinates $(x_{s,u}, x_{ss})$ near $\bar{\Gamma} \cup \bar{\Gamma}_1$.

Let $\mathcal{U}$ be a small tubular neighbourhood of $\bar{\Gamma} \cup \bar{\Gamma}_1$, such that the flow is directed into $\mathcal{U}$ at $\partial\mathcal{U}$. The existence of small neighbourhoods with this property is established in Homburg [1993]. In fact, $\mathcal{U}$ can be chosen diffeomorphic to a bundle over a two dimensional manifold $\mathcal{N}$ with fiber $(-1, 1)$. In the system of coordinates $(x_{s,u}, x_{ss})$ near $\overline{\Gamma} \cup \overline{\Gamma_1}$, $\mathcal{N}$ is diffeomorphic to $\{x_{ss} = 0\}$. Let C be the one dimensional manifold $\{x_{ss} = 0\} \cap \partial\mathcal{U}$, C is diffeomorphic to $\partial\mathcal{N}$.

Two cases occur, depending on whether the bundle of centre directions along $\overline{\Gamma} \cup \overline{\Gamma_1}$ is orientable or not. If it is, $\mathcal{N}$ can be seen to be diffeomorphic to a torus with a disk removed, otherwise $\mathcal{N}$ is diffeomorphic to a Klein bottle with a disk removed. In both cases C is diffeomorphic to a circle, C being the boundary of the removed disk, see Homburg [1993].

We claim, without giving the proof here, that $W^{s,u}_{loc}(\bar{\Gamma} \cup \bar{\Gamma}_1)$ can be obtained by saturating C, meaning

$$W^{s,u}_{loc}(\bar{\Gamma} \cup \bar{\Gamma}_1) = \overline{\bigcup_{t \geq 0} X_t(C)}.$$

To develop some geometric intuition, suppose X_μ on $\mathcal{U}$ is of the form:

$$Y_\mu(x_{s,u}) \frac{\partial}{\partial x_{s,u}} + \lambda^{ss} x_{ss} \frac{\partial}{\partial x_{ss}}.$$

One may then easily be convinced of the correctness of this construction. The justification for the general case proceeds by using graph transform techniques (Hirsch, Pugh and Shub [1977]).

For $\mu = 0$, the bundle of tangent spaces of $W^{s,u}_{loc}(\bar{\Gamma} \cup \bar{\Gamma}_1)$ along $\bar{\Gamma} \cup \bar{\Gamma}_1$ equals the bundle of centre directions. Of course, if the bundle of centre directions along $\bar{\Gamma} \cup \bar{\Gamma}_1$ is orientable resp. nonorientable, the centre manifold will have this property.

3 The orientable centre manifold

In this section we study a family of vector fields X_μ as in the previous section, such that the centre manifold $W^{s,u}_{loc}(\bar{\Gamma} \cup \bar{\Gamma}_1)$, whose existence is guaranteed by Theorem 2.1, is orientable.

We restrict X_μ to $W^{s,u}_{loc}(\bar{\Gamma} \cup \bar{\Gamma}_1)$. Let Σ be a section on $W^{s,u}_{loc}(\bar{\Gamma} \cup \bar{\Gamma}_1)$, transversally intersecting Γ for $\mu = 0$. We identify Σ with an interval in $\mathbb{R}$, such that $W^{ss,s}_{loc}(0)$ intersects Σ at $\{x = 0\}$. Consider a first return map π_μ on Σ. The following lemma describes some properties of π_μ, for the proof we refer the reader to Homburg [1993].

Lemma 3.1 *Let π_μ be a one parameter family of return maps on a section Σ, as above. Then, for a proper choice of Σ and, if necessary, after a reparametrization of the parameter, π_μ has the following properties for small parameter values.*

(P1) *π_μ is C^1 outside 0, where we have $0 < \pi_\mu' \leq h$, for some $h < 1$.*

(P2) *π_μ is discontinuous in 0, $\pi_\mu(0^+) = -\mu, \pi_\mu(0^-) = K(\mu)$ for some positive function $K(\mu)$.*

(P3) $\frac{\partial}{\partial \mu} \pi_\mu(x) < 0,\ x \neq 0.$ □

In this section we analyse π_μ and derive a bifurcation theorem. For $\mu < 0$, π_μ has a stable fixed point, corresponding to an attracting periodic orbit of X_μ. Observe that for definiteness, we suppose that the attracting periodic orbit of X_μ which disappears in the homoclinic orbit at $\mu = 0$, exists for $\mu < 0$. There is a first bifurcation at $\mu = 0$. Note that π_μ maps the interval $I = [-\mu, K(\mu)]$ into itself. In fact, π_μ restricted to $I - \{0\}$ is injective, see Figure 2. We shall consider π_μ for $\mu \geq 0$ on I, every other point will eventually end up in I.

We can make π_μ into a monotonic but discontinuous map on the circle by identifying the point $-\mu$ with $K(\mu)$. This makes π_μ discontinuous in $-\mu$. The fact that we can make π_μ into a circle map, suggests that we consider a rotation number. The map π_μ can be viewed as the inverse of an increasing continuous circle map (which is constant on a subinterval of I). For a continuous increasing circle map, the rotation number exists and is independent of x. This implies that a rotation number $\rho_\mu(x)$ for π_μ exists for almost all x and is independent of x. If desired, we can thus define $\rho_\mu(x)$ for all x. Because $\rho_\mu(x)$ does not depend on x, we

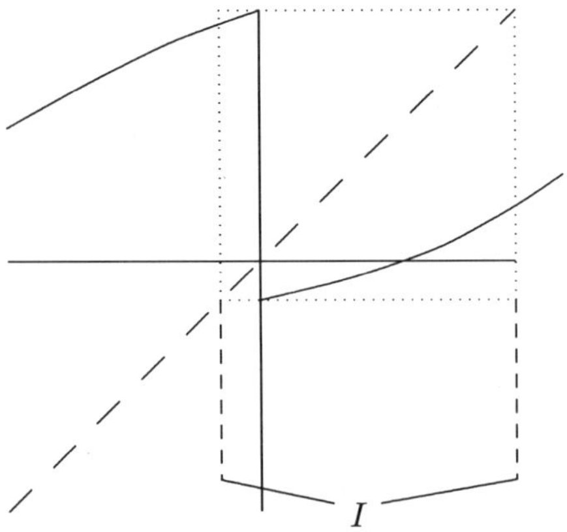

Figure 2 The return map π_μ, for $\mu > 0$.

drop the reference to x from $\rho_\mu(x)$. Furthermore, ρ_μ depends continuously on μ. By (P3), ρ_μ is a nonincreasing function of μ. The parameter values where $\rho_\mu \in \mathbb{Q}$, are intervals.

We can now easily translate the above information into knowledge on the dynamics of X_μ. At values of μ from the interior of an interval with $\rho_\mu \in \mathbb{Q}$, X_μ has a periodic attractor. At values of μ at the boundaries of such intervals, this periodic attractor disappears in a homoclinic bifurcation. In the closure of the homoclinic bifurcation values, π_μ has the bifurcations where ρ_μ is irrational. Observe that X_μ has a nonclosed recurrent orbit at such bifurcation values. Summarizing, we have the following theorem.

Theorem 3.2 *Let X_μ be a one parameter family of vector fields as in Section 2. Suppose the centre manifold $W^{s,u}_{loc}(\bar\Gamma \cup \bar\Gamma_1)$ is orientable.*

In a suitable and small neighbourhood of $\bar\Gamma \cup \bar\Gamma_1$ and, if necessary, after reversing the parameter, X_μ has the following bifurcations for small μ. The bifurcation set of X_μ is a Cantor set contained in $\mu \geq 0$. The left and right boundary points of the Cantor set are homoclinic bifurcation values.

For μ outside bifurcation values, the nonwandering set of X_μ is the union of the singularity and a periodic attractor. □

We mention that the bifurcation set has zero measure, see Homburg [1993].

4 The nonorientable centre manifold

In this section we study a family of vector fields X_μ as in Section 2, such that the centre manifold $W^{s,u}_{loc}(\bar\Gamma \cup \bar\Gamma_1)$, whose existence is guaranteed by Theorem 2.1, is nonorientable. In this section we derive a bifurcation theorem for such a situation.

Let π_μ be the return map on a section Σ, as in the previous section.

Lemma 4.1 *Let π_μ be a one parameter family of return maps on a section Σ, as above. Then, for a proper choice of Σ and, if necessary, after a reparametrization of the parameter, π_μ has the following properties for small parameter values.*

(P1) *π_μ is C^1 outside 0. For some $h < 1$ we have $0 < \pi_\mu'(x) \leq h$ if $x > 0$ and $-h \leq \pi_\mu'(x) < 0$ if $x < 0$.*

(P2) π_μ *is discontinuous in* 0, $\pi_\mu(0^+) = -\mu, \pi_\mu(0^-) = K(\mu)$ *for some positive function* $K(\mu)$.

(P3) $\frac{\partial}{\partial\mu}\pi_\mu(x) < 0,\ x \neq 0$.

For $\mu < 0$, π_μ has a stable fixed point. We shall consider π_μ for $\mu \geq 0$ restricted to the interval $I = [-\mu, \pi_\mu(-\mu)]$. On this interval (without 0), π_μ is injective. Every point eventually enters this interval, see Figure 2. To study the dynamics of this map, we consider the iterates of the interval $J = (-\mu, 0)$. Since every orbit passes through this interval, a description of the iterates of J yields a description of all the occuring dynamics. Let n be the smallest positive integer such that $\pi_\mu^n(J) \cap J \neq \emptyset$. There are two possibilities, either there exists a point x in J such that $\pi_\mu^n(x) = 0$ or such a point does not exist. In the former case, consider the intervals $J_1 = (-\mu, x)$ and $J_2 = (x, 0)$. Then $\pi_\mu^{n+1}(J_1) \subset J_1$ and $\pi_\mu^n(J_2) \subset J_2$. Since 0^+ and 0^- are not mapped to 0, X_μ has two periodic orbits. In the latter case $\pi_\mu^n(J) \subset J$, if also 0^+ or 0^- are not mapped to 0, X_μ has one periodic orbit. Finally, it is easily seen that if $\pi_\mu^n(0^-) = 0$ or $\pi_\mu^n(0^+) = 0$, X_μ has a homoclinic orbit and one periodic orbit.

The bifurcation set thus consists of a countable set of homoclinic bifurcations, with $\mu = 0$ as its only accumulation point. Outside bifurcation values, the nonwandering set Ω_μ is the union of p_μ and one or two periodic orbits. The vector field X_μ has alternatingly one or two periodic orbits.

Theorem 4.2 *Let* X_μ *be a one parameter family of vector fields as in Section* 2. *Suppose the centre manifold* $W^{s,u}_{loc}(\bar{\Gamma} \cup \bar{\Gamma}_1)$ *is nonorientable.*

In a suitable and small neighbourhood of $\bar{\Gamma} \cup \bar{\Gamma}_1$ *and, if necessary, after reversing the parameter,* X_μ *has the following bifurcations for small* μ. *The bifurcation set of* X_μ *is a countable set of homoclinic bifurcation values in* $\mu > 0$, *converging to* $\mu = 0$.

For μ *outside bifurcation values, the nonwandering set of* X_μ *is the union of the singularity and one or two periodic attractors.* □

5 Scaling properties

We now explain that the homoclinic bifurcation of X_μ at $\mu = 0$ gives rise to intermittency. Intermittent dynamics are dynamics where regular, for example almost periodic behavior, is from time to time interrupted by seemingly irregular dynamics which takes place in other parts of the phase space. One may think of a bifurcation in a one parameter family of vector fields in which a periodic attractor disappears and where after this bifurcation typical orbits still spend lots of time in a small neighbourhood of the former periodic orbit, but from time to time make an excursion through other parts of the state space. This is, as we shall see, actually the scenario occuring in our situation. It is in such cases interesting to have quantitative information on the lengths of the time intervals in which regular or irregular dynamics occurs. For more information on intermittency and definitions see Bergé, Pomeau and Vidal [1984] and Takens [1989].

Let $\mathcal{U}$ be a small neighbourhood of $\Gamma \cup \bar{\Gamma}_1$. Denote by Ω_μ the nonwandering set of X_μ in $\mathcal{U}$. Outside bifurcation values, Ω_μ is the union of the singularity and one or two periodic attractors. Consider the homoclinic bifurcation at $\mu = 0$. After a possible reparametrization of μ, we may assume that the periodic attractor which disappears in the homoclinic orbit at $\mu = 0$, exists for $\mu < 0$. We have $\lim_{\mu\uparrow 0}\Omega_\mu = \bar{\Gamma}$ while $\lim_{\mu\downarrow 0}\Omega_\mu = \bar{\Gamma} \cup \bar{\Gamma}_1$, which is strictly bigger. The limits here are in the Hausdorff sense. Denote by $\mathcal{V}$ the part of $\mathcal{U}$ which is near $\bar{\Gamma}$. For small $\mu < 0$, orbits in $\mathcal{U}$ all move into and then stay in $\mathcal{V}$. As a consequence of the previously derived bifurcation theorems, we see that for small positive μ, orbits spend most time in $\mathcal{V}$, but are outside $\mathcal{V}$ for bounded intervals in time.

Consider an orbit $(X_\mu)_t(x)$ in $\mathcal{U}$ for all $t \geq 0$. Adapting the terminology from Bergé, Pomeau and Vidal [1984], we speak of a laminar phase, if the orbit $(X_\mu)_t(x)$ is in $\mathcal{V}$, and we call the phase between two laminar phases the relaminarization phase. Clearly the duration of the relaminarization phases is almost constant. The proposition below contains a statement on the scaling of the duration of the laminar phases, as depending on μ.

Theorem 5.1 *Let X_μ be a one parameter family of vector fields as in Section 2. The duration of the laminar phase is bounded from below by $-C\ln(\mu)$, for some constant $C > 0$.*

Remark 5.2 An upper bound on the duration of the laminar phases, as function of μ, can not be given. For all $\mu > 0$, there are orbits which stay arbitrarily long in $\mathcal{V}$.

Proof We give the proof under the assumption of the existence of smooth local coordinates near 0 in which X_μ is a linear vector field. This is a generic assumption, as follows from a parameter dependent version of a result by Sternberg [1958].

So, we assume that in local coordinates (x_{ss}, x_s, x_u), X_μ is linear. Let S be the cross section $\{x_s = d\}$, transversally intersecting $\bar{\Gamma} \cup W^{ss,s}_{loc}(0)$ for $\mu = 0$. By a rescaling, we may assume $d = 1$. Consider a second section $T = \{x_u = 1\}$. An easy computation, using that X_μ is locally linear near 0, shows that the local transition map $S \to \pm T$, equals

$$(x_{ss}, 1, x_u) \quad \mapsto \quad (x_{ss}|x_u|^{-\lambda^{ss}/\lambda^u}, \pm|x_u|^{-\lambda^s/\lambda^u}, \pm 1). \tag{5.1}$$

The time required to go from a point $(x_{ss}, 1, x_u) \in S$ to $\pm T$ by the flow of X_μ is $\frac{-1}{\lambda^u}\ln(|x_u|)$.

Let Σ be the intersection of S with $W^{s,u}_{loc}(\bar{\Gamma} \cup \bar{\Gamma}_1)$ and let π_μ denote the Poincaré return map on Σ. As an immediate consequence of (5.1) and the fact that the centre manifold $W^{s,u}_{loc}(\bar{\Gamma} \cup \bar{\Gamma}_1)$ is C^1, after a C^1 reparametrization of μ, we can choose a coordinate x on Σ such that the following asymptotic description for π_μ holds:

$$\pi_\mu(x) = \begin{cases} -\alpha|x|^\vartheta + K + o(|x|^\vartheta), & \text{if } x < 0 \\ \beta x^\vartheta - \mu + o(x^\vartheta), & \text{if } x > 0. \end{cases} \tag{5.2}$$

Here α, β depend continuously on μ, and K, ϑ depend C^1 on μ. The functions β and K are positive, α is either positive or negative depending on whether the centre manifold $W^{s,u}_{loc}(\bar{\Gamma} \cup \bar{\Gamma}_1)$ is orientable or nonorientable. The o-terms depend C^1 on x and continuously on μ.

We may restrict to orbits on $W^{s,u}_{loc}(\bar{\Gamma} \cup \bar{\Gamma}_1)$. After the reparametrization of μ such that (5.2) holds, $W^u(0)$ intersects Σ in $-\mu$. So the transition time needed to go from this first intersection point of $W^u(0)$ with S to $\pm T$, equals $\frac{-1}{\lambda^u}\ln(-\mu)$. This clearly is a lower bound for the duration of the laminar phases. Using this parameter we have found the statement of the lemma. Because our change of parameter is C^1, the bound on the duration of the laminar phases follows. □

A second interesting scaling concerns the scaling of a particular sequence of bifurcation values. For each sufficiently large i, there are intervals in parameter space where X_μ has a periodic attractor which makes a passage through $\mathcal{U} - \mathcal{V}$ and then i tours in $\mathcal{V}$ before closing. Define μ_i to be the homoclinic bifurcation value where this periodic orbit disappears when μ decreases. For the proof of the following theorem, see Homburg [1993].

Theorem 5.3 *Let X_μ be a one parameter family of vector fields as in Section 2. Write $\vartheta = -\lambda^s/\lambda^u$. Let the bifurcation values μ_i be defined as above. Then*

$$\lim_{i\to\infty} \ln(\mu_{i+1})/\ln(\mu_i) = \vartheta. \quad \square$$

Observe that the sequence of bifurcation values μ_i converges to zero very fast. From $\lim_{i\to\infty} \ln(\mu_i)/\ln(\mu_{i-1}) = -\lambda^s/\lambda^u$ we further deduce $-\ln(\mu_i) = CM(i)^i$ for some constant C and some function $M(i)$ with $\lim_{i\to\infty} M(i) = -\lambda^s/\lambda^u$. Taking logarithms gives $\ln(-\ln(\mu_i)) = i\ln(M(i)) + \ln(C)$. Thus

$$\lim_{i\to\infty} \frac{i}{\ln(-\ln(\mu_i))} = \ln(-\lambda^s/\lambda^u).$$

In Homburg [1993] we study scaling properties in greater detail, specifically we do not restrict to the special sequence μ_i of bifurcation values (if the centre manifold $W^{s,u}_{loc}(\bar{\Gamma} \cup \bar{\Gamma}_1)$ is orientable, there are many more converging sequences of bifurcation values).

References

Bergé, P., Pomeau, Y. and Vidal, C. [1984] *L'ordre dans le chaos*, Hermann, Paris.

Fenichel, N. [1971] *Persistence and smoothness of invariant manifolds of flows,* Indiana Univ. Math. J., **21**, 193–226.

Gambaudo, J.-M. [1987] *Ordre, désordre, et frontière des systeme Morse-Smale,* thesis, University of Nice.

Glendinning, P.A. [1988] *Global bifurcations in flows*; in New Directions in Dynamical Systems, Cambridge University Press, Cambridge.

Hirsch, M. Pugh, C. and Shub, M. [1977] *Invariant manifolds*, LNM **583**, Springer Verlag, Berlin.

Homburg, A.J. [1993] *Some global aspects of homoclinic bifurcations,* thesis, Univ. of Groningen.

Kim, S. and Ostlund, S. [1989] *Universal scaling in circle maps*, Physica D, **39**, 365–392.

Palis, J. and de Melo, W. [1982] *Geometric theory of dynamical systems*, Springer Verlag, New York.

Palis, J. and Takens, F. [1993] *Homoclinic bifurcations: hyperbolicity, fractional dimensions and infinitely many attractors*, Cambridge Univ. Press.

Pomeau, Y. and Manneville, P. [1980] *Intermittent transition to turbulence in dissipative dynamical systems,* Comm. Math. Phys., **74**, 189–197.

Procaccia, I., Thomae, S. and Tresser, C. [1987] *First return maps as a unified renormalization scheme for dynamical systems*, Phys. Rev. A, **35**, 1884–1900.

Sotomayor, J. [1973] *Generic one parameter families of vector fields on two-dimensional manifolds,* Publ. Math. I.H.E.S. **43**, 5–46.

Shil'nikov, L.P. [1968] *On the generation of a periodic motion from trajectories doubly asymptotic to an equilibrium state of saddle type,* Math. USSR Sbornik, **6**, 427–437.

Sternberg, S. [1958] *On the structure of local homeomorphisms of Euclidean n-space, II,* American Journal of Math., **80**, 623–631.

Takens, F. [1989] *Intermittency; global aspects;* in Dynamical Systems, Valparaiso 1986, LNM, **1331**, Springer Verlag, Berlin.

Fields Institute Communications
Volume **4**, 1995

A Codimension 2 Bifurcation for Reversible Vector Fields

Gérard Iooss
Institut Universitaire de France
INLN, UMR 129 CNRS - UNSA
Sophia-antipolis
1361 route des Lucioles, F 06560 Valbonne

Abstract. For a family of reversible vector fields having a fixed point at the origin, we present the problem where, at criticality, the derivative at the origin has a multiple 0 eigenvalue with a 4×4 Jordan block. This is a codimension 2 singularity for reversible vector fields. This case happens in the water-wave problem for Bond number 1/3 and Froude number 1.

We study the persistence of all the known phenomena on the codimension one curves (in the parameter plane), especially concerning homoclinic orbits. One of these unfoldings is the 1:1 resonance Hopf bifurcation. The study strongly relies upon the knowledge of the reversible normal forms associated with the 4×4 Jordan block, and the unfolded situations, together with appropriate scalings.

1 Introduction

Let us consider a four-dimensional reversible vector field, having a fixed point at the origin, and such that 0 is a quadruple eigenvalue of the linear part, with a 4×4 Jordan block. This is a codimension 2 singularity for reversible vector fields, as we show in Section 2. For the unfolded vector field, we assume in what follows that 0 remains a fixed point. This is of course a restriction, but it leads to interesting cases, and simplifies the analysis. (It also implies the existence of another fixed point.) This situation can be found, for instance, in the water-wave problem, for Bond number 1/3 and Froude number 1, once the problem is reduced to a four-dimensional ordinary differential equation using a center manifold argument, where the evolution variable is the space variable (see Kirchgässner [1988]).

In the present work, we first give a normal form for the unfolded vector field. Then, we show how to pass to the limit for codimension one bifurcations which

1991 *Mathematics Subject Classification.* Primary 58F14; Secondary 76B15, 34C20.

Acknowledgement: The Fields Institute for Research in Mathematical Sciences (Canada) is gratefully acknowledged for their hospitality during which the author had time to complete this work.

Proof Let us choose the new variables p_1, p_2, p_3, x_2. The Jacobian of the transformation is $-6x_1^3$, and the inverse transformation is given by

$$x_1 = p_1,\ x_2 = x_2,\ x_3 = (2p_1)^{-1}[x_2^2 - p_2],\ x_4 = (3p_1^2)^{-1}(p_3 + \frac{1}{2}x_2^3 - \frac{3}{2}x_2 p_2).$$

It then results that any polynomial $P(x_1, x_2, x_3, x_4)$ may be written as a function $Q(p_1, p_2, p_3, x_2)$, where Q is a polynomial of p_2, p_3, x_2, with coefficients rational functions of p_1. Now, since $\partial^* P = 0$, we get $\partial Q / \partial x_2 = 0$, and then

$$P(x_1, x_2, x_3, x_4) = Q(p_1, p_2, p_3) = \sum_{\alpha,\beta,\gamma} a_{\alpha\beta\gamma} p_1^\alpha p_2^\beta p_3^\gamma$$

where the summation is *finite*, and taken over $\alpha \in \mathbb{Z}$ and $\beta, \gamma \in \mathbb{N}$. Let us now define the 'weight' of a monomial as follows: for x_1 we give the weight a, for x_2 we give $a+1$, for x_3 we give $a+2$, and for x_4 we give $a+3$. Then the differential operator ∂^* has the property to *conserve the degree of monomials and to decrease the weight by one.* It follows that one can consider *independently* monomials of different degrees and weights, and assume that all monomials in Q have the same degree and the same weight. Then, the monomials of Q have degree $\alpha + 2\beta + 3\gamma$ and weight $a(\alpha + 2\beta + 3\gamma) + 2\beta + 3\gamma$. This should be true for all values of a, hence we have α and $2\beta + 3\gamma$ which are the same for all our monomials, which may be written now as

$$Q(p_1, p_2, p_3) = p_1^\alpha \sum_{\beta,\gamma} b_{\beta\gamma} p_2^\beta p_3^\gamma$$

where α may be < 0. Now, if $\alpha < 0$, for $x_1 = 0$ we necessarily have

$$\sum_{\beta,\gamma} b_{\beta\gamma} x_2^{2\beta+3\gamma} = 0, \text{ hence} \sum_{\beta,\gamma} b_{\beta\gamma} = 0.$$

This leads to the property that

$$\sum_{\beta,\gamma} b_{\beta\gamma} p_2^\beta p_3^\gamma = \sum_{\beta,\gamma} b_{\beta\gamma} (p_2^3)^{\beta/3} (p_3^2)^{\gamma/2}$$

is homogeneous in (p_2^3, p_3^2) and equals zero if $p_2^3 - p_3^2 = 0$. Hence, $\sum_{\beta,\gamma} b_{\beta\gamma} p_2^\beta p_3^\gamma$ is divisible by $p_2^3 - p_3^2 = p_1^2 p_4$. This increases α by 2 and introduces p_4 into Q. It is then easy to iterate the reasoning until $\alpha \geq 0$. The conclusion of the lemma follows immediately from (2.10). □

Now, the reversibility operator S leaves invariant p_1, p_2, p_4 and changes p_3 into $-p_3$, hence from (2.6) and the above lemma we need to have

$$N_1(X) = p_3 P_1(p_1, p_2, p_4). \tag{2.11}$$

To make progress in constructing our normal form we need to introduce the following new polynomials q_j, r_j, s_j defined by

$$\partial^* q_j = p_j, \quad \partial^* r_j = q_j \quad \text{for } j = 1, 2, 3, \quad \text{and} \quad \partial^* s_j = r_j \quad \text{for } j = 1, 3. \tag{2.12}$$

We have:

$$\begin{aligned} q_1 &= x_2, \\ q_2 &= -3x_1x_4 + x_2x_3, \\ q_3 &= 3x_1x_2x_4 - 2x_1x_3^2 + x_2^2x_3, \end{aligned} \tag{2.13}$$

$$\begin{aligned} r_1 &= x_3, \\ r_2 &= -3x_2x_4 + 2x_3^2, \\ r_3 &= -3x_1x_3x_4 + 3x_2^2x_4 - x_2x_3^2, \end{aligned} \tag{2.14}$$

$$\begin{aligned} s_1 &= x_4, \\ s_3 &= 3x_2x_3x_4 - (4/3)x_3^3 - 3x_1x_4^2. \end{aligned} \tag{2.15}$$

Notice that s_2, such that $\partial^* s_2 = r_2$, is not a polynomial, so we do not want to introduce it. The change of X into SX leaves unchanged q_3, r_1, r_2, s_3 and changes q_1, q_2, r_3, s_1 into their negative. Now equation (2.7) for N_2 may be solved by adding a particular solution q_3P_1 to the general solution of the equation $\partial^* f = 0$, so by the lemma, $N_2 = q_3P_1(p_1,p_2,p_4) + Q_0(p_1,p_2,p_4) + p_3Q_1(p_1,p_2,p_4)$ and we have $Q_1 \equiv 0$, because of (2.6). Now N_2 may be written as

$$N_2 = q_3P_1(p_1,p_2,p_4) + p_1P_2(p_1,p_2,p_4) + p_2Q(p_2,p_4) + R(p_4) \tag{2.16}$$

where P_2, Q, and R are polynomials in their arguments, defined by $p_2Q + R = Q_0(0,p_2,p_4)$, and $R(p_4) = Q_0(0,0,p_4)$. Now equation (2.7) for N_3 leads by the same arguments to

$$N_3 = r_3P_1(p_1,p_2,p_4) + q_1P_2(p_1,p_2,p_4) + q_2Q(p_2,p_4) + \frac{q_1}{p_1}R(p_4) + \varphi$$

where φ satisfies $\partial^*\varphi = 0$. Here the particular solution of (2.7) is not a polynomial, due to $(q_1/p_1)R(p_4)$. It is clear that $\psi = p_1\varphi$ has to be a polynomial and that $q_1R(p_4) + \psi$ cancels for $x_1 = 0$ (φ itself is not necessarily a polynomial). Then observing that $p_{4|x_1=0} = x_2^2y_4$, where $y_4 = 3x_3^2 - 6x_2x_4$, this last property may be written as

$$x_2R(x_2^2y_4) + \psi(0, x_2^2, x_2^3, x_2^2y_4) \equiv 0.$$

The fact that x_2 and $x_2^2y_4$ are independent variables and a simple examination of the monomials shows that we necessarily have $R \equiv 0$, therefore ψ divisible by p_1, i.e. φ is a polynomial. Finally by the reversibility property (2.6), N_3 reads

$$N_3 = r_3P_1(p_1,p_2,p_4) + q_1P_2(p_1,p_2,p_4) + q_2Q(p_2,p_4) + p_3P_3(p_1,p_2,p_4), \tag{2.17}$$

and we get directly, following the same lines

$$\begin{aligned} N_4 = s_3P_1(p_1,p_2,p_4) + r_1P_2(p_1,p_2,p_4) + r_2Q(p_2,p_4) + q_3P_3(p_1,p_2,p_4) \\ + P_4(p_1,p_2,p_4). \end{aligned} \tag{2.18}$$

Now, collecting the results (2.11), (2.16), (2.17), (2.18) we obtain the desired normal form for our reversible vector field

$$\frac{dX}{dt} = L_0X + P_4(x_1,p_2,p_4)\begin{pmatrix}0\\0\\0\\1\end{pmatrix} + P_2(x_1,p_2,p_4)\begin{pmatrix}0\\x_1\\x_2\\x_3\end{pmatrix}$$
$$+Q(p_2,p_4)\begin{pmatrix}0\\p_2\\q_2\\r_2\end{pmatrix} + P_1(x_1,p_2,p_4)\begin{pmatrix}p_3\\q_3\\r_3\\s_3\end{pmatrix} + P_3(x_1,p_2,p_4)\begin{pmatrix}0\\0\\p_3\\q_3,\end{pmatrix}, \tag{2.19}$$

where $P_j, j = 1,2,3,4$ and Q are polynomials in their arguments, taking into account that the pure linear part is entirely contained into L_0X.

Assuming that 0 stays a fixed point of the unfolded vector fields, the principal part (linear and quadratic parts) of the right hand side of (2.19) is then generally given by

$$\begin{aligned} P_4(x_1,p_2,p_4) &= \mu_2 x_1 + ax_1^2 + bp_2 + O(3) \\ P_2(x_1,p_2,p_4) &= \mu_1 + cx_1 + O(2) \\ Q(p_2,p_4) &= d + O(2) \end{aligned} \tag{2.20}$$

where μ_1 and μ_2 are the small unfolding parameters, and where $O(n)$ gives the order in X of unwritten terms. Other coefficients depend on parameters but in principle they are of order one when the parameter (μ_1,μ_2) is 0. We shall see in what follows, that the *most important nonlinear coefficient is a*.

Notice that quadratic terms of the system corresponding to (2.19) for the water-wave problem, not in this normal form, were given by Kirchgässner [1988]. This actual normal form for quadratic terms was recently completed by Menasce [1992].

3 Local study near fixed points of the normal form

The fixed points for the vector field (2.19) are given by the system of equations

$$x_2 + p_3P_1(x_1,p_2,p_4) = 0,$$

$$x_3 + x_1P_2(x_1,p_2,p_4) + p_2Q(p_2,p_4) + q_3P_1(x_1,p_2,p_4) = 0,$$

$$x_4 + x_2P_2(x_1,p_2,p_4) + q_2Q(p_2,p_4) + r_3P_1(x_1,p_2,p_4) + p_3P_3(x_1,p_2,p_4) = 0,$$

$$P_4(x_1,p_2,p_4) + x_3P_2(x_1,p_2,p_4) + r_2Q(p_2,p_4) + s_3P_1(x_1,p_2,p_4)$$
$$+q_3P_3(x_1,p_2,p_4) = 0,$$

for which the first 3 equations can be solved with respect to (x_2,x_3,x_4) using the implicit function theorem. Then there remains one equation for x_1 depending on parameters,

$$(\mu_2 - \mu_1^2)x_1 + [a + (2b - 2c)\mu_1]x_1^2 + O(x_1^3) = 0, \tag{3.1}$$

which gives *two fixed points*: the first is the origin O, the second O', also invariant under S, is such that

$$\begin{aligned} x_1 &= a^{-1}(\mu_1^2 - \mu_2) + O(\mu_1^3, \mu_1\mu_2, \mu_2^2), \\ x_2 &= 0, \\ x_3 &= -a^{-1}\mu_1(\mu_1^2 - \mu_2) - ca^{-2}\mu_2^2 + O(\mu_1^4, \mu_1\mu_2^2, \mu_2\mu_1^2, \mu_2^3), \\ x_4 &= 0. \end{aligned} \tag{3.2}$$

The matrix of the derivative at the origin O is

$$\begin{pmatrix} 0 & 1 & 0 & 0 \\ \mu_1 & 0 & 1 & 0 \\ 0 & \mu_1 & 0 & 1 \\ \mu_2 & 0 & \mu_1 & 0 \end{pmatrix}, \tag{3.3}$$

and its eigenvalues λ are given by

$$\lambda^4 - 3\mu_1\lambda^2 + \mu_1^2 - \mu_2 = 0. \tag{3.4}$$

Notice that the linear operator (3.3) is reversible (anticommutes with S), so the set of eigenvalues is symmetric with respect to both axes in the complex plane. We give in Figure 1, the position of these four eigenvalues in the complex plane, as a function of (μ_1, μ_2).

The matrix of the derivative at the second fixed point O' is of the form:

$$\begin{pmatrix} 0 & 1+O(2) & 0 & O(2) \\ \begin{array}{c}\mu_1 + 2cx_1 - 2dx_3 \\ +O(2)\end{array} & 0 & \begin{array}{c}1-2dx_1 \\ +O(2)\end{array} & 0 \\ 0 & \mu_1 + cx_1 + dx_3 + O(2) & 0 & 1 - 3dx_1 + O(2) \\ \begin{array}{c}\mu_2 + 2ax_1 \\ +(c-2b)x_3 + O(2)\end{array} & 0 & \begin{array}{c}\mu_1 + (c-2b)x_1 \\ +4dx_3 + O(2)\end{array} & 0 \end{pmatrix} \tag{3.5}$$

where $O(2)$ means order 2 terms in x_j. The eigenvalues λ are solutions of the equation

$$\lambda^4 - [3\mu_1 + (2b - 4c)a^{-1}\mu_2 + \text{h.o.t.}]\lambda^2 - \mu_1^2 + \mu_2 + \text{h.o.t.} = 0. \tag{3.6}$$

We give in Figure 2 the position of these four eigenvalues in the complex plane, as a function of (μ_1, μ_2).

4 Computation of normal forms near curves $\Gamma_j (j = 0, 1, 2)$

Now we want to give the principal part of the normal forms near curves $\Gamma_j (j = 0, 1, 2)$ of the parameter plane (μ_1, μ_2), where codimension one bifurcations occur. We anticipate *singular coefficients* when approaching the origin of the parameter plane. This will give us an idea of the required scaling to be used on the normal form (2.19) for the codimension 2 problem.

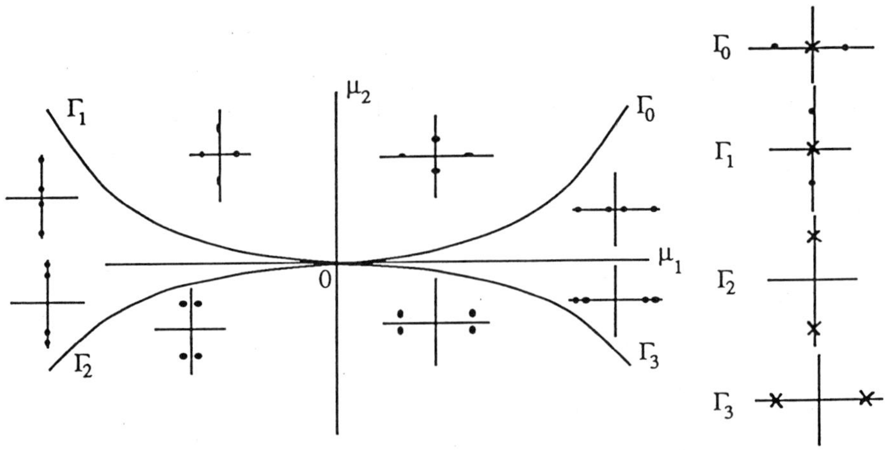

Figure 1 Eigenvalues of matrix (3.3)–linearization at 0. $\Gamma_0 : \mu_2 - \mu_1^2 = 0$, $\Gamma_1 : \mu_2 - \mu_1^2 = 0$, $\Gamma_2 : 4\mu_2 + 5\mu_1^2 = 0$, $\Gamma_3 : 4\mu_2 + 5\mu_1^2 = 0$.

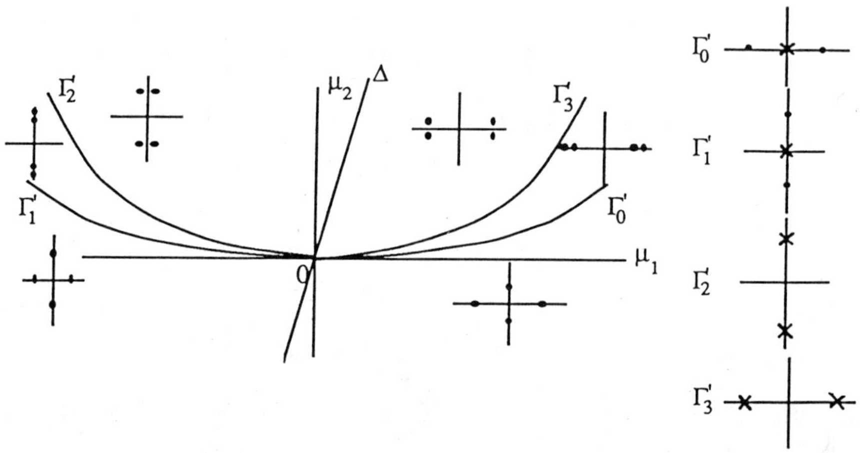

Figure 2 Eigenvalues of matrix (3.5)—linearization at $0'$. Γ_0' and $\Gamma_1' : \mu_2 - \mu_1^2 + \text{h.o.t.} = 0$, Γ_2' and $\Gamma_3' : 4\mu_2 - 13\mu_1^2 + \text{h.o.t.} = 0$, $\Delta : 3\mu_1 + (2b - 4c)a^{-1}\mu_2 + \text{h.o.t.} = 0$.

4.1 Normal form near Γ_1 Let us set $\mu_1 = \mu$ and $\mu_2 = \mu^2 + \epsilon$, $\mu < 0$. Then, in what follows we consider μ as fixed and ϵ is the bifurcation parameter. For $\epsilon = 0$, the eigenvalues are $\lambda_\pm = \pm i\sqrt{-3\mu}$ which are simple and 0 which is double. For $\epsilon > 0$, the double eigenvalue splits into two real eigenvalues $\pm\sqrt{-\epsilon/3\mu}$ + h.o.t., while for $\epsilon < 0$, they all are pure imaginary.

We need to introduce the eigenvectors $\xi_0, \zeta_0, \bar{\zeta}_0$ and the generalized eigenvector ξ_1

$$\xi_0 = \begin{pmatrix} 1 \\ 0 \\ -\mu \\ 0 \end{pmatrix} \quad \xi_1 = \begin{pmatrix} 0 \\ 1 \\ 0 \\ -2\mu \end{pmatrix} \quad \zeta_0 = \begin{pmatrix} 1 \\ \lambda_+ \\ 2\mu \\ \lambda_+\mu \end{pmatrix}, \tag{4.1}$$

which satisfy $S\xi_0 = \xi_0, S\xi_1 = -\xi_1, S\zeta_0 = \bar{\zeta}_0$.

Then the new coordinate system $(A, B,\ z,\ \bar{z})$ is defined by

$$X = A\xi_0 + B\xi_1 + z\zeta_0 + \bar{z}\bar{\zeta}_0 + \Phi(\epsilon, A, B, z, \bar{z}) \tag{4.2}$$

where the Taylor expansion of Φ may be computed in such a way that the system (2.19) takes the following normal form (see Iooss and Kirchgässner [1992] for the characterization and the computation of this normal form which is integrable, up to any order)

$$\begin{aligned} \tfrac{dA}{dt} &= B, \\ \tfrac{dB}{dt} &= (-3\mu)^{-1}\{\epsilon A + [a + 2\mu(b-c)]A^2 + \nu\,|z|^2\} + \dots, \\ \tfrac{dz}{dt} &= i\sqrt{-3\mu}z[1 + (18\mu^2)^{-1}\epsilon + \delta A + \dots], \end{aligned} \tag{4.3}$$

where $\nu = 2a + 2\mu(c - 7b) + 12\mu^2 d$, and $\delta = (9\mu^2)^{-1}[a + \mu(4c - b) - 12\mu^2 d]$.

We want to choose scales of A, B, z, t and ϵ, such that this system is no longer singular when μ goes to 0. The scales for A and ϵ are given by the z equation. The scale for t is also given by the last equation, in such a way that the eigenvalues do not meet at $\mu = 0$. Then the B scale is determined by keeping the form of the first equation, and it remains to check that it works in the B equation.

The following choice is then made

$$A = \mu^2\tilde{A}, \quad B = (-\mu)^{5/2}\tilde{B}, \quad z = \mu^2\tilde{z}, \quad t = (-\mu)^{-1/2}\tau, \quad \epsilon = \mu^2\tilde{\epsilon}. \tag{4.4}$$

Now, (4.1) and (4.2) show that we need to set, with $\mu_1 = \mu$ and $\mu_2 = \mu^2(1 + \tilde{\epsilon})$,

$$x_1 = \mu^2 y_1,\ x_2 = (-\mu)^{5/2} y_2,\ x_3 = -\mu^3 y_3,\ x_4 = (-\mu)^{7/2} y_4,\ t = (-\mu)^{-1/2}\tau. \tag{4.5}$$

4.2 Normal form near Γ_0 (reversible Takens-Bogdanov bifurcation) The computation is even simpler than in the previous case. We set again $\mu_1 = \mu$ and $\mu_2 = \mu^2 + \epsilon$, but now $\mu > 0$. The only relevant eigenvalue is 0 at $\epsilon = 0$. The two other eigenvalues are real $\lambda_\pm = \pm\sqrt{3\mu}$, and the corresponding modes are eliminated by a center manifold argument. So we have the vectors ξ_0 and ξ_1 as in (4.1), and a suitable Φ in

$$X = A\xi_0 + B\xi_1 + \Phi(\epsilon, A, B) \tag{4.6}$$

gives the two-dimensional set of amplitude equations (integrable up to any fixed order)

$$\begin{aligned} \tfrac{dA}{dt} &= B \\ \tfrac{dB}{dt} &= (-3\mu)^{-1}\{\epsilon A + [a + 2\mu(b-c)]A^2\} + \dots . \end{aligned} \tag{4.7}$$

Looking at this system is not sufficient to determine scales for the four-dimensional system (2.19). In fact, we also want to choose a scale in t such that the eigenvalues $\pm\sqrt{3\mu}$ become of order one (otherwise this introduces small denominators into the expression of the center manifold (4.6)). We then obtain the following choice of scales

$$A = \mu^2 \tilde{A}, \quad B = \mu^{5/2}\tilde{B}, \quad z = \mu^2 \tilde{z}, \quad t = \mu^{-1/2}\tau, \quad \epsilon = \mu^2 \tilde{\epsilon}, \tag{4.8}$$

which leads to: $\mu_1 = \mu$ and $\mu_2 = \mu^2(1+\tilde{\epsilon})$,

$$x_1 = \mu^2 y_1,\ x_2 = \mu^{5/2} y_2,\ x_3 = \mu^3 y_3,\ x_4 = \mu^{7/2} y_4,\ t = \mu^{-1/2}\tau. \tag{4.9}$$

Remark 4.1 Indeed, the coefficients of the terms in $A^3, \epsilon A^2, \epsilon^2 A$ in the second equation of (4.7) have order μ^{-3} which agrees with (4.8). (See Menasce [1991] for the computation of higher order terms for the water-wave problem.) In principle, it would be possible for the special form of (2.19) to lead to more regular coefficients than expected, so that 'exploding' the real eigenvalues $\pm\sqrt{3\mu}$ would not be necessary.

4.3 Normal form near Γ_2 (reversible 1:1 resonance Hopf bifurcation) Let us set $\mu_1 = \mu$ and $\mu_2 = (-5/4)\mu^2 - \epsilon$, $\mu < 0$, and proceed as above. For $\epsilon = 0$, the eigenvalues are pure imaginary and double: $\lambda_\pm = \pm i\sqrt{(-3/2)\mu}$. For $\epsilon > 0$ they split and escape from the imaginary axis in two complex simple pairs symmetric with respect to both axes, while for $\epsilon < 0$, they split in two pairs of pure imaginary simple eigenvalues. Here we need to introduce the eigenvectors $\zeta_0, \bar{\zeta}_0$ and the generalized eigenvectors $\zeta_1, \bar{\zeta}_1$

$$\zeta_0 = \begin{pmatrix} 1 \\ \lambda_+ \\ \mu/2 \\ -\lambda_+\mu/2 \end{pmatrix}, \quad \zeta_1 = \begin{pmatrix} 0 \\ 1 \\ 2\lambda_+ \\ 5\mu/2 \end{pmatrix}, \tag{4.10}$$

which satisfy

$$S\zeta_0 = \bar{\zeta}_0,\ S\zeta_1 = -\bar{\zeta}_1. \tag{4.11}$$

The new coordinates $(A, B, \bar{A}, \bar{B})$ are defined by

$$X = A\zeta_0 + B\zeta_1 + \bar{A}\bar{\zeta}_0 + \bar{B}\bar{\zeta}_1 + \Phi(\epsilon, A, B, \bar{A}, \bar{B}) \tag{4.12}$$

where Φ is such that the system (2.19) takes the following normal form (see for instance Iooss and Adelmeyer [1992] for the characterization and Iooss, Mielke and

Demay [1989] for the computation of such normal forms which appear to be *integrable up to any fixed order*)

$$\left.\begin{aligned}\tfrac{dA}{dt} &= i\sqrt{(-3/2)\mu}A + B - (6\mu\sqrt{-6\mu})^{-1}i\epsilon A + \dots \\ \tfrac{dB}{dt} &= i\sqrt{(-3/2)\mu}B - (6\mu\sqrt{-6\mu})^{-1}i\epsilon B \\ &\quad -(6\mu)^{-1}\epsilon A + (76a^2/243\mu^3)A\,|A|^2 + \dots\end{aligned}\right\}. \tag{4.13}$$

The choice of scales is then as follows: first the time scale has to place the double eigenvalue at distance $O(1)$ from O, then the scales for B/A and ϵ follow after a simple examination of the first equation, and then the scale of A by an examination of the second equation.

The following choice is then made

$$A = \mu^2\tilde{A},\ B = (-\mu)^{5/2}\tilde{B},\ t = (-\mu)^{-1/2}\tau,\ \epsilon = 5/4\mu^2\tilde{\epsilon}. \tag{4.14}$$

Now, (4.10) and (4.12) show that we need to set, with $\mu_1 = \mu$ and $\mu_2 = (-5/4)\mu^2(1+\tilde{\epsilon})$,

$$x_1 = \mu^2 y_1,\ x_2 = (-\mu)^{5/2}y_2,\ x_3 = -\mu^3 y_3,\ x_4 = (-\mu)^{7/2}y_4,\ t = (-\mu)^{-1/2}\tau, \tag{4.15}$$

which is identical to (4.5).

5 Rescaled codimension two normal forms: continuation results

Let us start with the scalings (4.5) (4.15), then the system (2.19) becomes

$$\begin{aligned}\tfrac{dY}{d\tau} &= L_0 Y + \mu^{-4}P_4(\mu^2 y_1, -\mu^5 p_2', -\mu^{11}p_4')\begin{pmatrix}0\\0\\0\\1\end{pmatrix}\\ &-\mu^{-1}P_2(\mu^2 y_1, -\mu^5 p_2', -\mu^{11}p_4')\begin{pmatrix}0\\y_1\\y_2\\y_3\end{pmatrix} + \mu^2 Q(-\mu^5 p_2', -\mu^{11}p_4')\begin{pmatrix}0\\p_2'\\q_2'\\r_2'\end{pmatrix}\\ &-\mu^5 P_1(\mu^2 y_1, -\mu^5 p_2', -\mu^{11}p_4')\begin{pmatrix}p_3'\\q_3'\\r_3'\\s_3'\end{pmatrix} + \mu^4 P_3(\mu^2 y_1, -\mu^5 p_2', -\mu^{11}p_4')\begin{pmatrix}0\\0\\p_3'\\q_3'\end{pmatrix},\end{aligned}$$

whose principal part reads

$$\frac{dY}{d\tau} = \Lambda_\epsilon Y + [ay_1^2 - b\mu p_2']\begin{pmatrix}0\\0\\0\\1\end{pmatrix} - c\mu y_1\begin{pmatrix}0\\y_1\\y_2\\y_3\end{pmatrix} + \mu^2 d\begin{pmatrix}0\\p_2'\\q_2'\\r_2'\end{pmatrix} + \begin{pmatrix}O(\mu^5)\\O(\mu^3)\\O(\mu^3)\\O(\mu^2)\end{pmatrix}, \tag{5.1}$$

where

$$\Lambda_\epsilon = \begin{pmatrix} 0 & 1 & 0 & 0 \\ -1 & 0 & 1 & 0 \\ 0 & -1 & 0 & 1 \\ \tilde{\mu}_2 & 0 & -1 & 0 \end{pmatrix},$$

and $\tilde{\mu}_2 = 1 + \tilde{\epsilon}$ near Γ_1, and $\tilde{\mu}_2 = (-5/4)(1+\tilde{\epsilon})$ near Γ_2.

In the case where we are near Γ_0, the scaling (4.9) gives in an analogous way

$$\frac{dY}{d\tau} = \Lambda'_\epsilon Y + [ay_1^2 + b\mu p'_2] \begin{pmatrix} 0 \\ 0 \\ 0 \\ 1 \end{pmatrix} + c\mu y_1 \begin{pmatrix} 0 \\ y_1 \\ y_2 \\ y_3 \end{pmatrix} + \mu^2 d \begin{pmatrix} 0 \\ p'_2 \\ q'_2 \\ r'_2 \end{pmatrix} + \begin{pmatrix} O(\mu^5) \\ O(\mu^3) \\ O(\mu^3) \\ O(\mu^2) \end{pmatrix}, \tag{5.2}$$

where

$$\Lambda'_\epsilon = \begin{pmatrix} 0 & 1 & 0 & 0 \\ 1 & 0 & 1 & 0 \\ 0 & 1 & 0 & 1 \\ \tilde{\mu}_2 & 0 & 1 & 0 \end{pmatrix}, \quad \tilde{\mu}_2 = 1 + \tilde{\epsilon}.$$

These rescaled normal forms (5.1),(5.2), are valid in the regions of the parameter plane (μ_1, μ_2) defined by $\mu_2 = \mu_1^2 \tilde{\mu}_2$ with $\tilde{\epsilon}$ small with respect to 1. These are little 'horns' centered on curves $\Gamma_j, j = 0,1,2$. (See Figure 3.)

We observe that, as expected, the eigenvalues of Λ_ϵ are

i) for $\tilde{\mu}_2 = 1 + \tilde{\epsilon}$: double 0 if $\tilde{\epsilon} = 0$, and $\pm i\sqrt{3}$

ii) for $\tilde{\mu}_2 = (-5/4)(1+\tilde{\epsilon})$: double $\pm i\sqrt{3/2}$ if $\tilde{\epsilon} = 0$

iii) of Λ'_ϵ for $\tilde{\epsilon} = 0$: double 0 and $\pm\sqrt{3}$.

The computation made above essentially shows that the coefficients of systems (5.1), (5.2) are regular functions of μ. Now, if we consider higher order terms, not written in the normal form (2.19), it should be clear that they are also regular functions of μ, since, to have that, it would be sufficient to consider the normal form up to quadratic terms.

Now the computation of the normal forms corresponding to the above three cases give coefficients which are *regular functions of the two parameters* $\tilde{\epsilon}$ *and* μ. The computations made in Section 4 show that the new normal forms are now:

i) for $\tilde{\mu}_2 = 1 + \tilde{\epsilon}$, near $\Gamma_1 (\mu < 0)$: (we still denote the amplitudes by A, B, z)

$$\begin{aligned} \tfrac{dA}{d\tau} &= B, \\ \tfrac{dB}{d\tau} &= (3)^{-1}\{\tilde{\epsilon}A + [a + 2\mu(b-c)]A^2 + \nu |z|^2\} + \dots, \\ \tfrac{dz}{d\tau} &= i\sqrt{3}z[1 + (18)^{-1}\tilde{\epsilon} + \delta' A + \dots], \end{aligned} \tag{5.3}$$

where $\nu = 2a + 2\mu(c - 7b) + 12\mu^2 d$, and $\delta' = (9)^{-1}[a + \mu(4c - b) - 12\mu^2 d]$ and where unwritten terms are at least of order μ^2. (We could not guarantee this in Section 4.)

ii) for $\tilde{\mu}_2 = 1 + \tilde{\epsilon}$, near $\Gamma_0(\mu > 0)$: (we still denote the amplitudes by A, B)

$$\frac{dA}{d\tau} = B,$$
$$\frac{dB}{d\tau} = (-3)^{-1}\{\tilde{\epsilon}A + [a + 2\mu(b-c)]A^2\} + \dots, \tag{5.4}$$

where unwritten terms are at least of order μ^2. (We could not guarantee this in Section 4.)

iii) for $\tilde{\mu}_2 = (-5/4)(1 + \tilde{\epsilon})$, near $\Gamma_2(\mu < 0)$: (we still denote the amplitudes by A and B)

$$\frac{dA}{d\tau} = i\sqrt{(3/2)}A + B + 5(24\sqrt{6})^{-1}i\tilde{\epsilon}A + \dots,$$
$$\frac{dB}{d\tau} = i\sqrt{(3/2)}B + 5(24\sqrt{6})^{-1}i\tilde{\epsilon}B + (5/24)\tilde{\epsilon}A \tag{5.5}$$
$$-(76a^2/243)A\,|A|^2 + \dots$$

where unwritten terms are of higher order. (We could not guarantee this in Section 4.)

The phase portraits of the normal forms of the reversible vector fields given by (5.3, 5.4, 5.5) are well known. However, the corresponding phase portraits of the full vector fields are only partially known. For instance, for the case (5.3) and (5.4), the detailed study is made in Iooss and Kirchgässner [1992].

In particular, we see on (5.3), i.e. near Γ_1, that $K = |z|^2$ is a first integral of the normal form, up to any order, and the trace of the phase portrait on the plane A, B is given in Figure 4. In such a case, for $K = 0$, there is a homoclinic orbit, starting and ending at the origin, for $\tilde{\epsilon} > 0$. It is known that this homoclinic orbit may not persist for the full vector field (only one stable and one unstable direction, in a four-dimensional space). There is a model equation, with the same linear part, for which the homoclinic orbit is proved not to persist (Eckhaus [1991], Hammersley and Mazzarino [1989], Pomeau, Ramani, and Grammaticos [1988]), and this leads to the fact that here it might not persist in a wide region of the parameter space. For $K \neq 0$, one has circles of solutions curves homoclinic to periodic orbits ($B = 0, A$ and $|z|$ fixed). It is shown by Iooss and Kirchgässner [1992] and Lombardi [1992] for very small periodic orbits, that in each of these families, the two reversible homoclinic orbits persist for the full vector field, and an analogous result holds for quasi-periodic solutions.

For (5.4), i.e. near Γ_0, the situation is simpler since it reduces to dimension 2. The phase portrait is the same as for $K = 0$ on Figure 4 except that we have to change $\tilde{\epsilon}$ into $-\tilde{\epsilon}$, hence for $\tilde{\epsilon} < 0$ such orbits homoclinic to 0 for the full original systems (2.1), exist in one side of the 'horn' centered on Γ_0, in the parameter plane.

In the case of the normal form (5.5), which corresponds to the reversible 1:1 resonance Hopf bifurcation, the detailed study is made in Iooss and Pérouème [1993]. We observe that the coefficient of $A\,|A|^2$ is negative, so we give in Figure 5 the projection of the phase portrait on the plane r_0, r_1 where $A = r_0 e^{i\Psi_0}, B = r_1 e^{i\Psi_1}$ and where we fix the first integral $K = A\bar{B} - B\bar{A}$ to be 0, i.e. we make $\Psi_0 = \Psi_1$ and let r_0 be ≥ 0 and r_1 take real values. It is proved in Iooss and Pérouème [1993] that the two reversible homoclinic orbits of each of these families persist for the full

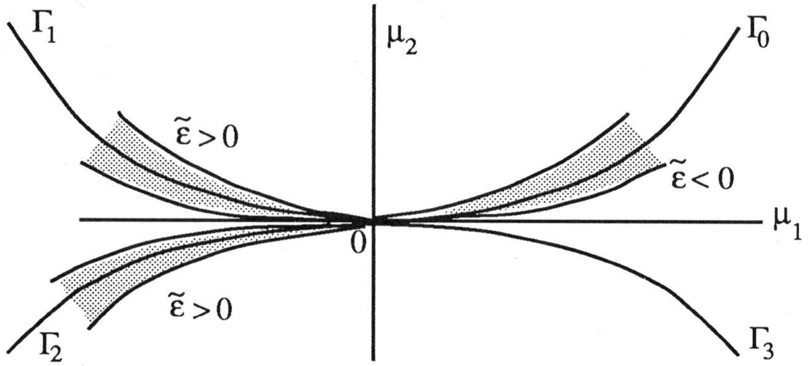

Figure 3 Three horns where the scaled normal forms are valid. We indicate for each normal form the sides where homoclinics to 0 exist.

vector fields. This implies that there are also two reversible orbits, homoclinic to 0, for the full system (2.1), in the side of the 'horn' centered on Γ_2 which is such that $\tilde{\epsilon} > 0$. These results hold provided the coefficient a is non-zero. This coefficient was recently computed in the water-wave problem, it is indeed non-zero and can be deduced from the computations made in Dias and Iooss [1993] where (4.13) is precisely given.

5.1 Simplified model and open problems The complete study of the phase portraits in dimension four is a wide open problem. However, we notice from the above computations, that the really important nonlinear coefficient is a. If we wildly truncate the problem, just keeping this nonlinear term in (2.19), we obtain an apparently simple 4th order differential equation for $x_1(t)$ denoted hereafter by $x(t)$:

$$x^{(4)} - 3\mu_1 x'' + (\mu_1^2 - \mu_2)x - ax^2 = 0. \tag{5.6}$$

This equation contains all the statements shown in the previous sections, concerning what we know in the horns in the (μ_1, μ_2) plane. Despite its apparent simplicity, and the fact that *one can integrate twice this equation* [set u(x)=$(x')^2(t)$ and integrate once], one should note a particular study relative to the neighbourhood of Γ_1, made by several authors on equations equivalent to (5.6) (see Eckhaus [1991], Amick and McLeod [1991], Hammersley and Mazzarino [1989], Pomeau, Ramani, and Grammaticos [1988]) showing that there is no reversible orbit homoclinic to 0, even though at any order the normal form of (5.6) possesses such orbits. Another interesting problem (work in progress) is what happens to the two homoclinics existing near Γ_2, when we escape from the corresponding little horn and go towards the horn where there exists only one homoclinic orbit near Γ_0. For instance, crossing of the curve Γ_3 shows a 'homoclinic multiplying' reversible bifurcation (see Champneys and Toland [1993]), among the new homoclinic orbits, one is clearly related with one of the two homoclinic orbits existing near Γ_2, while the other one is just the continuation of the basic one near Γ_0.

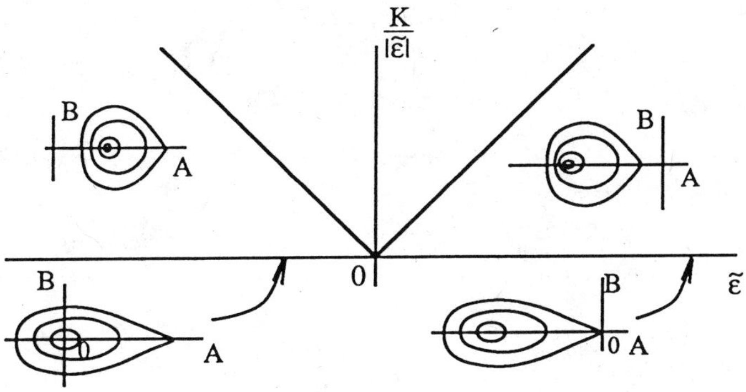

Figure 4 Phase portrait in the (A, B) plane for parameters in the horn near Γ_1. (See (5.3).)

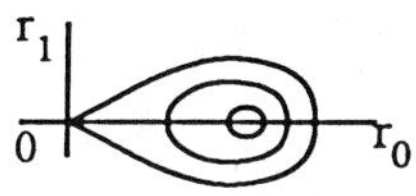

Figure 5 Phase portrait in the (r_0, r_1) plane for parameters in the horn near Γ_2, for $\tilde{\epsilon} > 0$.

References

Amick, C. and McLeod, J. [1991] *A singular perturbation problem in water waves,* Stab. and Appl. Anal. of Cont. media., **1**, 127–148.

Arnold, V. [1983] *Geometrical methods in the theory of ordinary differential equations,* Springer Verlag, New York.

Belitskii, G. [1981] *Normal forms relative to a filtering action of a group,* Trans. Mosc. Math. Soc., **40**, 1–39.

Cerezo, A. [1988] *Sur les invariants algébriques du groupe engendré par une matrice nilpotente,* Mémoire d'Habiltation, Université de Nice.

Champneys, A.R. and Toland, J.F. [1993]. *Bifurcation of a plethora of large amplitude homoclinic orbits for Hamiltonian systems,* Nonlinearity **6**.

Cushman, R. and Sanders, J.A. [1990] *A survey of invariant theory applied to normal forms of vector fields with nilpotent linear part,* Invariant theory and Tableaux. D. Stanton ed., IMA, **19**, 82–106, Springer Verlag, New York.

Dias, F. and Iooss, G. [1993] *Capillary-gravity solitary waves with damped oscillations,* Physica D, **65**, 399–423.

Eckhaus, W. [1991] *Singular perturbations of homoclinic orbits in* $\mathbb{R}^4$, Preprint 642, Dept. of Math, Univ. of Utrecht.

Elphick, C., Tirapegui, E., Brachet, M., Coullet, P. and Iooss, G. [1987] *A simple global characterization for normal forms of singular vector fields,* Physica D, **29**, 95–127.

Hammersley, J. and Mazzarino, G. [1989] *Computational aspects of some autonomous differential equations,* Proc. Roy. Soc. London A, **424**.

Iooss, G. and Adelmeyer, M. [1992] *Topics in bifurcation theory and applications,* Adv. Ser. Nonlinear Dynamics., **3**, World Scientific.

Iooss, G. and Kirchgässner, K. [1992] *Water waves for small surface tension: An approach via normal form,* Proc. Roy. Soc., Edinburgh A, **122A**, 267-299.

Iooss, G., Mielke, A. and Demay, Y. [1989] *Theory of steady Ginzburg-Landau equation in hydrodynamic stability problems,* Eur. J. Mech. B/Fluids., **8**, 229–268.

Iooss, G. and Pérouème, M.C. [1993] *Perturbed homoclinic solutions in reversible 1:1 resonance vector fields,* J. Diff. Eq., **102**, 1, 62-88.

Kirchgässner, K. [1988] *Nonlinearly resonant surface waves and homoclinic bifurcation,* Adv. Appl. Mech., **26**, 135–181.

Lombardi, E. [1992] *Bifurcation d'ondes solitaires à oscillations de faible amplitude à l'infini, pour un nombre de Froude proche de 1,* C.R. Acad. Sci. Paris., **314 I**, 493–496.

Menasce, D. [1991] *Mémoire de Maîtrise MIM,* Université de Nice.

Menasce, D. [1992] *Mémoire de DEA 'Turbulence et Systèmes Dynamiques',* Université de Nice.

Pomeau, Y., Ramani, A. and Grammaticos, B. [1988] *Structural stability of the Korteweg-De Vries solutions under a singular perturbation,* Physica D, **31**, 127–134.

Fields Institute Communications
Volume **4**, 1995

Nonasymptotically Stable Attractors in O(2) Mode Interactions

Martin Krupa
Department of Mathematics
University of Groningen
PO Box 800
9700 AV Groningen
The Netherlands

Ian Melbourne
Department of Mathematics
University of Houston
Houston, Texas
77204-3476, USA

Abstract. Heteroclinic cycles are a natural source of nonasymptotically stable attractors in systems with symmetry. In this paper, stability properties are completely classified for a large class of heteroclinic cycles. In particular, we establish the existence of several nonasymptotically stable attractors in codimension two mode interactions with $\mathbf{O}(2)$ symmetry, in the process explaining the results of some numerical experiments. In the Hopf/Hopf mode interaction we show that two heteroclinic cycles can coexist as nonasymptotically stable attractors.

1 Introduction

It is now well known, following the work of Field [1980] and in particular Guckenheimer and Holmes [1988], that structurally stable heteroclinic cycles occur naturally in low codimension bifurcation theory when there is a group of symmetries present. The majority of these heteroclinic cycles may be asymptotically stable. For example, codimension two mode interactions with $\mathbf{O}(2)$ symmetry provide a rich supply of asymptotically stable heteroclinic cycles between equilibria and/or periodic solutions. (See Armbruster et al. [1988], Proctor and Jones [1988] and Melbourne et al. [1989].)

1991 *Mathematics Subject Classification.* Primary 58F12; Secondary 34C37.

The research of I.M. was supported in part by NSF Grant DMS-9101836 and by the Texas Advanced Research Program (003652037). Both authors would like to thank The Fields Institute for their support and hospitality during visits in November 1992 and February 1993.

Unpublished numerical experiments by the authors Melbourne et al. [1989] indicated that certain heteroclinic cycles are numerically observable even when they are not asymptotically stable. Motivated by these observations, Melbourne [1991] gave an example of a heteroclinic cycle in three dimensions that is not asymptotically stable but is an attractor. This example is somewhat idealized, though it occurs naturally in work of Lauterbach and Roberts [1992]. The examples of Melbourne et al. [1989] have complicating factors such as high dimension and continuous group actions which serve to disguise the underlying ideas in Melbourne [1991]. However the framework introduced in Krupa and Melbourne [1992] for analyzing asymptotic stability of heteroclinic cycles is designed to circumvent such difficulties. In this paper, we show that the combination of the ideas in Melbourne [1991] and the framework in Krupa and Melbourne [1992] leads to a proof that the numerics described above are indeed explained by the existence of a nonasymptotically stable attractor (at the same time improving upon the results in Melbourne [1991]).

These nonasymptotically stable heteroclinic cycles have very strong attractivity properties which we now describe. Although there is not an open basin of attraction, the cycles satisfy a measure-theoretic notion of attractor (see Milnor [1985]) where nearby trajectories remain close and are asymptotic to the attractor provided that they lie in sets of positive measure. In fact, we claim much more — the measure of these sets of initial conditions becomes arbitrarily close to full measure in small enough neighborhoods of the cycle. This property is called essential asymptotic stability in Melbourne [1991] and is formalized in the following definition.

Definition 1.1 *A flow-invariant set X is said to be* essentially asymptotically stable *if there is a set D such that for any open neighborhood U of X and any $\epsilon > 0$, there exists a (smaller) open neighborhood V such that:*

(a) *trajectories starting in $V - D$ remain in U for all forward time and are asymptotic to X.*

(b) $\lambda(V - D)/\lambda(V) > 1 - \epsilon$, *where λ is Lebesgue measure.*

Note that if we choose $D = \emptyset$ then we recover the notion of asymptotic stability. In the examples in Melbourne [1991] and in this paper, the invariant set X is a heteroclinic cycle and D is a cuspoidal region abutting one or more (but not all) of the heteroclinic connections that make up the cycle. Recently, it has been shown by Alexander et al. [1992] and Buescu and Stewart [1993] that flow-invariant subspaces (forced for example by symmetry) may contain essentially asymptotically stable (but not asymptotically stable) invariant sets. Here, the set D is considerably more complicated than in our examples.

Usually, an invariant set is defined to be unstable if there are trajectories that start arbitrarily close but do not remain close. In the light of our examples this definition is very weak since unstable sets can be essentially asymptotically stable. These considerations suggest the utility of a stronger notion of instability. We propose a definition that demands instability on the same scale that we demanded asymptotic stability in our definition of essential asymptotic stability.

Definition 1.2 *A flow-invariant set X is* almost completely unstable *if there is a set D and an open neighborhood U of X such that for any $\epsilon > 0$ there exists an open neighborhood V of X such that:*

(a) *no trajectory starting in $V - D$ remains in U for all forward time,*

(b) $\lambda(V - D)/\lambda(V) > 1 - \epsilon$.
The set X is completely unstable *if D is a set of measure zero (in which case condition (b) is automatic).*

Standard examples of completely unstable invariant sets are provided by saddle points and sources. (We do not distinguish between saddles and sources; for a source $D = \emptyset$, for a saddle D is nonempty but of measure zero.) More generally, any invariant set that has a hyperbolic structure is either asymptotically stable or completely unstable.

Our main result in this paper is to show for a large class of heteroclinic cycles that essential asymptotic stability and almost complete instability are typically the only possibilities. It is of interest to isolate the asymptotically stable cycles (rather less to isolate the completely unstable cycles) so the cycles fall into one of the following three mutually exclusive categories:

- asymptotically stable,
- unstable but essentially asymptotically stable,
- almost completely unstable.

We note that our conditions for a heteroclinic cycle to be a nonasymptotically stable attractor are an improvement on these in Melbourne [1991]. (Conditions (i) and (ii) in Theorem 1.2 of Melbourne [1991] are necessary and sufficient, condition (iii) is superfluous.) In addition, the fact that there is no middle ground between essential asymptotic stability and almost complete instability for the heteroclinic cycles under consideration was not at all apparent from the treatment in Melbourne [1991]. Kirk and Silber [1993] consider examples, some of which do not fall into the class of heteroclinic cycles described in this paper (specifically, hypothesis (S2) below is invalid) and show that such examples may lead to heteroclinic cycles that are neither essentially asymptotically stable nor almost completely unstable.

In Krupa and Melbourne [1992], necessary and sufficient conditions for asymptotic stability were derived for heteroclinic cycles satisfying certain hypotheses ((H1)–(H3) in Krupa and Melbourne [1992]). If the conditions for asymptotic stability fail the cycle is unstable. The conditions are stated in terms of the real parts of eigenvalues of the linearized vector field at the equilibria (or relative equilibria) on the heteroclinic cycle. In this paper, we consider the unstable case in more detail and demonstrate that the cycle is either essentially asymptotically stable or almost completely unstable. Again, conditions are given in terms of the real parts of eigenvalues.

In Section 2 we state our main results. The conditions for asymptotic stability in Krupa and Melbourne [1992] are obtained by considering the significant components of a Poincaré map defined around the heteroclinic cycle. Essentially this reduces to the analysis of the stability properties of the origin for a class of mappings of $\mathbf{R}^2$. In Section 3 we concentrate on these stability issues when the origin is unstable. Then in Section 4 we apply our results to the examples that arise in $\mathbf{O}(2)$ mode interactions. Finally, in Section 5 we show that two of the cycles discussed in Section 4 can simultaneously exist and be essentially asymptotically stable. Moreover, for this to happen each cycle must be unstable.

2 Eigenvalue data for heteroclinic cycles

For a general vector field without symmetry, a heteroclinic cycle is necessarily structurally unstable. However, symmetry may force the flow-invariance of certain subspaces and this may permit structural stability to occur. In this section, we describe the setup surrounding such heteroclinic cycles, and define the eigenvalue data associated with these cycles. Once we have done this we are able to state our main results.

This section is similar to Sections 2 and 3 in Krupa and Melbourne [1992]. (All numbered references to Krupa and Melbourne [1992] are to the revised version.) However there are several differences due to the fact that the interest in Krupa and Melbourne [1992] centered exclusively around asymptotic stability. We shall emphasize these differences and refer to Krupa and Melbourne [1992] and references therein for details on such objects as isotropy subgroups and their fixed-point spaces, isotypic decompositions, relative equilibria, normal vector fields and so on.

Let Γ be a compact Lie group acting on $\mathbf{R}^n$ and suppose that $f : \mathbf{R}^n \to \mathbf{R}^n$ is a Γ-equivariant vector field. Suppose that ξ_j, $j = 1, \dots, m$ are hyperbolic relative equilibria with stable and unstable manifolds $W^s(\xi_j)$ and $W^u(\xi_j)$. The set of group orbits of heteroclinic connections

$$X = \{W^u(\gamma\xi_j) \cap W^s(\delta\xi_{j+1}),\ j = 1, \dots, m,\ \delta, \gamma \in \Gamma\},$$

forms a *heteroclinic cycle* provided

$$(W^u(\xi_j) - \{\xi_j\}) \cap W^s(\gamma\xi_{j+1}) \neq \emptyset \text{ for some } \gamma \in \Gamma.$$

This is more general than the corresponding definition in Krupa and Melbourne [1992] where we demanded that $W^u(\xi_j) - \{\xi_j\}$ be completely contained in $\bigcup_{\gamma\in\Gamma} W^s(\gamma\xi_{j+1})$. Clearly, the two definitions coincide when X is asymptotically stable.

We shall make several hypotheses which we label (S1)–(S4) to distinguish them from (H1)–(H3) and (H3)′ in Krupa and Melbourne [1992]. The first hypothesis guarantees that the heteroclinic cycle is robust (cf. Proposition 2.5 in Krupa and Melbourne [1992]).

(S1) There is an isotropy subgroup Σ_j with fixed-point subspace $P_j = \mathrm{Fix}(\Sigma_j)$ such that $W^u(\xi_j) \cap P_j \subset W^s(\xi_{j+1})$ and ξ_{j+1} is a sink in P_j.

Remark 2.1 In our definition of heteroclinic cycle and in hypothesis (S1) we have relaxed the assumptions on $W^u(\xi_j)$ in Krupa and Melbourne [1992]. It should be noted that any heteroclinic cycle that is asymptotically stable satisfies the definition of heteroclinic cycle in Krupa and Melbourne [1992]. Also, (S1) together with asymptotic stability implies (H1) in Krupa and Melbourne [1992]. (Kirk and Silber [1993] have studied an example where (S1) fails. In their example, $W^u(\xi_j) \cap P_j$ is a one-dimensional manifold and they demand only that one branch of this manifold is contained in $W^s(\xi_{j+1})$.)

It is possible also to relax the assumption in hypothesis (S1) that ξ_{j+1} is a sink in P_j. Then the cycle could still be robust provided the intersection of $W^u(\xi_j) \cap P_j$ with $W^s(\xi_{j+1}) \cap P_j$ is a transverse intersection of manifolds in P_j. However, if ξ_{j+1} is not a sink, it is clear that the cycle is almost completely unstable.

Let f_N denote the normal vector field of f defined in a neighborhood of the relative equilibria ξ_j. Choose $x_j \in \xi_j$ and consider the linearization $(df_N)_{x_j}$. The results in this paper and in Krupa and Melbourne [1992] are stated in terms of the real parts of the eigenvalues of $(df_N)_{x_j}$, $j = 1, \ldots, m$. Recall that the real parts are independent of the choice of $x_j \in \xi_j$ (Field [1980]).

As in Krupa and Melbourne [1992] the geometry of a heteroclinic cycle satisfying hypothesis (S1) allows us to divide the eigenvalues into four classes, *radial*, *contracting*, *expanding* and *transverse*. Let $L_j = P_{j-1} \cap P_j$. We divide up the eigenvalues as follows.

Eigenvalues	Description
radial	eigenvalues in L_j
contracting	nonradial eigenvalues in P_{j-1}
expanding	nonradial eigenvalues in P_j
transverse	remaining eigenvalues

Note that the collections of radial, contracting and expanding eigenvalues are always nonempty, that the radial and contracting eigenvalues have negative real part and that there is at least one expanding eigenvalue with positive real part. Set $-c_j$, e_j and t_j to be the maximum real parts of the contracting, expanding and transverse eigenvalues respectively. Hence we have selected the weakest contracting, strongest expanding and most unstable transverse eigenvalues. If there are no transverse eigenvalues, set $t_j = -\infty$. Observe that $c_j > 0$, $e_j > 0$, $t_j \neq 0$.

Remark 2.2 In Krupa and Melbourne [1992], the transverse eigenvalues are assumed to have negative real part, this being a necessary condition for asymptotic stability of the cycle. This assumption is relaxed here and the transverse eigenvalues may have positive real part. It is this greater generality that leads to the occurrence of nonasymptotically stable attractors.

Corresponding to each isotropy subgroup Σ_j in (S1) is the isotypic decomposition $\mathbf{R}^n = W_0 \oplus \cdots \oplus W_q$ of $\mathbf{R}^n$ into isotypic components. We may choose $W_0 = P_j$. Let $N(\Sigma_j)$ denote the normalizer of Σ_j in Γ.

(S2) The eigenspaces corresponding to c_j, t_j, e_{j+1} and t_{j+1} lie in the same Σ_j-isotypic component.

(S3) $\dim W^u(\xi_j) \cap P_j = \dim (N(\Sigma_j)/\Sigma_j) + 1$.

(S4) All transverse eigenvalues of ξ_j with positive real part lie in the same Σ_j-isotypic component.

Hypotheses (S2) is identical to hypotheses (H2) in Krupa and Melbourne [1992]. In addition, (S3) reduces to (H3)$'$ in Krupa and Melbourne [1992] when there are no transverse eigenvalues with positive real part. Of course (S4) is vacuous in the context of Krupa and Melbourne [1992].

Set $\rho_j = \min(c_j/e_j, 1 - t_j/e_j)$ and define $\rho = \rho_1 \cdots \rho_m$. We make the nondegeneracy assumptions $c_j \neq e_j - t_j$, $t_j \neq e_j$ and $\rho \neq 1$. Then Theorem 3.3 in Krupa and Melbourne [1992] states that, provided $t_j < 0$ for each j, generically a heteroclinic cycle satisfying hypotheses (S1)–(S4) is asymptotically stable if and only if $\rho > 1$. In fact it follows easily from the proof in Krupa and Melbourne [1992] that the cycle is completely unstable when $\rho < 1$. (Complete instability is established on cones that can be made arbitrarily wide).

In this paper, we prove the following result.

Lemma 2.3 *Suppose that X is a heteroclinic cycle satisfying hypotheses (S1)–(S4). Generically X is essentially asymptotically stable if and only if $\rho > 1$ and $t_j < e_j$, $j = 1, \ldots, m$. Otherwise X is almost completely unstable.*

Combining Lemma 2.3 with Theorem 3.3 in Krupa and Melbourne [1992] yields the following.

Theorem 2.4 *Suppose that X is a heteroclinic cycle satisfying hypotheses (S1)–(S4). Then generically the stability of X is described by precisely one of the following possibilities.*

(a) *asymptotically stable ($\rho > 1$ and $t_j < 0$ for each j),*

(b) *unstable but essentially asymptotically stable ($\rho > 1$, $t_j < e_j$ for each j and $t_j > 0$ for some j),*

(c) *almost completely unstable ($\rho < 1$ or $t_j > e_j$ for some j).*

Moreover, if $\rho < 1$ the cycle is completely unstable.

Remark 2.5 It follows from Theorem 2.4 that if $t_j > 0$ for each j then X is completely unstable. For then $\rho_j < 1$ for each j, and hence $\rho < 1$.

3 Stability of a class of planar maps

In Krupa and Melbourne [1992], stability of structurally stable heteroclinic cycles under an equivariant vector field is related to the stability of certain dynamically invariant regions under a Poincaré map g. Moreover, it is shown that certain components of this Poincaré map govern the stability. Ultimately, this leads to the analysis of a rather unusual class of planar maps and an investigation of the stability of the origin under these maps. The mappings are not necessarily defined in an open neighborhood of the origin, but are defined at least on an open dense full measure subset of a neighborhood of the origin.

Suppose that X is a heteroclinic cycle satisfying hypotheses (S1)–(S4) with eigenvalue data $c_j > 0$, $e_j > 0$, and $t_j \neq 0$ for $j = 1, \ldots, m$. To prove Lemma 2.3 we may reduce as in Krupa and Melbourne [1992] to proving a similar result about the stability of the origin under the mapping $g = g_m \circ \cdots \circ g_1$ where g_j has to lowest order the form

$$g_j(w, z) = (A_j w^{c_j/e_j} + B_j w^{-t_j/e_j} z, C_j w^{c_j/e_j} + D_j w^{-t_j/e_j} z).$$

For completeness, we describe quickly how this is arrived at. The details can be found in Krupa and Melbourne [1992].

The map g_j represents the significant components of the part of the Poincaré map that begins and ends in neighborhoods of ξ_j and ξ_{j+1}. In the notation of Krupa and Melbourne [1992], $g_j = \psi_j \circ \phi_j$ where ϕ_j is a first hit map defined in a small neighborhood of ξ_j and ψ_j is a diffeomorphism connecting the neighborhoods of ξ_j and ξ_{j+1}. The coordinates w and z correspond to the most expanding and largest transverse directions. Finally, the constants A_j, B_j, C_j, D_j are drawn from the linearization of ψ_j (taking only the significant components into account) and

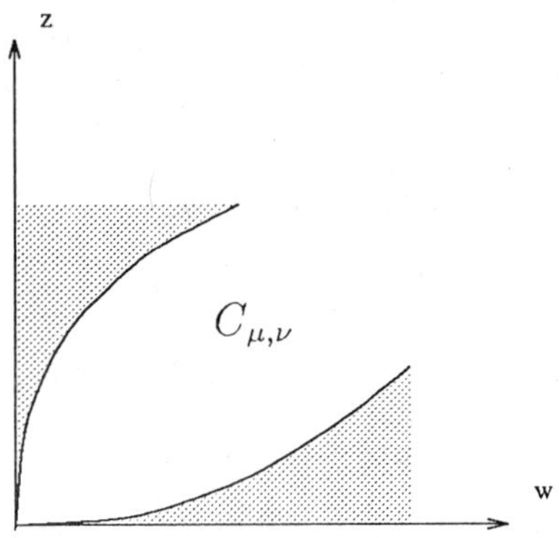

Figure 1 The cuspoidal cone in the first quadrant.

are generically nonzero by hypothesis (S2), see Remark 4.11 in Krupa and Melbourne [1992]). Hypothesis (S4) implies that we can neglect transverse directions corresponding to the other unstable transverse eigenvalues (if any).

As before, we set $\rho_j = \min(c_j/e_j, 1 - t_j/e_j)$ and $\rho = \rho_1 \cdots \rho_m$. Also, we assume the nondegeneracy conditions $c_j \neq e_j - t_j$, $t_j \neq e_j$ and $\rho \neq 1$. We must prove the following result.

Lemma 3.1 *Generically, 0 is essentially asymptotically stable under g if and only if $\rho > 1$ and $t_j < e_j$ for each j. Otherwise 0 is almost completely unstable under g.*

To simplify the notation it is convenient to generalize the class of mappings and allow g_j to have to lowest order the form

$$g_j(w, z) = (A_j w^{a_j} z^{b_j} + B_j w^{c_j} z^{d_j}, C_j w^{a_j} z^{b_j} + D_j w^{c_j} z^{d_j}),$$

where A_j, B_j, C_j, D_j are nonzero constants and $a_j + b_j \neq c_j + d_j$. Lemma 3.1 is then a special case of Corollary 3.3 below.

Without loss of generality we may assume that $a_j + b_j < c_j + d_j$. Accordingly, we set $\rho_j = a_j + b_j$ and $\rho = \rho_1 \cdots \rho_m$. Choose constants $\beta > \alpha > 0$ such that

$$4\alpha|A_j| \leq |C_j| \leq \frac{1}{4}\beta|A_j|, \quad 4\alpha|B_j| \leq |D_j| \leq \frac{1}{4}\beta|B_j|.$$

For $\mu > 1$ and $0 < \nu < 1$ define the *cuspoidal cone*

$$\mathcal{C}_{\mu,\nu} = \{\alpha|w|^\mu < |z| < \beta|w|^\nu\}.$$

The portion of the cuspoidal cone that lies in the first quadrant is illustrated in Figure 1.

Observe that if B_ϵ is the open ball of radius ϵ and center 0 in $\mathbf{R}^2$ and λ is Lebesgue measure, then $\lambda(B_\epsilon \cap \mathcal{C}_{\mu,\nu})/\lambda(B_\epsilon) \to 1$ as $\epsilon \to 0$. (The easiest way to check this is to use square neighborhoods and explicitly integrate.) We shall take D to be the complement of such a cuspoidal cone in Definitions 1.1 and 1.2.

Lemma 3.2 *There are numbers $\mu_0 > 1$, $\nu_0 \in (0,1)$, an open neighborhood V of 0 and positive constants k, K such that for $1 < \mu < \mu_0$, $\nu_0 < \nu < 1$,*

(a) $g_j(\mathcal{C}_{\mu,\nu} \cap V) \subset \mathcal{C}_{\mu,\nu}$.

(b) *If* $b_j \geq 0$

$$k|w|^{a_j+b_j\mu} \leq |g_j(w,z)| \leq K|w|^{a_j+b_j\nu} \text{ for } (w,z) \in \mathcal{C}_{\mu,\nu} \cap V.$$

A similar inequality holds for $b_j < 0$ with μ and ν interchanged.

Proof For ease of notation we suppress the index j throughout the proof.
(a) For $(w,z) \in \mathcal{C}_{\mu,\nu}$, we have

$$|w^a z^b| \geq \begin{cases} \alpha^b|w|^{a+b\mu}; & b \geq 0 \\ \beta^b|w|^{a+b\nu}; & b < 0 \end{cases}.$$

Similarly, we have

$$|w^c z^d| \leq \begin{cases} \beta^d|w|^{c+d\nu}; & d \geq 0 \\ \alpha^d|w|^{c+d\mu}; & d < 0 \end{cases}.$$

Hence for μ and ν close enough to 1, we have $|w^c z^d| = o(|w^a z^b|)$. To lowest order, $g(w,z) = (Aw^a z^b, Cw^a z^b)$.

Write g in components, $g = (g^w, g^z)$. For (w,z) close enough to 0 we compute that

$$\alpha|g^w(w,z)|^\mu \leq 2\alpha|Aw^a z^b| \leq \frac{1}{2}|Cw^a z^b| \leq |g^z(w,z)|.$$

A similar computation shows that $|g^z(w,z)| \leq \beta|g^w(w,z)|^\nu$ so that $g(w,z) \in \mathcal{C}_{\mu,\nu}$ as required.
(b) Proceeding as in (a) we have $g(w,z) = (Aw^a z^b, Cw^a z^b)$ at lowest order. Using the estimates in (a) for $|w^a z^b|$ we take $k = \frac{1}{2}\min(|A|,|C|)\min(\alpha^b,\beta^b)$. This yields the required lower bound. A similar argument gives the upper bound. □

Corollary 3.3 *Suppose that $\rho \neq 1$ and that $\rho_j \neq 0$ for each j. Then 0 is essentially asymptotically stable if and only if $\rho > 1$ and $\rho_j > 0$ for each j. Otherwise 0 is almost completely unstable.*

Proof Choose μ and ν close enough to 1 so that parts (a) and (b) of Lemma 3.2 are valid. Part (a) of the lemma implies that points in $\mathcal{C}_{\mu,\nu}$ remain in $\mathcal{C}_{\mu,\nu}$ under iteration by g for as long as they are close to 0. Hence we may consider the stability of 0 under the mapping g restricted to $\mathcal{C}_{\mu,\nu}$. Note that asymptotic stability in $\mathcal{C}_{\mu,\nu}$ implies essential asymptotic stability in $\mathbf{R}^2$. Similarly, complete instability in $\mathcal{C}_{\mu,\nu}$ corresponds to almost complete instability in $\mathbf{R}^2$.

Suppose first that $\rho_j < 0$ for some j. Then $a_j + b_j < 0$ and we can choose μ and ν close enough to 1 so that $a_j + b_j\mu < 0$ and $a_j + b_j\nu < 0$. Then it follows from the lower estimates in Lemma 3.2(b) that for V a small enough neighborhood of 0, $g_j(V \cap \mathcal{C}_{\mu,\nu}) \cap V = \emptyset$. Hence 0 is completely unstable in $\mathcal{C}_{\mu,\nu}$.

Now suppose that $\rho_j > 0$ for each j. This time the upper estimates imply (for ν close enough to 1) that each g_j extends to a continuous map on $\overline{\mathcal{C}_{\mu,\nu}}$ with 0 a fixed point. Combining the estimates in Lemma 3.2(b) for each of the g_j we obtain an estimate for g

$$k^m|w|^{\rho+\epsilon} \leq |g(w,z)| \leq K^m|w|^{\rho-\epsilon},$$

for $(w, z) \in \mathcal{C}_{\mu,\nu}$, where $\epsilon > 0$ can be made as small as we wish by choosing μ and ν close enough to 1. It follows that g is uniformly contracting in $\mathcal{C}_{\mu,\nu}$ if $\rho > 1$ and uniformly expanding in $\mathcal{C}_{\mu,\nu}$ if $\rho < 1$. Hence 0 is asymptotically stable in $\mathcal{C}_{\mu,\nu}$ or completely unstable in $\mathcal{C}_{\mu,\nu}$ respectively as required. □

4 Cycles in O(2) mode interactions

Codimension two mode interactions with $\mathbf{O}(2)$ symmetry have provided a rich source of examples of structurally stable heteroclinic cycles. (See Armbruster et al. [1988], Proctor and Jones [1988] and Melbourne et al. [1989].) Necessary and sufficient conditions for asymptotic stability of many of these heteroclinic cycles were obtained in Section 6 of Krupa and Melbourne [1992].

For convenience we give a quick (but incomplete) description of the heteroclinic cycles that occur in $\mathbf{O}(2)$ mode interactions. There are three codimension two mode interactions.

- steady state/steady state (Armbruster et al. [1988], Proctor and Jones [1988])
- steady state/Hopf (Melbourne et al. [1989])
- Hopf/Hopf (Melbourne et al. [1989])

The cycles occur on center manifolds whose dimension is determined by the multiplicity of the eigenvalues passing through the imaginary axis. Generically these eigenvalues are simple or have multiplicity two corresponding to the absolutely irreducible representations of $\mathbf{O}(2)$. The existence of structurally stable heteroclinic cycles requires all eigenvalues to have multiplicity two and the center manifolds for the three mode interactions have dimension 4, 6 and 8 respectively.

4.1 Steady state/steady state interaction In the steady state/steady state interaction the center manifold is four-dimensional and coordinates can be chosen on $\mathbf{C}^2$ so that the action of $\mathbf{O}(2)$ is given by

$$\theta \cdot z = (e^{ki\theta} z_1, e^{\ell i\theta} z_2),$$

$$\kappa \cdot z = (\bar{z}_1, \bar{z}_2),$$

where k and ℓ are positive coprime integers and $k \leq \ell$.

Structurally stable heteroclinic cycles occur only when $k = 1$, $\ell = 2$, and consist of two equilibria and two heteroclinic connections between these equilibria. The pair of equilibria are related by elements of the group, as are the heteroclinic connections. The equilibria have isotropy subgroup $\mathbf{D}_2$ generated by rotation through π and the reflection κ. The heteroclinic connections lie in the two-dimensional flow-invariant subspaces $Fix(\kappa)$ and $Fix(\pi\kappa)$. In fact, there is a continuous group orbit of such heteroclinic cycles.

Since the equilibria on the cycle are related by the group of symmetries we may view this as a cycle with $m = 1$. There are no transverse eigenvalues and the eigenvalue data consists of $c > 0$ and $e > 0$. It was shown in Krupa and Melbourne [1992] and in Proctor and Jones [1988] that generically the heteroclinic cycle is asymptotically stable if $c > e$ and unstable if $c < e$. It follows from Theorem 2.4 that when $c < e$ the cycle is completely unstable. In particular the cycle is never a nonasymptotically stable attractor.

4.2 Steady state/Hopf interaction We can choose coordinates on the six-dimensional center manifold $\mathbf{C}^3$ so that the $\mathbf{O}(2)$ action is given by

$$\phi \cdot (z_0, z_1, z_2) = (e^{ki\phi} z_0, e^{\ell i\phi} z_1, e^{-\ell i\phi} z_2),$$

$$\kappa. \cdot (z_0, z_1, z_2) = (\bar{z}_0, z_2, z_1),$$

where k and ℓ are positive coprime integers and $k \leq \ell$. There is also an $\mathbf{S}O(2)$ phase shift symmetry

$$\theta \cdot (z_0, z_1, z_2) = (z_0, e^{i\theta} z_1, e^{i\theta} z_2).$$

Structurally stable heteroclinic cycles occur when $k = \ell = 1$.

A schematic diagram of the heteroclinic cycle is given in Figure 2. There is an equilibrium (1) lying in an invariant line and a periodic solution (2) in an invariant plane. These are joined by heteroclinic connections in the three-dimensional invariant subspaces (3) and (4). The equilibrium has eigenvalue data $c_1 > 0$, $e_1 > 0$ and $t_1 = -\infty$ (no transverse eigenvalues). The periodic solution has a single transverse eigenvalue and hence data $c_2 > 0$, $e_2 > 0$ and $t_2 \neq 0$.

Theorem 4.1 *Generically, the heteroclinic cycle in the steady state/Hopf interaction is a nonasymptotically stable attractor if and only if the following conditions are valid.*

(a) $0 < t_2 < e_2$,

(b) $c_1 \min(c_2, e_2 - t_2) > e_1 e_2$.

More precisely, the cycle is unstable yet essentially asymptotically stable if these conditions are satisfied.

If the second inequality in (a) or the inequality in (b) is reversed, the cycle is almost completely unstable.

If the first inequality in (a) is reversed, but the others maintained, the cycle is asymptotically stable.

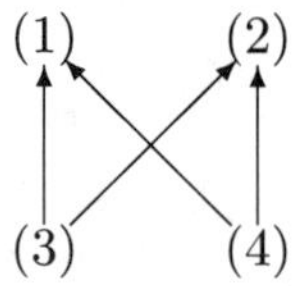

Figure 2 The cycle in the steady state/Hopf interaction.

4.3 Hopf/Hopf interaction Effectively, the symmetry group is $\mathbf{O}(2) \times T^2$, the T^2-symmetry arising from the simultaneous Hopf bifurcations. We can choose coordinates so that the action of $\mathbf{O}(2) \times T^2$ is as follows:

$$\begin{aligned}
\phi \cdot z &= (e^{ik\phi} z_1, e^{-ik\phi} z_2, e^{i\ell\phi} z_3, e^{-i\ell\phi} z_4), \quad \phi \in \mathbf{S}O(2), \\
(\psi_1, \psi_2) \cdot z &= (e^{i\psi_1} z_1, e^{i\psi_1} z_2, e^{i\psi_2} z_3, e^{i\psi_2} z_4), \quad (\psi_1, \psi_2) \in T^2, \\
\kappa \cdot z &= (z_2, z_1, z_4, z_3),
\end{aligned}$$

where k and ℓ are positive coprime integers and $k \leq \ell$.

Heteroclinic cycles occur for all values of k and ℓ, but the cycles for $\ell > 1$ violate hypothesis (S2), see Krupa and Melbourne [1992]. Hence we restrict attention to the case $k = \ell = 1$.

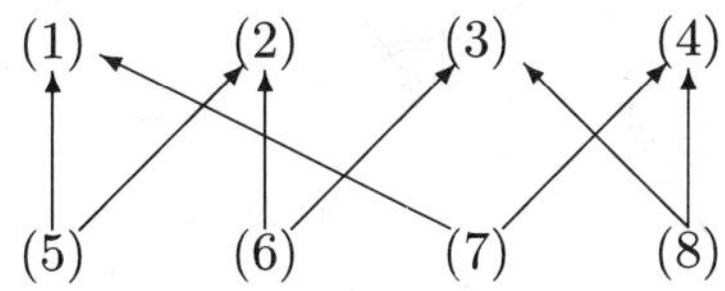

Figure 3 The cycle of rotating and standing waves in the Hopf/Hopf interaction.

There are three heteroclinic cycles to consider, one that connects a pair of rotating waves, another connecting a pair of standing waves and a third connecting all four of these solutions. Schematically, the first two cycles look like the one in the steady state/Hopf interaction. (See Figure 2.) The third cycle is shown schematically in Figure 3. In the notation of Melbourne et al. [1989] modes (1) and (4) correspond to the rotating waves, and modes (2) and (3) to the standing waves. Each of the modes (j), $j = 1, \cdots, 4$ has a pair of transverse eigenvalues and hence data $c_j > 0$, $e_j > 0$ and $t_j \neq 0$.

Theorem 4.2 *Generically, the cycles in the Hopf/Hopf interaction are unstable but essentially asymptotically stable if and only if*

(a) $t_j < e_j$, *for each* j,

(b) $t_j > 0$ *for at least one* j,

(c) $\prod_{j=1}^{m} \min(c_j, e_j - t_j) > \prod_{j=1}^{m} e_j$ *($m = 2$ for the first two cycles and $m = 4$ for the third).*

If an inequality in (a) or (c) is reversed, then the cycle is almost completely unstable. If (a) and (c) are valid but $t_j < 0$ for each j then the cycle is asymptotically stable.

5 Bistability in the Hopf/Hopf interaction

In Section 4 we saw that there are three distinct heteroclinic cycles that may occur in the Hopf/Hopf mode interaction with $\mathbf{O}(2)$ symmetry when $k = \ell = 1$. These cycles connect the rotating waves (1) and (4), the standing waves (2) and (3), and all four periodic solutions (1)–(4) respectively. It is natural to ask whether these cycles can coexist and be asymptotically stable or at least essentially asymptotically stable together.

In fact it is clear that if the cycle connecting all four modes is even essentially asymptotically stable then the other two cycles are automatically almost completely unstable. To see this observe that if two cycles have a mode in common, then the expanding eigenvalues for one cycle are transverse eigenvalues for the other cycle and vice versa. Hence one of the cycles must have a transverse eigenvalue dominating the expanding eigenvalues so that the cycle is almost completely unstable.

The only possibility for bistability lies with the cycle joining rotating waves and the cycle joining standing waves. We show that this possibility is realized but only with both cycles being unstable (but essentially asymptotically stable).

The vector field on the eight-dimensional manifold $\mathbf{C}^4$ can be written in the form $f = (f_1, f_2, f_3, f_4)$ where

$$\begin{aligned} f_1(z) &= P^1 z_1 + R^1 z_2 z_3 \bar{z}_4, \\ f_2(z) &= f_1(z_2, z_1, z_4, z_3), \\ f_3(z) &= P^3 z_3 + R^3 z_1 \bar{z}_2 z_4, \\ f_4(z) &= f_3(z_2, z_1, z_4, z_3). \end{aligned}$$

Here, P^1, P^3, R^1 and R^3 are complex valued functions of parameters $\lambda, \mu \in \mathbf{R}$ and the invariants

$$\rho_i = z_i \bar{z}_i,\ i = 1, \dots, 4, \quad \Re\alpha, \quad \Im\alpha,$$

where $\alpha = z_1 \bar{z}_2 \bar{z}_3 z_4$. Let p^i denote the real part of P^i, and r^i the real part of R^i for $i = 1, 3$. Existence results for the heteroclinic cycles are stated in Melbourne et al. [1989]. Following the treatment there we set $p^1_\mu(0) = p^3_\mu(0) = 0$. From now on we shall take it as given that the various Taylor coefficients are to be evaluated at 0. We write p^1_μ instead of $p^1_\mu(0)$ and so on. By scaling we may assume that $p^1_\lambda = p^3_\lambda = 1$.

Each of the modes (1)–(4) has four eigenvalues up to multiplicities forced by the group action. We are interested in the lowest order coefficient in the real part of each eigenvalue. These are determined by the Taylor coefficients of the real parts p^1, p^3 up to cubic order. Following Melbourne et al. [1989] we list these lowest order coefficients corresponding to the mode (j) as ϵ_{ij}, $i = 1, \dots, 4$.

$$\epsilon_{11} = 2p^1_{\rho_1} \qquad \epsilon_{12} = 2(p^1_{\rho_1} + p^1_{\rho_2}) \qquad \epsilon_{13} = 2(p^3_{\rho_3} + p^3_{\rho_4}) \qquad \epsilon_{14} = 2p^3_{\rho_3}$$

$$\epsilon_{21} = p^1_{\rho_2} - p^1_{\rho_1} \qquad \epsilon_{22} = 2(p^1_{\rho_1} - p^1_{\rho_2}) \qquad \epsilon_{23} = 2(p^3_{\rho_3} - p^3_{\rho_4}) \qquad \epsilon_{24} = p^3_{\rho_4} - p^3_{\rho_3}$$

$$\epsilon_{31} = p^3_{\rho_1} - p^1_{\rho_1} \qquad \epsilon_{32} = p^3_{\rho_1} + p^3_{\rho_2} + r^3 - p^1_{\rho_1} - p^1_{\rho_2} \qquad \epsilon_{33} = p^1_{\rho_3} + p^1_{\rho_4} + r^1 - p^3_{\rho_3} - p^3_{\rho_4} \qquad \epsilon_{34} = p^1_{\rho_4} - p^3_{\rho_3}$$

$$\epsilon_{41} = p^3_{\rho_2} - p^1_{\rho_1} \qquad \epsilon_{42} = p^3_{\rho_1} + p^3_{\rho_2} - r^3 - p^1_{\rho_1} - p^1_{\rho_2} \qquad \epsilon_{43} = p^1_{\rho_3} + p^1_{\rho_4} - r^1 - p^3_{\rho_3} - p^3_{\rho_4} \qquad \epsilon_{44} = p^1_{\rho_3} - p^3_{\rho_3}$$

We also define numbers ϵ_5, ϵ_6, ϵ_7 and ϵ_8 as follows.

$$\epsilon_5 = \frac{p^1_{\rho_3}}{p^3_{\rho_3}} + \frac{p^3_{\rho_1}}{p^1_{\rho_1}} \qquad \epsilon_7 = \frac{p^3_{\rho_1} + p^3_{\rho_2} + r^3}{p^1_{\rho_1} + p^1_{\rho_2}} + \frac{p^1_{\rho_3} + p^1_{\rho_4} + r^1}{p^3_{\rho_3} + p^3_{\rho_4}}$$

$$\epsilon_6 = \frac{p^1_{\rho_4}}{p^3_{\rho_3}} + \frac{p^3_{\rho_2}}{p^1_{\rho_1}} \qquad \epsilon_8 = \frac{p^3_{\rho_1} + p^3_{\rho_2} - r^3}{p^1_{\rho_1} + p^1_{\rho_2}} + \frac{p^1_{\rho_3} + p^1_{\rho_4} - r^1}{p^3_{\rho_3} + p^3_{\rho_4}}$$

According to Theorem 4.1 in Melbourne et al. [1989] the cycle between rotating waves exists provided

(i) $\epsilon_{11} < 0$, $\epsilon_{14} < 0$,

(ii) $\epsilon_{31} > 0$, $\epsilon_{34} > 0$, $\epsilon_{41} < 0$, $\epsilon_{44} < 0$,

(iii) $\epsilon_5 > -2$, $\epsilon_6 > -2$.

In our terminology, the data in (i) corresponds to the radial eigenvalues, the data in (ii) to the contracting and expanding eigenvalues. The cycle exists also if the signs in (ii) are reversed, interchanging the contracting and expanding eigenvalues. The conditions (iii) ensure boundedness of trajectories in the invariant subspaces so that the required connections are made.

We shall let c_j^R, e_j^R and t_j^R, $j = 1, 2$ denote the eigenvalue data for the cycle connecting rotating waves. Then

$$c_1^R = -\epsilon_{41},\ e_1^R = \epsilon_{31},\ t_1^R = \epsilon_{21},\ c_2^R = -\epsilon_{44},\ e_2^R = \epsilon_{34},\ t_2^R = \epsilon_{24}.$$

Similarly, by Theorem 4.4 in Melbourne et al. [1989] the cycle between standing waves exists provided

(i) $\epsilon_{12} < 0,\ \epsilon_{13} < 0,$

(ii) $\epsilon_{32} > 0,\ \epsilon_{43} > 0,\ \epsilon_{33} < 0,\ \epsilon_{42} < 0,$

(iii) $\epsilon_7 > -2,\ \epsilon_8 > -2.$

If c_j^S, e_j^S and t_j^S, $j = 1, 2$ denotes the eigenvalue data for the cycle connecting standing waves, we have

$$c_1^S = -\epsilon_{42},\ e_1^S = \epsilon_{32},\ t_1^S = \epsilon_{22},\ c_2^S = -\epsilon_{33},\ e_2^S = \epsilon_{43},\ t_2^S = \epsilon_{23}.$$

Observe that the transverse eigenvalues of the two cycles are related: $t_j^S = -2t_j^R$. It follows that if one of the cycles is asymptotically stable with negative transverse eigenvalues, then the other cycle has both transverse eigenvalues positive. That cycle is then almost completely unstable by Remark 2.5.

We show that the two cycles exist and are simultaneously unstable but essentially asymptotically stable for a nonempty open set of codimension two bifurcation problems with $\mathbf{O}(2)$ symmetry. Since the conditions for existence and stability are given by finitely many inequalities, it is sufficient to find one set of Taylor coefficients for which the inequalities are satisfied simultaneously. Consider the following values for the Taylor coefficients.

$$\begin{array}{lllll} p^1_{\rho_1} = -1.0 & p^1_{\rho_2} = 0.0 & p^1_{\rho_3} = -4.0 & p^1_{\rho_4} = 0.0 & r^1 = -5.0 \\ p^3_{\rho_1} = 5.1 & p^3_{\rho_2} = -7.0 & p^3_{\rho_3} = -1.0 & p^3_{\rho_4} = -2.0 & r^3 = 2.0\ . \end{array}$$

We compute that

$$\begin{array}{llll} \epsilon_{11} = -2.0 & \epsilon_{12} = -2.0 & \epsilon_{13} = -6.0 & \epsilon_{14} = -2.0 \\ \epsilon_5 = -1.1 & \epsilon_6 = 7.0 & \epsilon_7 = 2.9 & \epsilon_8 = 107/30 \\ c_1^R = 6.0 & c_2^R = 3.0 & c_1^S = 2.9 & c_2^S = 6.0 \\ e_1^R = 6.1 & e_2^R = 1.0 & e_1^S = 1.1 & e_2^S = 4.0 \\ t_1^R = 1.0 & t_2^R = -1.0 & t_1^S = -2.0 & t_2^S = 2.0\ . \end{array}$$

Note that conditions (i)–(iii) for existence of each cycle are satisfied. Moreover, each cycle has an unstable transverse eigenvalue (t_1^R and t_2^S) and hence is unstable. Finally it is easily checked that the condition $\rho > 1$ for essential asymptotic stability in Theorem 2.4 is satisfied for the two cycles.

References

Alexander, J.C., Kan, I., Yorke, J.A. and You, Z. [1992] *Riddled basins*, Int. J. of Bif. and Chaos, **2**, 795–813.

Armbruster, D., Guckenheimer, J. and Holmes, P. [1988] *Heteroclinic cycles and modulated waves in systems with* $\mathbf{O}(2)$ *symmetry*, Phys. D, **29**, 257–282.

Buescu, J. and Stewart, I.N. [1993] In preparation.

Field, M. [1980] *Equivariant dynamical systems*, Trans. Amer. Math. Soc., **259**, 185–205.

Guckenheimer, J. and Holmes, P. [1988] *Structurally stable heteroclinic cycles*, Math. Proc. Cambridge Philos. Soc., **103**, 189–192.

Kirk, V. and Silber, M. [1993] In preparation.

Krupa, M. and Melbourne, I. [1992] *Asymptotic stability of heteroclinic cycles in systems with symmetry*, University of Houston Research Report UH/MD-123, 1991, revised December 1992. Ergodic Theory Dyn. Syst., submitted.

Lauterbach, R. and Roberts, M. [1992] *Heteroclinic cycles in dynamical systems with broken spherical symmetry*, J. Diff. Eq., **100**, 22–48.

Melbourne, I. [1991] *An example of a non-asymptotically stable attractor*, Nonlinearity, **4**, 835–844.

Melbourne, I., Chossat, P. and Golubitsky, M. [1989] *Heteroclinic cycles involving periodic solutions in mode interactions with* $\mathbf{O}(2)$ *symmetry*, Proc. Roy. Soc. Edinburgh, **113A**, 315–345.

Milnor, J. [1985] *On the concept of attractor*, Commun. Math. Phys., **99**, 177–195.

Proctor, M.R.E. and Jones, C.A. [1988] *The interaction of two spatially resonant patterns in thermal convection, Part 1. Exact 1:2 resonance*, J. Fluid Mech., **188**, 301–335.

Fields Institute Communications
Volume 4, 1995

Determining the Global Topology of Resonance Surfaces for Periodically Forced Oscillator Families

Richard P. McGehee
School of Mathematics
University of Minnesota
Minneapolis, MN 55455

Bruce B. Peckham
Department of Mathematics and Statistics
University of Minnesota at Duluth
Duluth, Minnesota 55812

Abstract. Maps of the plane can be generated by sampling the flow of periodically forced planar oscillators at the period of forcing. Numerical studies of the bifurcations present in a two-parameter family of such maps, obtained by varying the forcing frequency and amplitude, have revealed a rich structure. Resonance regions in the parameter space, corresponding to maps having periodic orbits of a certain period, are always a part of the bifurcation picture.

Much insight has been gained into the bifurcation structure by viewing resonance regions as projections to the two-dimensional parameter space of "resonance surfaces" from the four-dimensional phase × parameter space. Here we continue the study of these surfaces by presenting an algorithm to determine their global topology from the bifurcation diagram in the parameter plane and knowledge of generic codimension-one and -two bifurcations.

1 Background: Forced oscillators and resonance surfaces

Differential equations which can be classified as periodically forced planar oscillators are abundant in science and engineering. These oscillators are often studied by doing a bifurcation analysis on the maps generated, as explained in Section 1.1, by sampling the flow at the time period of forcing. As is typical in studies of periodically forced oscillators, we use the frequency of forcing and amplitude of forcing as our two parameters.

1991 *Mathematics Subject Classification.* Primary 70K30; Secondary 34C23, 34C30.

Work was partially supported by NSF grant DMS-9020220. Figures 4a and 4b were obtained with the aid of "geomview," an interactive geometry viewer developed by The Geometry Center, Minneapolis, MN, a Science and Technology Center supported by the National Science Foundation.

Period-q "resonance regions" (also called Arnold tongues, horns, or entrainment regions), defined as the regions of parameter space where the corresponding maps have a period-q orbit, are always a prominent feature of the bifurcation diagrams. In previous studies (Aronson et al. [1986], Peckham [1988], Peckham [1990]), it proved useful to consider sets of period-q points in the phase $\times$ parameter space and view the resonance regions as projections of these sets to the parameter space. Because the parameter space is two-dimensional, the period-q sets are two-dimensional manifolds which we call "resonance surfaces".

In this paper, we look specifically at the question of determining the global topology of various period-q resonance surfaces from the bifurcation diagram in the parameter plane. This is similar to the problem of determining the topology of a surface in three dimensions from its shadow on a plane, but the fact that the ambient space is four-dimensional rather than three-dimensional makes the problem more challenging. We rely on the local features of the resonance surfaces, determined by normal forms and universal unfoldings of bifurcations, to arrive at their global topology. We hope that an understanding of the topology of the resonance surfaces and their relationship to corresponding bifurcation diagrams will aid in understanding of the underlying bifurcation theory.

1.1 The periodically forced oscillator model A standard model of a periodically forced planar oscillator with parameters α (forcing amplitude) and ω (forcing frequency) is given by

$$\frac{d\mathbf{x}}{dt} = \mathbf{V}(\mathbf{x}) + \alpha\mathbf{W}(\mathbf{x}, \omega t) \tag{1.1}$$

where $\mathbf{x} \in \mathbf{R}^2; \omega, \alpha, t \in \mathbf{R}$; and $\mathbf{W}$ is periodic with period one in its second variable. The vector fields $\mathbf{V}$ and $\mathbf{W}$ are assumed to be smooth. We assume that for $\alpha = 0$ the nonautonomous flow of (1.1) has a repelling equilibrium point $\mathbf{c}_0$ inside a normally hyperbolic attracting limit cycle C_0 with frequency $\omega_0 > 0$. We call C_0 the *unforced oscillator* and the frequency ω_0 the *natural frequency* of the system. For simplicity we assume that the unforced flow travels counterclockwise around C_0, C_0 is globally attracting except for the repelling equilibrium point c_0, and $\omega > 0$. See Figure 1.

In performing a bifurcation study of (1.1), it turns out to be more convenient to use ω_0/ω than ω as our first parameter, and to restrict α to $[0, \infty)$, so we define $\mu := (\omega_0/\omega, \alpha) \in (0, \infty) \times [0, \infty)$. We define $\phi_\mu(\mathbf{x}, t)$ to be the flow of (1.1) satisfying $\phi_\mu(\mathbf{x}, 0) = \mathbf{x}$. We can then reduce the study of (1.1) to a two-parameter family $\mathbf{f}_\mu$ of diffeomorphisms of the plane by considering time $1/\omega$ maps of the flows:

$$\mathbf{f}_\mu(\mathbf{x}) := \phi_\mu(\mathbf{x}, 1/\omega). \tag{1.2}$$

We assume that for $\alpha > 0$ the family $\mathbf{f}_\mu$ is *generic* (in the space of smooth two-parameter families of smooth diffeomorphisms of the plane) so that only codimension-one and -two bifurcations need to be considered. (At $\alpha = 0$, $\mathbf{f}_\mu$ is a time $1/\omega$ map of an autonomous flow, which is not generic: restricted to the invariant circle C_0, $\mathbf{f}_{(a,0)}$ is conjugate to a rigid rotation of $\omega_0/\omega = a$ times around the circle; in particular, if ω_0/ω equals the rational number p/q, every point on the circle C_0 is a period-q point.)

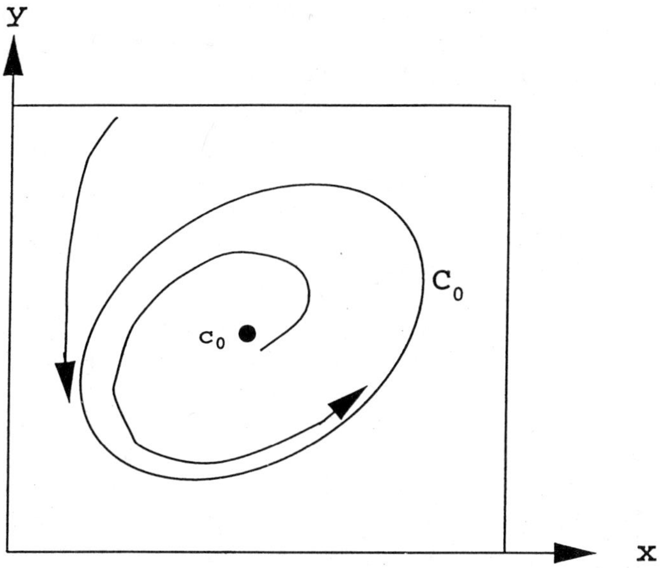

Figure 1 Phase portrait of the 'unforced' oscillator .

1.2 Small forcing amplitude The normal hyperbolicity of C_0 at $\alpha = 0$ ensures that for small $\alpha \geq 0$, $\mathbf{f}_\mu$ will continue to have an attracting invariant curve near C_0. Since all recurrence (other than the fixed point which persists from the unforced vector field's repelling equilibrium $\mathbf{c}_0$) must be on this invariant curve, a bifurcation analysis can be made via circle map theory. The assumptions that $\mathbf{f}_\mu$ is a rigid rotation for $\alpha = 0$ and that $\mathbf{f}_\mu$ is generic for $\alpha > 0$ imply that *resonance regions* (horns, tongues), nonoverlapping for small α, generically open into the first quadrant of the parameter plane from every point on the $\alpha = 0$ axis where $\omega_0/\omega = p/q$ is rational (Arnold [1982], Hall [1984]). A period-q resonance region is defined as the set of parameter values for which the corresponding map has a "least-period-q orbit". (The persistence of a period-q orbit implies a constant ratio of the *response* frequency to the *forcing* frequency ω for the original differential equation in (1.1). The two frequencies are said to be *entrained* or *locked* or *in resonance.*) See Figure 2 for the low forcing amplitude portion of a single (period-2) resonance region and some phase portraits for the *second* iterate of the map which might correspond to the indicated parameter values.

Inside the "period-q resonance region", the corresponding phase portraits generically include an attracting invariant "circle in resonance" with at least one pair of period-q orbits, one of the pair attracting and one of the pair repelling when restricted to the circle in resonance. Phase portraits B and C of Figure 2 include examples of circles in resonance. The two sides of the period-q resonance region are saddle-node bifurcation curves for the qth iterate of the map. As the parameters are varied to approach a saddle-node curve from inside a period-q resonance region, a pair of orbits coalesces into a single period-q orbit, neutrally stable when restricted to the circle. Outside the period-q resonance region, as in phase portraits A and

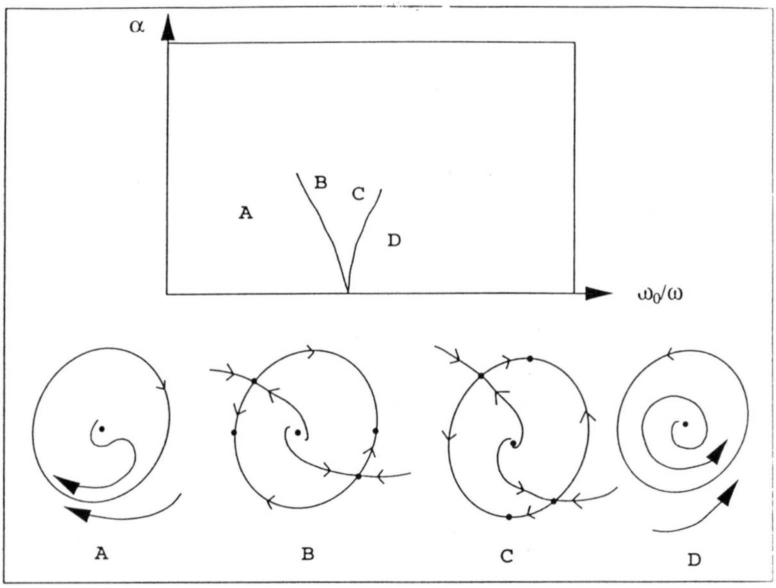

Figure 2 The low forcing amplitude part of a period-two resonance horn in the parameter space and accompanying phase portraits for the second iterate of the map.

D of Figure 2, there are no period-q orbits. Except for the number of orbits which exist at points inside the resonance regions, the collection of all resonance horns, one for each rational point on the ω/ω_0 axis, is the complete bifurcation story for small α.

1.3 Higher forcing amplitudes As we increase the forcing amplitude, the invariant circles which necessarily persist for small forcing amplitude, can, and in many examples do, break apart. Circle map theory is therefore no longer sufficient to explain the bifurcations.

However, the period-q points whose existence determines the parameter values that are inside a resonance region can persist whether or not the original invariant circle persists. The same is true for the saddle-node bifurcation curves that form the boundaries of the period-q resonance regions for small α. Many researchers, in fact, were able to numerically "continue" to higher forcing amplitude these saddle-node curves (Kai and Tomita [1979], Aronson et al. [1986], Kevrekidis, Aris and Schmidt [1986], McKarnin, Schmidt and Aris [1988], Peckham [1988], Schreiber et al. [1988], Vance and Ross [1989]). The numerical work leads to some fascinating bifurcation diagrams. One common feature of these bifurcation diagrams is that, at least for $q \geq 3$, the two saddle-node curves which begin as the two sides of a resonance region almost always come back together at some higher forcing amplitude. As is the case for our application in Section 2, the compact region of parameter space bounded by the saddle-node curves is often the projection of a compact surface from the phase $\times$ parameter space to the parameter space. We define these surfaces in the next subsection.

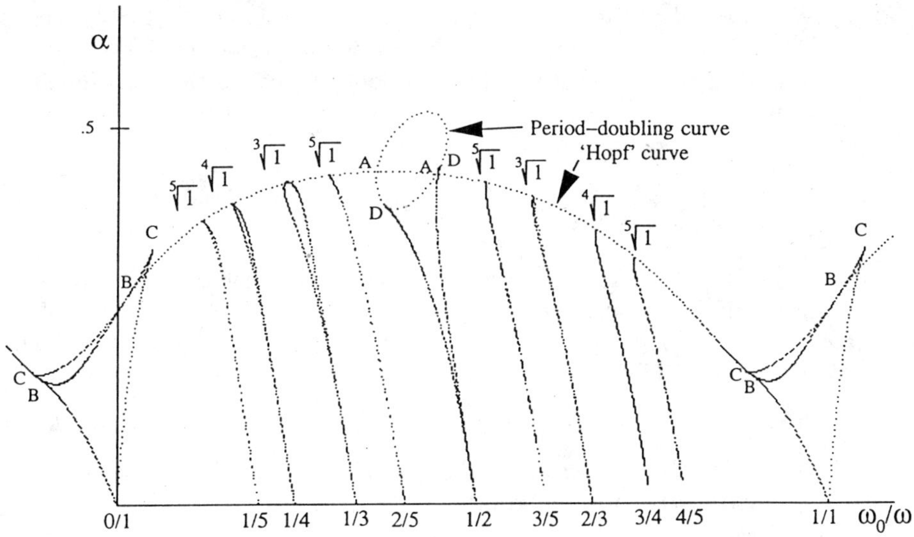

Figure 3 (a) Partial parameter space bifurcation diagram for the application of Section 2.3:
A - Double -1 point
B - Takens-Bogdanov point (double +1 point)
C - Cusp point
D - Degenerate period-doubling point
qth root of one points - Resonant Hopf points "Hopf curve" - defined by product of eigenvalues = 1 Period-doubling curve - defined by an eigenvalue = -1. All other curves are saddle-node curves defined by an eigenvalue of the qth iterate of the map = 1. (The appropriate value of q is given by the denominator of the label at the bottom tip of the respective resonance regions.)

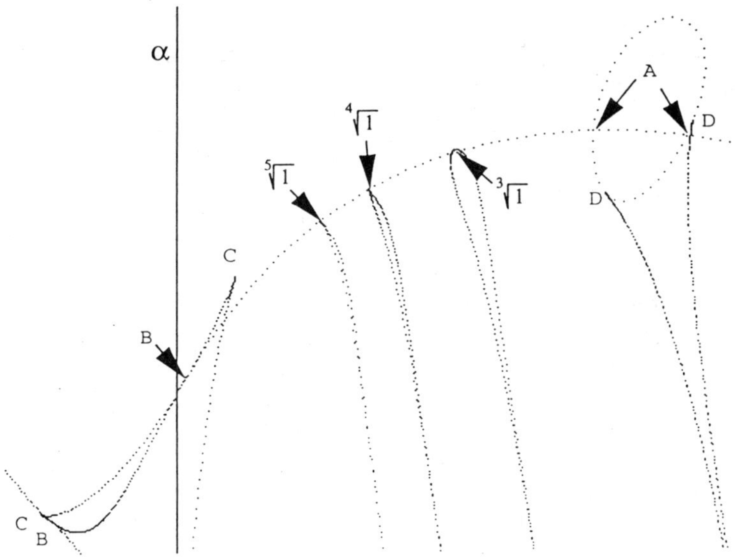

Figure 3 (b) Enlargement of a part of Figure 3a.

1.4 Resonance surfaces As suggested in Aronoson et al. [1986], because a period-q resonance region is defined via the existence of period-q points, it turns out to be more natural to consider the sets of period-q points in the four-dimensional phase $\times$ parameter space than just the corresponding set of parameter values. So we make the following definitions:

$$\Gamma(q) := \begin{array}{l}\{(\mathbf{x},\mu) \in \mathbf{R}^2 \times ((0,\infty)\times[0,\infty)) : \mathbf{x} \text{ is a least period } q \\ \text{point of } \mathbf{f}_\mu\} \ .\end{array}$$

Because the parameter space is two-dimensional, we expect these sets to be two-dimensional manifolds or surfaces. We actually need to take the *closure* of $\Gamma(q)$ to fill in some "missing points" on the corresponding surfaces for $q \geq 2$. (We are using the subspace topology for $\mathbf{R}^2 \times (0,\infty) \times [0,\infty) \subset \mathbf{R}^4$.)

Definition 1.1 *A* **period-q resonance surface** *is a component of $cl(\Gamma(q))$.*

Definition 1.2 *The projection to the parameter plane of a period-q resonance surface is a* **period-q resonance region**.

Because we are most interested in the periodic point surfaces which emanate from the "unforced oscillator", as described in the Small Forcing Amplitude subsection above, we make the following definitions as well.

Definition 1.3 *The p/q* **resonance surface** : = *the period-q resonance surface containing $C_0 \times (p/q, 0), q \geq 2$.*

Definition 1.4 *The projection to the parameter plane of the p/q resonance surface is the p/q* **resonance region**.

Remarks 1.5

1. The definition of p/q resonance surfaces forces p and q to be relatively prime or the rigid rotation of C_0 under $\mathbf{f}_{(p/q,0)}$ would not have *least* period q, as is required for points on a period-q resonance surface.

2. Surfaces can be identified with a generalized rotation number [Peckham, 1990] instead of merely a period. Among other things, this means that a p/q surface and an r/q surface cannot connect if $p \neq r$.

3. The fact that a period-q surface "continues" from $C_0 \times (p/q, 0)$, as suggested in the definition of the p/q resonance surface, is justified in the Lemma in the following section.

2 Identifying resonance surfaces

2.1 Local topology The first order of business is to justify that our resonance surfaces are in fact surfaces. This justification comes from a combination of the implicit function theorem and bifurcation theory; we provide a brief outline of the arguments below. See Peckham [1990] for more details.

Note first that all points on a period-q resonance surface (the period-q points *and* any points in their closure) satisfy $\mathbf{F}(\mathbf{x},\mu) := \mathbf{f}_\mu{}^q(\mathbf{x}) - \mathbf{x} = \mathbf{0}$. The implicit function theorem guarantees that the period-q surface has a unique local continuation whenever the 2×4 Jacobian matrix $D\mathbf{F}(\mathbf{x},\mu)$ has rank 2. In order to be sure the surface has a local homeomorphic projection to the parameter plane, however, we require the more restrictive condition: the 2×2 Jacobian matrix $D_\mathbf{x}\mathbf{F} = D\mathbf{f}_\mu{}^q(\mathbf{x}) - I$ must be nonsingular. This is equivalent to requiring that $D\mathbf{f}_\mu{}^q(\mathbf{x})$ *not* have an eigenvalue of one. Consequently, we define the **period-q resonance surface projection singularities** (singular with respect to projection to the parameter plane), $Z(q)$, as

$$Z(q) := \{(\mathbf{x}, \mu) \in \text{a period-}q \text{ resonance surface} : D_{\mathbf{x}}\mathbf{f}_{\mu}{}^{q}(\mathbf{x}) \text{ has an eigenvalue equal to one}\}.$$

In the context of dynamical systems, the surface projection singularities are map bifurcation points, most of which have been extensively studied in the literature. We assume that by restricting ourselves to generic periodically forced oscillators, we can restrict our list of possible projection singularities (corresponding to positive forcing amplitudes, at least) to those of codimension-one and -two. The list is even further restricted because the continuity of a "self rotation number", defined in Peckham [1990], allows only fixed points to be added to the least-period-q points by the closure operation in the definition of the resonance surfaces. We then rely on existing normal forms and universal unfoldings to see that the period-q surfaces are manifolds near all relevant codimension-one and -two bifurcation points. We list the universal unfoldings of the relevant bifurcations, and explicit formulas for the local pieces of period-q resonance surfaces, in the Appendix.

Because of the special form at zero forcing amplitude of maps generated by periodically forced oscillators (as described in Section 1.1), these points cannot be treated with generic map bifurcation theory. The implicit function theorem can still be used, however, to ensure that the period-q surfaces extend locally from the "unforced oscillator" to a topological cylinder. Because this result is not as well-known, we state it more formally.

Lemma 2.1 *Let the family $\mathbf{f}_{\mu}(\mathbf{x})$ be defined from a periodically forced planar oscillator as in Section 1.1. (This implies $\mathbf{f}_{\mu}(\mathbf{x})$ is C^{∞} as a function from $\mathbf{R}^2\times$ (parameter space) to $\mathbf{R}^2$.) Then for each rational $p/q > 0$, the set of all least-period-q points near the unforced oscillator, $C_0 \times (p/q, 0)$, and having $\alpha \geq 0$, is a C^{∞} topological cylinder with $C_0 \times (p/q, 0)$ as a boundary component.*

Proof By the assumptions in Section 1.1, $\mathbf{f}_{(p/q,0)}$ restricted to C_0 is conjugate to a rigid rotation by p/q times around C_0. In particular, all points on $C_0 \times (p/q, 0)$ are least-period-q points and satisfy $\mathbf{F}(\mathbf{x}, \mu) := \mathbf{f}_{\mu}{}^{q}(\mathbf{x}) - \mathbf{x} = \mathbf{0}$. We make a local change of phase variables near C_0 by choosing a θ variable in $\mathbf{T} := \mathbf{R}$ (mod 1) along C_0 and an r variable in $(1-\epsilon_1, 1+\epsilon_1)$ (for some $\epsilon_1 > 0$) transverse to C_0. Assume C_0 is described by $r = 1$. Let $(R_{\mu}(r,\theta), \Theta_{\mu}(r,\theta))$ be the new coordinates of $\mathbf{f}_{\mu}{}^{q}(\mathbf{x})$ where (r, θ) are the new coordinates of $\mathbf{x}$. Then $\mathbf{F}(\mathbf{x}, \mu) = \mathbf{0}$ is equivalent to the two scalar equations

$$R_{\mu}(r,\theta) - r = 0,$$
$$\Theta_{\mu}(r,\theta) - \theta = 0.$$

Our assumption that the unforced oscillator was a normally hyperbolic attracting limit cycle implies

$$\frac{\partial R_{\mu}(r,\theta)}{\partial r} < 1$$

at

$$(\mu_1, \mu_2, r, \theta) = (p/q, 0, 1, \theta)$$

for any θ. The use of ω_0/ω for our first parameter μ_1 implies that $\frac{\partial \Theta_{\mu}(r,\theta)}{\partial \mu_1} < 0$ at $(\mu_1, \mu_2, r, \theta) = (p/q, 0, 1, \theta)$ for any θ. These two partial derivatives imply, via the implicit function theorem, that solutions to the system of equations $\mathbf{F}(\mathbf{x}, \mu) = \mathbf{0}$ can be locally continued from the unforced oscillator $C_0 \times (p/q, 0)$, parametrized by the forcing amplitude α (i.e., μ_2) and the angular phase variable θ. Since our map was assumed to be C^∞ in both its phase and parameter variables, the implicit function theorem guarantees that the set of least-period-q points near the unforced oscillator is locally a C^∞ two-manifold parametrized by $(\alpha, \theta) \in (-\epsilon, \epsilon) \times \mathbf{T}$, for some small ϵ. By restricting to $\alpha \geq 0$, we are left with the C^∞ cylinder with boundary corresponding to $\alpha = 0$. This boundary is the unforced oscillator $C_0 \times (p/q, 0)$. □

2.2 The algorithm for determining global topology We now present an algorithm to identify the global topology of an individual p/q resonance surface with $q \geq 2$. The algorithm will be modified only slightly for the fixed point surface. Note that the completion of step 1, determining *all* bifurcations associated with a p/q resonance surface, may be impossible to justify in applications. Assuming step one, however, the remaining steps can be accomplished. We include step 1 in our "algorithm" because in practice, this is the first step one would perform.

1. **Determine the bifurcation diagram** in the parameter space for the period-q resonance surface in question. In practice, this is done by numerically continuing the two period-q saddle-node orbits which project to the left and right side boundaries near the zero-forcing-amplitude tip of the corresponding period-q resonance region. Any other projection singularities which connect to the continuation of the saddle-node curves are also continued. If other components of projection singularities exist on the surface, we usually locate them (or rule out the likelihood of their existence) with additional numerics (computing various "constant α cross sections", for example).
2. **Divide the surface into pieces** by "cutting" the surface along the curves of these singularities. If necessary to ensure each resulting surface piece projects injectively to the parameter plane (it will be necessary for the period-3 surfaces), make additional cuts which are not necessarily along projection singularities. (The number of cuts on the resonance surface corresponding to each bifurcation curve in the parameter plane is determined from bifurcation theory; the number of resulting pieces from the cuts is also determined from bifurcation theory.) "Flatten" each of the surface pieces by projecting into the parameter plane.
3. For each surface piece which emanates from the unforced oscillator $C_0 \times (p/q, 0)$ **add an appropriate part of the unforced oscillator** which the Lemma of Section 2.1 guarantees to be the boundary of the p/q surface.
4. **Label edges** so that pieces which were cut apart can later be glued back in place. (Familiarity with the "local" surface features, determined by local bifurcation theory partially summarized in the Appendix, and by the Lemma in Section 2.1, is essential for this identification.)
5. **Reassemble the surface** by rearranging, flipping, and stretching the flat pieces so the identified edges can be reglued.
6. **Identify the topology of the surface.**

2.3 An application We use for illustration a two-parameter family of maps which can be thought of as a model of a planar oscillator with periodic "impulse"

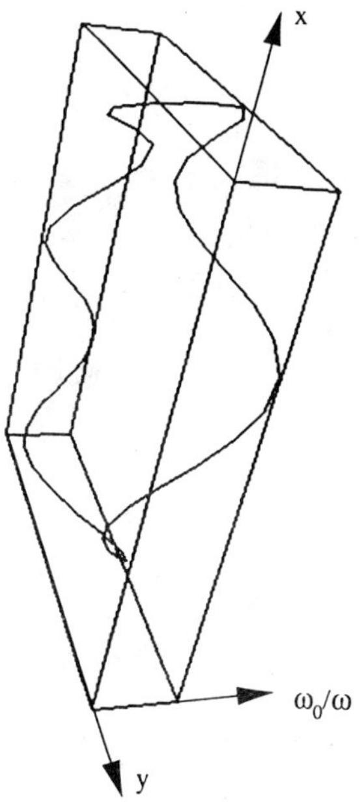

Figure 4 (a) An $\alpha = .38$ cross section of the 1/5 surface.

forcing. This model was preferable to using a standard planar oscillator with periodic forcing as in equation (1.1) because it enabled us to use an explicit formula for the maps, instead of computing each iteration of the map by integrating the differential equation in the form of equation (1.1) for the time period of forcing. The computational savings of this approach are considerable. Additionally, the model was chosen to guarantee the maps had a unique globally attracting fixed point for high enough value of a parameter which can be interpreted as the amplitude of forcing. Consequently, all resonance regions eventually terminate as the forcing amplitude increases.

We define our family of maps, $\mathbf{H}_\mu$, as the composition $\mathbf{g}_\alpha \circ \mathbf{h}_{\omega_0}$, where $\mathbf{h}_{\omega_0}$ is defined as the time one map of the (integrable) flow described by

$$\frac{dr}{dt} = \frac{r(1-r^2)}{1+r^2}, \quad \frac{d\theta}{dt} = 2\pi\omega_0 + \frac{1-r^2}{1+r^2}, \quad \omega_0 \in \mathbf{R}$$

and $\mathbf{g}_\alpha \equiv (1-\alpha)((x_1, x_2) - (1,0)) + (1,0), \alpha \in [0,1]$. Note that this caricature allows us to vary ω_0/ω by varying ω_0 instead of ω. ($\omega = 1$ for all our maps since we always take the time one map.) Consequently, we can allow the parameter ω_0/ω to take on any real value. The parameter α plays the role of the forcing amplitude.

Guided by the knowledge of the small amplitude of forcing theory, which assures us that a horn-shaped period-q resonance region emanated from each "rational" point on the ω_0/ω axis (of form $(p/q, 0)$), we numerically followed both (saddle-node) sides of each resonance horn for $1 \le q \le 5$ and $0 \le p/q < 1$. When the saddle-node curves intersected other codimension-one bifurcation curves (such as a fixed-point "Hopf" curve or a fixed-point period-doubling curve), we continued these codimension-one curves as well. The result was Figure 3a. An enlargement of a portion of Figure 3a is shown in Figure 3b. We put the word Hopf in quotes

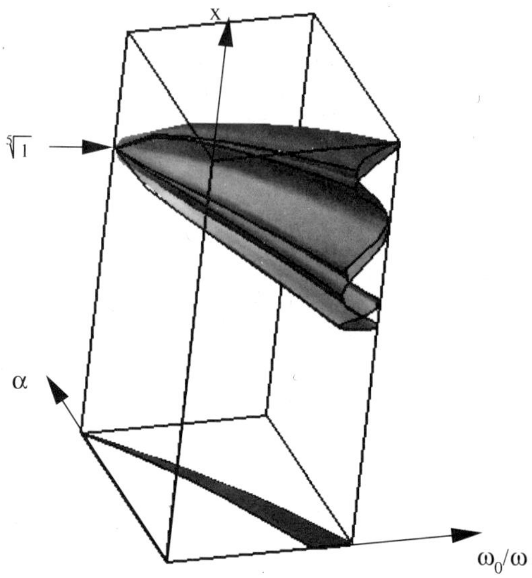

Figure 4 (b) The projection of the 1/5 resonance surface near the $\sqrt[5]{1}$ point from the four-dimensional $(x, y, \omega_0/\omega, \alpha)$ space to the three-dimensional $(x, \omega_0\ \omega, \alpha)$ space and its "shadow", part of the 1/5 resonance region, a further projection to $(\omega_0/\omega, \alpha)$ space. The five curves along the surface "folds" are period-five saddle-node curves; they all project to the left side of the 1/5 resonance region.

because the "Hopf" curve was defined numerically by requiring the product of the eigenvalues of the fixed point to be one. This condition allows the inclusion of three segments of our "Hopf" curve in Figure 3a which are actually saddles: between the two pairs of "B" points and the one pair of "A" points.

2.4 Applying the identification algorithm We now walk through the algorithm for several of the resonance surfaces for our example family of maps.

The 1/5 resonance surface

1. The two period-5 saddle-node curves which marked the boundary of the period-5 resonance region at small forcing amplitude were continued numerically from the parameter value $(1/5, 0)$. They joined a resonant Hopf bifurcation (a $\sqrt[5]{1}$ point). This formed the skinny 1/5 resonance region illistrated in Figure 3a. An enlargement including the part of the 1/5 region with the $\sqrt[5]{1}$ point is shown in Figure 3b. Since several "$\alpha =$ constant" cross sections of the 1/5 resonance surface all turned out to be qualitatively like the one shown in Figure 4a, we are convinced that there are no other projection singularities on the 1/5 surface. More specifically, there are exactly ten singularities in the projection of the slice in Figure 4a to the ω_0/ω axis: five corresponding to one period-five saddle-node orbit (all having a common parameter value on the left side of the 1/5 resonance region) and five corresponding to another period-five saddle-node orbit (all having a common parameter value on the right side of the 1/5 resonance region).

2. Because a pair of period-5 orbits is born as the parameters are varied from outside to inside the 1/5 resonance region by crossing the saddle-node boundary curve, there are ten saddle-node cuts to be made on the 1/5 surface: five projecting to the left side of the 1/5 region and five projecting to the right side. This results in ten pieces of the 1/5 surface.
3. We add a boundary piece at the "bottom" of each of the ten pieces. The ten boundary pieces make up the unforced oscillator $C_0 \times (1/5, 0)$. This gives us the shape of the ten pieces we start with in Figure 5a. We have labelled the surfaces with "S" or "N," according to whether they are saddles or nodes in the corresponding phase portraits for small forcing amplitudes, although that designation is not necessarily constant on the whole piece.

   Technical Note: Adding the boundary piece to get the shapes of the ten surface pieces in Figure 5a suggests that as we follow a saddle-node curve in the phase $\times$ parameter space with parameter values approaching the tip, $(p/q, 0)$, the corresponding phase coordinates approach a fixed phase point on C_0. This limiting behavior *is* suggested by our numerical studies (not presented here) and we believe it to be generic. In any case, the Lemma in Section 2.1 justifies the topology of Figure 5a, even if the saddle-node sides of the surface pieces *did not* limit to a fixed phase point as the parameters approached the tip.
4. The edge identifications near the $\sqrt[5]{1}$ point and the fact that the ten saddle-node curves on the 1/5 surface come together at the $\sqrt[5]{1}$ point are justified by the local bifurcation analysis of a $\sqrt[5]{1}$ point Arnold [1982]. The "cyclic" identification of the saddle and node pieces is corroborated by our cross section of Figure 4a and by the fact that the given identifications join the ten added boundary pieces to make a single topological circle. After filling in some details, it is also corroborated by the formula in the Appendix for the period-5 surface near the $\sqrt[5]{1}$ point: a small circle around the origin in the phase space travels back and forth five times across the 1/5 resonance region. For further illustration, we show some projections of a small part of the 1/5 surface near the $\sqrt[5]{1}$ point in Figure 4b.
5. Reassemble as indicated in the rest of Figure 5a.
6. The 1/5 resonance surface is a topological disk.

The 1/3 resonance surface

1. The left-hand side period-3 saddle-node curve which starts from parameter value $(1/3, 0)$ continues up, around, and back down to the right hand side saddle-node curve. No codimension-two points are encountered on the way. However, an isolated $\sqrt[3]{1}$ point lies inside (in the parameter space) the curve of saddle-nodes. The existence of this isolated surface singularity point is guaranteed by Theorem 2 in Peckham [1990]; we were able to easily locate it because it had to be on the fixed-point Hopf bifurcation curve; that it is an isolated surface singularity is consistent with the unfolding of the $\sqrt[3]{1}$ point in the Appendix. See the enlargement in Figure 3b where it is apparent that the period-three saddle-node curve extends above the Hopf bifurcation curve; again, this is consistent with the unfolding. As with the 1/5 resonance surface, various cross sections convinced us that there are no other projection singularities on the 1/3 surface.

2. Because of the singular point on the interior, it is still possible that after cutting along the three saddle-node curves, the surface pieces do not project injectively to the parameter plane. In fact, guided by the local universal unfolding of the $\sqrt[3]{1}$ point, we know that the period-3 surface near the $\sqrt[3]{1}$ point, whose formula is given in the Appendix, projects in a 3-1 fashion to a deleted neighborhood (after removing the $\sqrt[3]{1}$ point) of the parameter plane; all period-3 points in the deleted neighborhood of the $\sqrt[3]{1}$ point are saddles. So we make three extra cuts on the saddle surface, all three projecting to the same curve in the parameter space from the $\sqrt[3]{1}$ point to an arbitrarily chosen point on the period-3 saddle-node curve, as indicated in Figure 5b. This leaves us with three "saddle" pieces to accompany the three "node" pieces. All six pieces project to the parameter plane in the "inverted teardrop" shape of the 1/3 resonance region of Figure 3a, the saddle pieces each having an extra cut.
3. Add the six pieces corresponding to the boundary circle in a procedure similar to that for the 1/5 surface.
4. The cyclical identification of saddle pieces to each other along the "extra" cut is justified by the period-3 surface formula near the $\sqrt[3]{1}$ point given in the Appendix: a closed loop on the period-three surface parametrized by a small circle around the origin in the phase plane projects to a loop which travels three times around the $\sqrt[3]{1}$ point in the parameter plane. This forces the rest of the edge labels given in the first step of Figure 5b. As with the 1/5 surface, the labelling is corroborated because the six boundary pieces are joined to form a single topological circle.
5. Reassemble as indicated in the rest of Figure 5b.
6. The 1/3 surface is a topological disk. Notice that, similar to the 1/5 surface, the $\sqrt[3]{1}$ point can still be thought of as the center of the disk, but, unlike the 1/5 surface, the saddle-node curves don't meet this point.

The 1/2 resonance surface

1. The period-two saddle-node curves on the sides of the 1/2 horn for small forcing amplitude each terminate in a degenerate period-doubling bifurcation, where the two points on the period-two orbit come together to form a fixed point. The local bifurcation analysis of such a point (Peckham and Kevrekidis [1991]) shows a period-doubling curve passing through such a point. Continuation of this curve reveals that both degenerate period-doubling points turn out to be on the same period-doubling curve, which forms a circle in the parameter space. This gives us the "ice cream cone" shape for the 1/2 resonance region in Figure 3a. As for our other surfaces, additional cross sections convinced us that there are no more projection singularities on the 1/2 surface.
2. After cutting along the saddle-node curves and the period-doubling curves of the 1/2 surface, we are left with four pieces: two which project to the whole 1/2 resonance region, and two which project only to the bottom triangular "cone" part of the region.
3. Add the four pieces corresponding to the boundary circle in a procedure similar to that for the 1/5 surface.

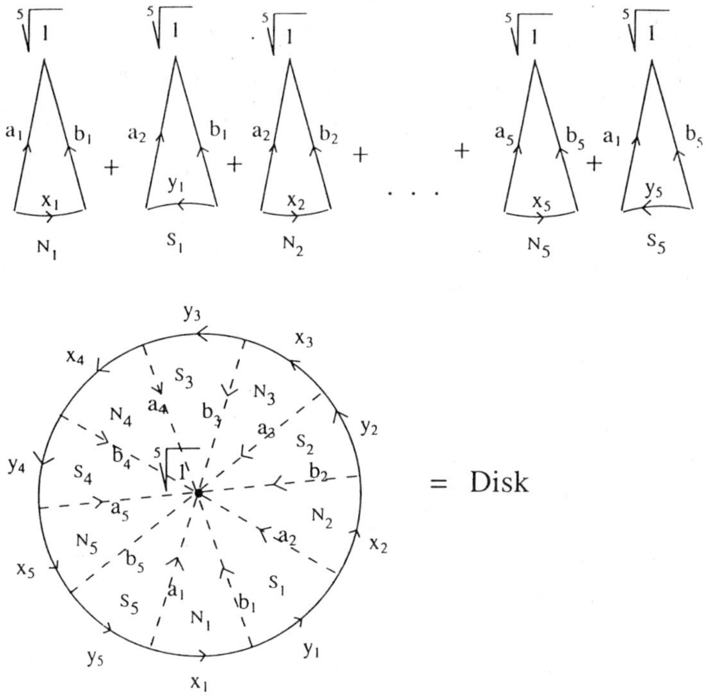

Figure 5 (a) The 1/5 surface. Lines a_i and b_i are period-five saddle-nodes.

4. The edge identifications are forced by the local bifurcation analysis of the degenerate period-doubling points, where all four surfaces come together, and by the fact that the four added boundary pieces must form a single topological circle. Which pieces are labelled saddles and which are labelled nodes can be determined from corresponding phase portraits, but this does not affect the topology. This is the starting point of Figure 5c.
5. Reassemble as indicated in the rest of Figure 5c.
6. The surface is a mobius strip!! Note that the period-doubling circle is a "generator" of the mobius strip.

The fixed point resonance surface

1. The continuation of the two saddle-node curves which emanate from $(p/1, 0)$ in the parameter plane for any integer p results in a triangular region in the parameter space. Triangles for $p = 0$ and $p = 1$ are shown in Figure 3a. The two top "corners" of the $p = 0$ triangle, shown in the enlargement of Figure 3b, are cusp bifurcations. As for the other surfaces, additional numerical cross sections convinced us there are no other fixed point projection singularities.

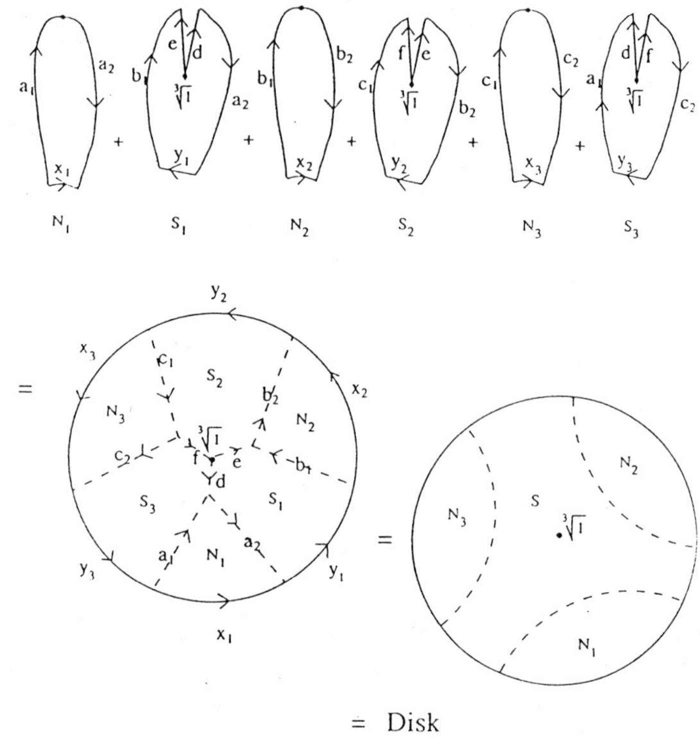

Figure 5 (b) The 1/3 surface. Lines a_i, b_i, and c_i are period-three saddle-nodes; lines d, e, and f are "extra" cuts.

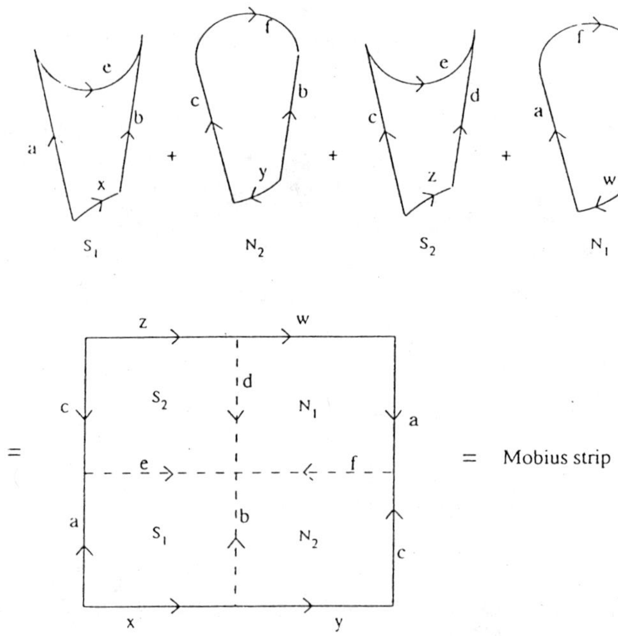

Figure 5 (c) The 1/2 surface. Lines a, b, c, and d are period-two saddle-nodes; lines e and f are period doublings.

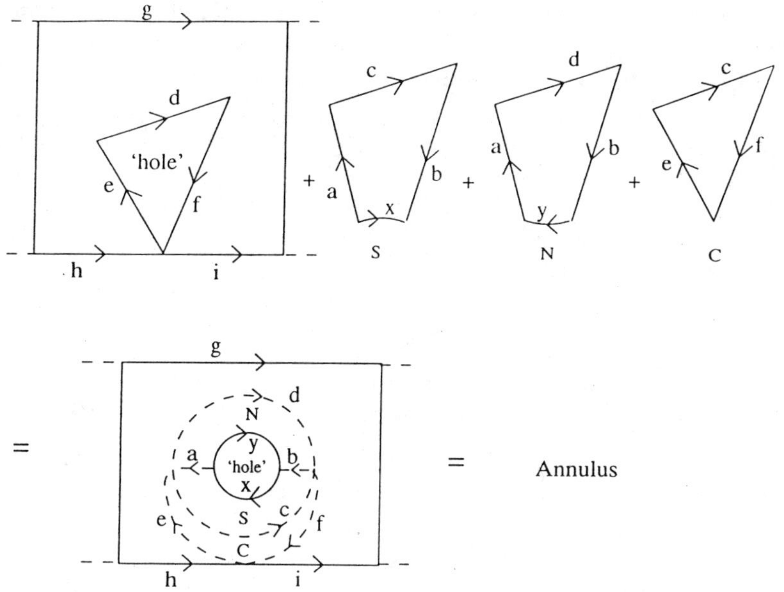

Figure 5 (d) The fixed-point surface restricted to $-1/2 \leq \omega_0/\omega \leq 1/2$. Lines a, b, and d are saddle-nodes; lines c, e, and f are "extra" cuts.

2. We make cuts along all the saddle-node curves. We note that there are two fixed points born as we enter any one of the triangular regions across a saddle-node bifurcation, but there is a third one which is involved in the cusp bifurcations. Thus there are three fixed points for parameter values inside the triangular regions and one fixed point outside. The "outside" point can be thought of as the continuation of the repelling "center" fixed point which came from the repelling equilibrium point of the original unforced oscillator. For increased clarity, we choose to make three extra cuts on the surface for each triangular region, one corresponding to the parameter values along each (saddle-node) leg of the triangle but for the surface which is not involved in the saddle-node bifurcation. This leaves us with three pieces which cover each triangular region and a single piece which covers the rest of the parameter space $(-\infty, \infty) \times [0, 1]$.

3. For each triangular region, add the two boundary pieces corresponding to the boundary circle to two of the three triangular surface pieces. (The "center" fixed point is not part of the unforced oscillator, so the boundary pieces are added to the saddle and node pieces only.)

4. We restrict our description and figures to $-1/2 \leq \omega_0/\omega \leq 1/2$, so that we are dealing with only one of the triangular regions. The saddle piece and node piece are identified along the edges corresponding to the two "legs" of the triangle. The local unfolding of the cusp points implies that the saddle-node pairings of the two surface pieces changes at a cusp point. So the identification at the top of the triangle must be between the third piece (the "center" piece) and one of the other two. As can be verified from phase portraits, the saddle piece is the one which pairs with the "center" piece.

The surface not involved in the saddle-node bifurcation must be the one identified with the sides of the "hole" in the surface which represents the unique fixed point for parameter values outside the triangle. This provides the starting point for Figure 5d. The dotted lines indicate the continuation of the fixed point surface to $|\omega_0/\omega| > 1/2$. Note also that the "center" piece of fixed points has boundaries at $\alpha = 0$ and $\alpha = 1$, where our caricature's parameter space was cut off. This boundary is unique to the fixed point surface because for all other p/q surfaces boundary components exist only at $C_0 \times (p/q, 0)$.

5. Reassemble the four pieces as in the rest of Figure 5d.

6. The surface restricted to $-1/2 \leq \omega_0/\omega \leq 1/2$ is an annulus. That is, topologically, each triangular region corresponds to a hole (a boundary component) in the single surface of fixed points. The whole fixed point surface, then, is a strip with an infinite number of holes.

Remarks 2.2

1. Because the normal forms for any $\sqrt[q]{1}$ point with $q \geq 5$ indicate pairings of saddle surfaces and node surfaces analogous to the pairings near a $\sqrt[5]{1}$ point, the strategy to determine the topology of surfaces with $q \geq 5$ is analogous to the strategy we used for the $1/5$ surface. Likewise, we expect the topology of all period-3 surfaces to be determined as was our $1/3$ surface, the topology of all period-2 surfaces to be determined as was our $1/2$ surface, and the topology of all fixed-point surfaces to be determined as was our fixed-point surface. Depending on the relative magnitudes of the normal form coefficients A and B (as in the Appendix) near a $\sqrt[4]{1}$ point, the period-4 surfaces could either be like a $1/5$ surface, with a $\sqrt[4]{1}$ point connected to the saddle-node curves, or like a $1/3$ surface, with the $\sqrt[4]{1}$ point isolated from the other singular points. The $1/4$ surface and $3/4$ surfaces of our caricature both appear to be of the former variety.

2. Although we have labelled most surfaces with S (respectively N) for saddle (respectively node), it is not necessary that the continuation of these surfaces stay saddles (respectively nodes) in all applications. Period doubling bifurcations could cause a change from one to the other.

3 Comments

This paper deals mainly with the topology of the resonance surfaces, but only hints at the geometry of the surfaces as they are situated in the four-dimensional phase $\times$ parameter space. Work in progress presents much more of the geometry, including various three-dimensional projections such as Figure 4b (from the four-dimensional phase $\times$ parameter space) of global resonance surfaces which have been numerically computed (McGehee, R.P. and Peckham, B.B. [1994 and to appear]). A better "feel" for the geometry of these surfaces is provided by a movie illustrating these numerically computed surfaces (McGehee and Peckham [1992]). Better yet is the understanding obtained by using the computer to interactively rotate the surfaces through various three- and four-dimensional projections.

In dealing with other forced oscillator families we note that for $q \geq 3$ the p/q resonance surfaces are generically orientable two-manifolds with one boundary component. Thus they must be disks with some number of handles. So the topology of resonance surfaces for other forced oscillators can differ from ours only in the number of handles it may have. (This statement holds for $q = 2$ as well, if the conjecture that all period-two surfaces are nonorientable is true.) Although we have seen no "natural" example of a forced oscillator family having a handle on any resonance surface, we can construct families having resonance surfaces with handles by transforming the parameter space. Other parameter space diagrams of forced oscillators (Schreiber et al. [1988]) appear to have sets of projection singularities more complicated than in our application, but such complications do not correspond to handles; they can be thought of merely as extra folds with respect to projection to the parameter plane. An investigation of surface folds appears in another paper (McGehee and Peckham, submitted); surface handles will be treated in future work.

We point out that although this study targets bifurcations for periodically forced oscillator systems, it is really applicable to generic two-parameter families of maps of the plane. Forced oscillator families differ from the generic case only because there is a known starting point: forcing amplitude (α) equals zero. Because the corresponding map behavior at zero forcing amplitude is not generic, the "boundary circles" we described do not occur in generic families of maps. The other "local" descriptions of periodic surfaces, however, still apply. For example, a period-5 resonance surface which is "born" in a Hopf bifurcation at a $\sqrt[5]{1}$ point can "die" at a second such point. This would result in a period-5 surface which would be a topological sphere. This is a typical scenario for resonance horns corresponding to "secondary" Hopf bifurcations as studied in Peckham, Frouzakis and Kevrekidis [1995].

We also acknowledge that our study is far from the final word on the study of bifurcations of forced oscillator systems. For example, we have said nothing about the global bifurcations (invariant manifold crossings), the relationship of the resonance surfaces to the global bifurcations, or the relationship of the resonance surfaces to associated phase portraits. A complete bifurcation study does not yet exist. However, see Vance and Ross [1989] and Peckham [1988] for more complete numerical studies of forced oscillator systems, and Aronson et al. [1983] for additional bifurcation features which may be present inside any given resonance region.

In addition to providing a philosophical plug for resonance surfaces as more natural building blocks for bifurcation analysis than the bifurcation regions obtained from their projections to the parameter plane, we note that the use of these surfaces may also provide some practical benefits for numerical computation of bifurcation diagrams. For example, it may be easier numerically to locate the saddle-node boundaries of a resonance surface by computing the whole resonance surface than by computing the saddle-node boundaries directly. In order to do so, however, we must first know the topology of the surface, as studied in this paper, so that we may succesfully parametrize it. Our real goal is to extend the same philosophy to the computation of global bifurcations: computing whole surfaces of homoclinic points may turn out to be a more tractable task than computing the curves of homoclinic tangencies that project to the boundaries of homoclinic regions in the parameter space. The further expectation that homoclinic surfaces can be viewed in many contexts as limits of periodic point surfaces is even more motivation for understanding the topology of resonance surfaces as presented in this paper.

Acknowledgements Discussions over the last several years with D. G. Aronson, G. R. Hall, and R. Moeckel have been instrumental in the development of the ideas presented in this paper.

4 Appendix

We list below a table of local representation of period-q surfaces (actually the closure of the least-period-q points) which exist in the neighborhood of the described fixed-point bifurcation.

Bifurcation:	q	Universal unfolding	Period-q Surface	Reference
Saddle-node (ev*=1)	1	$(x,y) \to (x+\epsilon_1 \pm x^2, by)$	$\epsilon_1 = \mp x^2, y=0$	Ar,GH
Cusp (ev*=1 and hod*)	1	$(x,y) \to (\epsilon_1 + (\epsilon_2+1)x \pm x^3, by)$	$\epsilon_1 = -(\epsilon_2 x \pm x^3)$	Ar, GH
Takens-Bogdanov (both ev*=1)	1	$(\dot{x},\dot{y}) = (y, \epsilon_1 + \epsilon_2 y + x^2 \pm xy)$	$\epsilon_1 = -x^2, y=0$	Ar,Bo GH, Ta
Period-doubling (ev*=-1)	2	$(x,y) \to ((\epsilon_1 - 1)x \mp x^3, -ay)$	$\epsilon_1 = \mp x^2, y=0$	Ar,GH
Degenerate period doubling (ev*=-1 and hod*)	2	$(x,y) \to ((\epsilon_1 - 1)x + \epsilon_2 x^3 \pm x^5, -ay)$	$\epsilon_1 = -(\epsilon_2 x^2 \pm x^4), y=0$	PK
Double -1 (both ev*=-1)	2	$(\dot{x},\dot{y}) = (y, \epsilon_1 x + 2\epsilon_2 y \pm x^3 - 2x^2 y)$	$\epsilon_1 = \mp x^2, y=0$	Ar, Ta
$\sqrt[q]{1}$ point	3	$\dot{\mathbf{z}} = \epsilon\mathbf{z} + A\mathbf{z}\lvert\mathbf{z}\rvert^2 + B\bar{\mathbf{z}}^2$	$\epsilon = -A\lvert\mathbf{z}\rvert^2 - B\bar{\mathbf{z}}^2/\mathbf{z}$	Ar, Ta
(ev* = $e^{\pm 2\pi i p/q}$)	4	$\dot{\mathbf{z}} = \epsilon\mathbf{z} + A\mathbf{z}\lvert\mathbf{z}\rvert^2 + B\bar{\mathbf{z}}^3$	$\epsilon = -A\lvert\mathbf{z}\rvert^2 - B\bar{\mathbf{z}}^3/\mathbf{z}$	Ar, Ta
	≥ 5	$\dot{\mathbf{z}} = \epsilon\mathbf{z} + \mathbf{z}^2\bar{\mathbf{z}}A(\lvert\mathbf{z}\rvert^2) + B\bar{\mathbf{z}}^{q-1}$	$\epsilon = -\lvert\mathbf{z}\rvert^2 A(\lvert\mathbf{z}\rvert^2) - B\bar{\mathbf{z}}^{q-1}/\mathbf{z}$	Ar, Ta

* ev = eigenvalue(s) of the fixed point, hod = higher order degeneracy

Comments:

1. For the universal unfoldings given as differential equations, the time one map of the differential equation agrees with the qth iterate of the original map (after a coordinate change) up to arbitrarily high order.
2. The period-q surface formulas are determined directly from the universal unfoldings. For $q \geq 2$, the fixed point at the origin of the phase plane must be "divided out" to leave us with a representation of the closure of the least-period-q surface.

References

Arnold V.I. [1982] *Geometrical Methods in the Theory of Ordinary Differential Equations*, Springer Verlag, New York.

Aronson D.G., McGehee R.P., Kevrekidis I.G. and Aris R. [1986] *Entrainment regions for periodically forced oscillators*, Phys. Rev. A, **33**, (3), 2190–2192.

Aronson D.G., Chory M.A., Hall G.R. and McGehee R.P. [1983] *Bifurcations from an invariant circle for two-parameter families of maps of the plane: a computer assisted study*, Comm. Math. Phys., **83**, 303–354.

Bogdanov R.I. [1976] *Versal deformation of a singularity of a vector field in the plane in the case of zero eigenvalues,* Trudy Seminara Imeni I. G. Petrovskogo, **2**, 23–36. English transl. in Selecta Math. Sovietica, **1**, No. 4, 389–421.

Guckenheimer, J. and Holmes, P. [1983] *Nonlinear Oscillations, Dynamical Systems and Bifurcations of Vector Fields*, Applied Mathematical Sciences, **42**, Springer Verlag, New York.

Hall G.R. [1984] *Resonance zones in two-parameter families of circle homeomorphisms,* SIAM J. Math. Anal., **15** (6), 1075–1081.

Kai, T. and Tomita, K. [1979] *Stroboscopic phase portrait of a forced nonlinear oscillator*, Progr. Theor. Phys., **61**(1), 54–73.

Kevrekidis I.G., Aris R. and Schmidt L.D. [1986] *The stirred tank forced,* Chem. Eng. Sci., **41**(6), 1549–1560.

McGehee R.P. and Peckham B.B., *Arnold Flames and Resonance Surface Folds*, submitted to the International Journal of Bifurcation and Chaos, 1994.

McGehee R. P. and Peckham B. B., *Resonance surfaces for forced oscillators*, Geometry Center Research Report GCG70 (1994) and to appear in *Experimental Mathematics.*

McGehee R. P. and Peckham B. B. [1992] *Resonance City,* a 15 minute VCR movie made at the Geometry Center, Minneapolis, MN. Available on request from the Geometry Center or the authors.

McKarnin M., Schmidt L., and Aris R. [1988] *Forced oscillations of a self-oscillating bimolecular surface reaction model,* Proc. R. Soc. Lond. A, **417**, 363–388.

Peckham B.B. [1988] *The Closing of Resonance Horns for Periodically Forced Oscillators*, thesis, University of Minnesota.

Peckham, B.B. [1990] *The necessity of the Hopf bifurcation for periodically forced oscillators with closed resonance regions,* Nonlinearity, **3**, 261–280.

Peckham B.B. and Kevrekidis I.G. [1991] *Period doubling with higher order degeneracies*, SIAM J. Math. Anal., **22**, No. 6, 1552–1574.

Peckham B.B., Frouzakis C.E. and Kevrekidis I.G. [1995] *Bananas and Banana Splits: A parametric degeneracy in the Hopf bifurcation for maps*, SIAM J. of Math. Anal., Vol. 26, No. 1.

Schreiber I., Dolnik M., Choc P. and Marek M. [1988] *Resonance behaviour in two-parameter families of periodically forced oscillators,* Physics Letters A, **128, 1.2**, 66–70.

Takens F. [1974] *Forced oscillations and bifurcations,* Applications of Global Analysis, Communications of the Mathematical Instituet Rijksuniversiteit Utrecht, Vol. **3**, 1–59.

Vance W. and Ross J. [1989] *A detailed study of a forced chemical oscillator: Arnold tongues and bifurcation sets,* J. Chem. Phys., **91**, 7654–7670.

Fields Institute Communications
Volume **4**, 1995

Normal Forms and Homoclinic Chaos: Application to Structural Systems

N. Sri Namachchivaya and Naresh Malhotra
Nonlinear Systems Group
Department of Aeronautical and Astronautical Engineering
University of Illinois at Urbana-Champaign
Urbana, IL
61801

Abstract. In this paper, we examine the nonlinear behavior of two different structural systems. First, we study the complex coupling between the bending and torsional motion of a simply supported beam transversely loaded by a follower force. Here, the presence of double Hopf bifurcation with non-semisimple 1:1 resonance is investigated with the help of three physical parameters using the theory developed by van Gils et al. [1990]. It is shown that the model describing the two degree of freedom bending-torsional motion of the beam is degenerate. However, an additional symmetry due to reversibility reveals that it is indeed a generic behavior for follower force systems (O'Reilly et al. [1994]). We next examine the global dynamics associated with an externally excited shallow arch structure under the conditions of external subharmonic resonance and one-one internal resonance. By using a perturbation technique developed by Kovačič and Wiggins [1992], we show that for some parameter values the shallow arch system may exhibit complicated dynamics. Using this technique we can predict an explicit range of parameter values where Silnikov type homoclinic orbits may occur and lead to chaotic behaviour.

1 Introduction

In this study we apply the ideas of normal forms and homoclinic chaos to examine the dynamical behavior of structural systems under different load conditions. The main objective is to investigate the local and global bifurcations in the motion of these systems in the presence of internal resonance. Previous studies have indicated that such systems are capable of exhibiting dynamical responses such as amplitude modulation and chaotic motion due to internal resonance even under simple harmonic excitation. In this study we examine the nonlinear response of two different structural systems using two-degree-of-freedom models.

1991 *Mathematics Subject Classification.* Primary 70K30; Secondary 70K40, 34CXX.

This research has been partially supported by AFOSR through Grant 93-0041 and the National Science Foundation through Grant MSS 90-57437 PYI.

First we investigate the flexural-torsional motion of a thin beam with rectangular cross section and a transverse follower force acting at its center. This system was previously examined by Fu and Nemat-Nasser [1972] who studied the stability of the nontrivial steady state response of the nonlinear beam structure under small nonautonomous follower force. For this case, they showed the existence of steady-state solutions in the averaged equations. In this investigation we apply the results of van Gils et al. [1990] to investigate the complex dynamics of the beam structures subjected to nonconservative autonomous loads in the presence of 1:1 internal resonance. Under critical conditions, the model describing the two degree of freedom bending-torsional motion of the beam can possess a pair of purely imaginary nonsemisimple eigenvalues. The local dynamics of this system can be analyzed by using the nonsemisimple Hopf bifurcation theory developed by van Gils et al. [1990] provided the model is non-degenerate. In this system, the dimensionality of the governing equations can be reduced by one, due to the presence of an S^1 symmetry.

Next we consider the dynamics of a shallow arch structure under harmonic excitation. Shallow arch structures subjected to various load conditions have been investigated by many researchers in the past. Dynamic behavior of the arch structures under impulsive loading was examined by Mettler [1962], Lock [1965], Humphreys [1966], and Hsu [1969]. Recently, Namachchivaya and Doyle [1988], and Tien et al. [1994] studied the bifurcation behavior and the chaotic dynamics of a shallow arch subjected to small amplitude periodic excitation. In this analysis we are interested in examining the global bifurcation behavior in the oscillations of an initially deformed shallow arch system and identifying the parameter regimes where this system may display chaotic dynamics. In real life systems such behavior most frequently occurs due to the presence of homoclinic orbits. Conventionally, one uses Melnikov type perturbation methods to analytically investigate the global dynamics. However such methods cannot be used in these resonance problems without further refining the scaling of various dynamics that co-exist. In this example we use a perturbation technique developed by Kovačič and Wiggins [1992] which uses such scaling and provides the criterion for the existence of a Silnikov type homoclinic orbit to a saddle-focus type fixed point that may be created near resonance under small perturbations. This may lead to a well described type homoclinic chaos provided certain conditions on the stability properties of the fixed point are satisfied. The result of this analysis is an analytical criterion in terms of system parameters for the existence of complex dynamical behaviour.

2 Flexural and torsional motion of beam structures

In this section, we examine the interactions between the flexural and torsional motion of a narrow rectangular beam. This beam is simply supported at both ends and is subjected to a transverse follower force at the center (Figure 1). The follower force changes its direction with the deformation, and thus always remains in the plane of the cross-section. Under critical load conditions (and in the absence of any damping), the equations governing the bending-torsional motion of the beam can be brought to an appropriate form where the analysis of van Gils et al. [1990] can be used to explore the complicated dynamics of this mechanical system.

The equations governing the flexural-torsional motion of the beam are derived using Hamilton's energy principle. While deriving these equations of motion, an

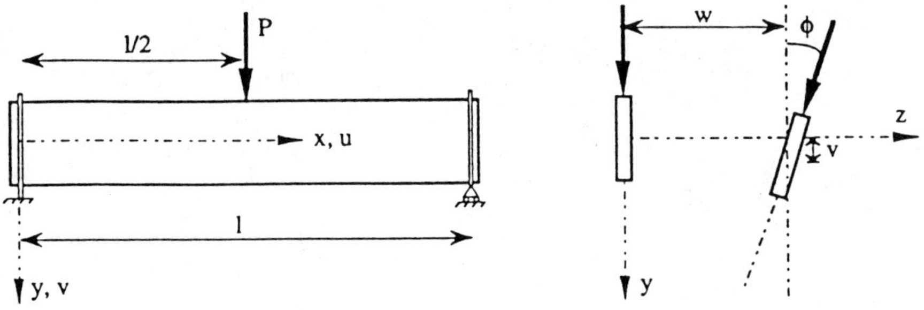

Figure 1 Narrow rectangular beam subjected to an autonomous follower force.

appropriate ordering scheme was used and all the nonlinear terms up to third order were retained. For a thin rectangular beam, the deflections in the x and y directions (u and v respectively) are assumed to be small. In this case, the partial differential equations describing the lateral deflection $w(x,t)$ and the angle of twist $\phi(x,t)$, are expressed as:

$$
\begin{aligned}
m w_{tt} + \beta_1 w_t + EI_z w_{xxxx} + (M_z \phi)_{xx} - E\ A\ h(w,t)\ w_{xx} + P\ \delta(l/2 - x)\ \phi = 0 \\
m\ r^2\ \phi_{tt} + \beta_2\ \phi_t + EI_w \phi_{xxxx} + M_z w_{xx} - G\ I_d\ w_{xx} + G\ I_p\ h(w,t)\ \phi_{xx} = 0
\end{aligned}
\tag{2.1}
$$

where

$$
M_z = \begin{cases} \frac{1}{2} P x & 0 \le x \le \frac{1}{2} \\ \frac{1}{2} P(l - x) & \frac{1}{2} \le x \le l \end{cases}
$$

$$
h(w,t) = \frac{1}{2l} \int_0^l w_x^2\, dx.
$$

In these equations, $w(x,t)$ represents the deflection in the x-z plane and $\phi(x,t)$ is the angle of twist in y-z plane (see Fig. 1). EI_y, GI_d and EI_w denote the bending, torsional and warping rigidities respectively. I_p is the polar moment of inertia, m is mass per unit length, r is the polar radius of inertia and A represents the cross sectional area of the beam. P is the autonomous follower force, β_1 and β_2 are the damping coefficients, and the terms involving h(w,t) are the nonlinear terms. In order to be consistent with the ordering scheme used in deriving the equations of motion (2.1) we can argue that the nonlinear term in the ϕ equation is much smaller than the corresponding nonlinear term in the transverse deflection equation, and thus we can drop the nonlinear term from the second equation. The assumption of simply supported boundary conditions at both the ends implies that:

$$w(0,t) = w(l,t) = w_{xx}(0,t) = w_{xx}(l,t);$$
$$\phi(0,t) = \phi(l,t) = \phi_{xx}(0,t) = \phi_{xx}(l,t).$$

Now, we consider the first mode expansion of the deflection (w) and the angle of twist (ϕ) that satisfies the boundary conditions at both the ends:

$$\begin{aligned} w(x,t) &= q_1(t)\ r\ \sin(\frac{\pi x}{l}); \\ \phi(x,t) &= q_2(t) \sin(\frac{\pi x}{l}). \end{aligned} \tag{2.2}$$

At this stage, we apply the Ritz-Galerkin method to reduce the equations of motion to a set of two second order nonlinear ordinary differential equations describing the evolution of $q_1(t)$ and $q_2(t)$. Replacing β_1 and β_2 by $(m \cdot \beta_1)$ and $(mr^2\beta_2)$, respectively, we get,

$$\begin{aligned} \ddot{q}_1 + \beta_1\dot{q}_1 + \Omega_1^2 q_1 + b_1 q_2 + a\ q_1^3 &= 0 \\ \ddot{q}_2 + \beta_2\dot{q}_2 + \Omega_2^2 q_2 + b_2 q_1 &= 0 \end{aligned} \tag{2.3}$$

where

$$\Omega_1^2 = \frac{1}{m}\ E\ I_y\ (\frac{\pi}{l})^4, \qquad \Omega_2^2 = \frac{1}{mr^2}\ \{EI_w(\frac{\pi}{l})^4 + GI_d(\frac{\pi}{l})^2\}$$
$$b_1 = \frac{P}{2mlr}\ (7 - \frac{\pi^2}{4}), \qquad b_2 = -\frac{P}{2mlr}\ (1 + \frac{\pi^2}{4}), \qquad a = \frac{1}{4m}\ E\ A\ r^2\ (\frac{\pi}{l})^4.$$

Here Ω_1 and Ω_2 are the two natural frequencies of the system, b_1 and b_2 represent the linear coupling between the amplitude of deflection and the angle of twist of the beam, and a is the nonlinear coefficient. In the phase space, these equations of motion can be expressed as:

$$\dot{x} = Ax + f(x) \tag{2.4}$$

where

$$x = [x_1, x_2, x_3, x_4]^T = [q_1, \dot{q}_1, q_2, \dot{q}_2]^T$$

$$A = \begin{bmatrix} 0 & 1 & 0 & 0 \\ -\Omega_1^2 & -\beta_1 & -b_1 & 0 \\ 0 & 0 & 0 & 1 \\ -b_2 & 0 & -\Omega_2^2 & -\beta_2 \end{bmatrix}$$

$$f(x) = [0,\ -a\ x_1^3,\ 0,\ 0]^T.$$

Under critical conditions the linear operator of this system assumes the nonsemisimple form. These critical conditions occur when $P = P_{cr}$ and $\beta_{1,2} = 0$, where

$$P_{cr} = \frac{4\ |\Omega_1^2 - \Omega_2^2|\ m\ l\ r}{\sqrt{(4+\pi^2)(28-\pi^2)}}. \tag{2.5}$$

This represents a double Hopf bifurcation with 1:1 resonance. At critical conditions, the equations of motion reduce to the following form:

$$\dot{z} = Bz + f(z, \bar{z}) \qquad z \in C^2 \tag{2.6}$$

where

$$z = \begin{bmatrix} z_1 & z_2 \end{bmatrix}^T$$

$$B = \begin{bmatrix} i\omega & 1 \\ 0 & i\omega \end{bmatrix}$$

$$\omega = \sqrt{\frac{\Omega_1^2 + \Omega_2^2}{2}}.$$

This gives a 4-dimensional autonomous system that has a nonsemisimple linear operator and cubic nonlinearities. The normal form for the generalized Hopf bifurcation with 1:1 internal resonance can be expressed as:

$$\begin{pmatrix} \dot{z}_1 \\ \dot{z}_2 \end{pmatrix} = \begin{pmatrix} i\omega & 1 \\ 0 & i\omega \end{pmatrix} \begin{pmatrix} z_1 \\ z_2 \end{pmatrix} + \{a_1 z_1 \bar{z}_1 + a_2 \left(z_1 \bar{z}_2 - \bar{z}_1 z_2\right)\} \begin{pmatrix} z_1 \\ z_2 \end{pmatrix} + \{b_1 z_1 \bar{z}_1 + b_2 \left(z_1 \bar{z}_2 - \bar{z}_1 z_2\right)\} \begin{pmatrix} 0 \\ z_1 \end{pmatrix} \tag{2.7}$$

where $a_k = c_k + id_k, b_k = d_k + ie_k,\ k = 1, 2$. The other two equations are simply the complex conjugates of these equations. The nonlinear coefficients in the normal form are expressed in terms of the nonlinear coefficients of the original system. One is referred to, Namachchivaya et al. [1994], for the explicit expressions of these nonlinear coefficients.

In this case, the 1:1 internal resonance singularity is unfolded with the help of three unfolding parameters (α_1, μ_1 and μ_2) which can be expressed in terms of physical quantities p, δ_1 and δ_2 where p is the deviation from the critical loading ($i.e., p = (P - P_{cr})$ and δ_1 and δ_2 are the small damping coefficients for the bending and torsional modes respectively.

$$\alpha_1 = -\frac{1}{4}(\delta_1 + \delta_2),$$

$$\mu_1 = \frac{1}{4} \frac{(\Omega_1^2 - \Omega_2^2)^2}{(\Omega_1^2 + \Omega_2^2)} \frac{p}{P_{cr}},$$

$$\mu_2 = \frac{1}{4\sqrt{2}} \frac{(\Omega_1^2 - \Omega_2^2)}{(\Omega_1^2 + \Omega_2^2)^{\frac{1}{2}}} (\delta_2 - \delta_1).$$

After an appropriate scaling of the variables and the unfolding parameters, the truncated normal form equations are expressed in (ρ, θ, u, v) coordinates as follows:

$$\begin{aligned} \rho' &= 2\rho(\beta + u) + O(\epsilon) \\ u' &= \nu_1 + (v^2 - u^2) + c_2\ \rho + O(\epsilon) \\ v' &= \nu_2 - 2\ u\ v + d_2\ \rho + O(\epsilon) \\ \theta' &= 1 + v + O(\epsilon) \end{aligned} \tag{2.8}$$

where $\rho = z_1\bar{z}_1$, $\theta = \frac{1}{2i}\ln(z_1/\bar{z}_1)$, $u = \frac{1}{2}(z_2/z_1 + \bar{z}_2/\bar{z}_1)$ and $v = \frac{1}{2i}(z_2/z_1 - \bar{z}_2/\bar{z}_1)$. The prime indicates differentiation with respect to slow time, and β, ν_1 and ν_2 are the scaled unfolding parameters.

Van Gils et al. [1990] assume the two nonlinear coefficients (c_2 and d_2) to be nonzero in their general analysis of autonomous systems, but for present case one of the nonlinear coefficients turns out to be degenerate ($d_2 = 0$) due to the nature

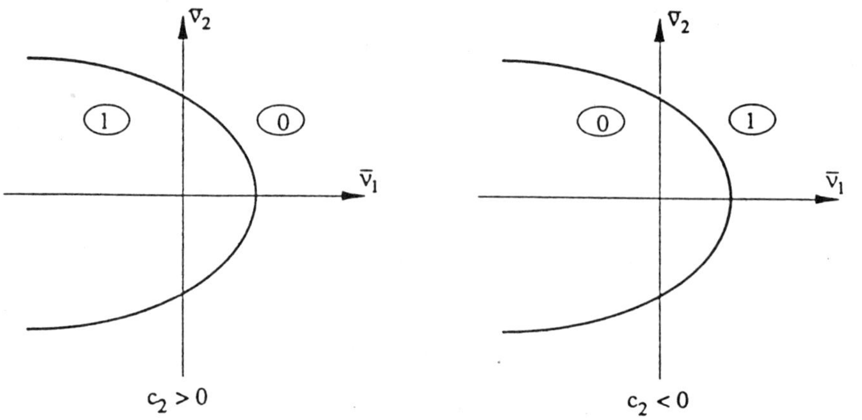

Figure 2 Number of permissible periodic orbits in the parameter space for different values of nondegenerate nonlinear coefficient.

of the nonlinearities present in the system. The other coefficient (c_2) is expressed in terms of system parameters as:

$$c_2 = -\frac{3}{16m}\frac{\Omega_1^2 - \Omega_2^2}{\Omega_1^2 + \Omega_2^2} EA\, r^2 \left(\frac{\pi}{l}\right)^4. \tag{2.9}$$

Most of the results of the general study cannot be applied to this example due to the degeneracy present in one of the nonlinear coefficients. Due to the presence of S^1 symmetry in the system, the θ equation decouples from the rest, and thus we only need to consider the first three equations. For $\beta \neq 0$, the equilibrium solutions are given as:

$$\bar{\rho} = \frac{1 - \bar{\nu}_1 - \bar{\nu}_2^2}{c_2/\beta^2}, \qquad \bar{u} = -\beta, \qquad \bar{v} = -\beta\bar{\nu}_2 \tag{2.10}$$

where $\bar{\nu}_1 = \nu_1/\beta^2$, and $\bar{\nu}_2 = \nu_2/(2\beta^2)$. For $c_2 > 0$, there exists only one real and positive root of ρ for the parameters values lying inside the parabolic region, while no nontrivial equilibrium solution is permissible outside the parabolic boundary (Figure 2). For $c_2 < 0$, we observe exactly the opposite scenario, i.e., no equilibrium is possible inside, while one steady state exists outside the parabolic region. A steady state of the normal form equations corresponds to a periodic orbit in the original system. Thus, in this case there may exist 0 or 1 small amplitude periodic orbits in the system of equations describing the beam dynamics. For secondary bifurcation to occur from a solution with $\rho > 0$, the Jacobian J_0 corresponding to the first three coupled equations, must have a zero or a pure imaginary pair of eigenvalues. J_0 is given as:

$$J_0 = \begin{bmatrix} 0 & 2\bar{\rho} & 0 \\ c_2 & 2\beta & -2\beta\bar{\nu}_2 \\ 0 & 2\beta\bar{\nu}_2 & 2\beta \end{bmatrix}. \tag{2.11}$$

J_0 is non-singular ($\mathrm{Det}(J_0) = -4\, c_2\, \beta\, \rho \neq 0$ since $\bar{\rho} > 0, \beta \neq 0$ and $c_2 \neq 0$ in general), and thus J_0 does not have a zero eigenvalue. We obtain the following conditions that must be satisfied for a pure imaginary pair of eigenvalues to exist:

$$\bar{\rho} c_2 \leq 2\beta^2(1+\nu_2^2) \quad \text{and} \quad \bar{\rho} c_2 = 4\beta^2(1+\nu_2^2).$$

These conditions can not be satisfied simultaneously, irrespective of the value of c_2 and hence there is no secondary Hopf bifurcation present in this case.

An extended version of the autonomous case recognizing the fact that the follower force systems are reversible systems, is presented in O'Reilly et al. [1994]. Reversible systems are a class of dynamical systems that remain invariant under a linear state space transformation along with the reversal of time. In this work, they show that $d_2 = 0$ is a generic property of the reversible systems and these systems are integrable on a three-dimensional reduced phase space. In the absence of any dissipative forces, the follower force systems are reversible. In such non-conservative systems that are originally undamped, the introduction of small dissipation may have a destabilizing effect. O'Reilly et al. [1994] examine the dynamics of a general class of reversible systems on a four-dimensional center manifold with and without 1:1 resonance. They also illustrate their results of reversible double Hopf bifurcation with 1:1 resonance on the beam model described in this section.

In a general case when the follower force is non-autonomous, the corresponding degeneracy condition becomes the simultaneous vanishing of both the nonlinear coefficients (i.e., $c_2 = d_2 = 0$) and thus, the beam dynamics can be studied in that framework. One is referred to Namachchivaya and Malhotra [1992] for a detailed analysis of the corresponding non-autonomous systems.

3 Chaotic motion of shallow arch structures

In this section, we investigate the dynamic response of an initially deformed shallow arch subjected to a lateral loading $p(x,t)$, as shown in Figure 3. Our approach is to first obtain the nonlinear perturbation equations near the time dependent single mode solutions. The external forcing appears as a parametric term in these perturbation equations. We explore the dynamical behavior of the averaged system in the presence of sub-harmonic external resonance and 1:1 internal resonance.

The equation of motion governing the lateral deflection $w(x,t)$ of this arch can be derived using the energy method, and is given as:

$$m w_{tt} + \beta w_t + EI w_{xxxx} - E\ A\ h(x,t)\ (\bar{w}_{xx} + w_{xx}) = p(x,t) \tag{3.1}$$

where

$$h(w,t) = \frac{1}{l}[u(l,t) + \int_0^l (\frac{1}{2} w_x^2 + \bar{w}_{xx} w_{xx})\, dx\ ]$$

$$\bar{w}_{xx} = -q_0 \sin\frac{\pi x}{l}$$

$$p(x,t) = (p_0\ + \rho \sin \nu t) \sin\frac{\pi x}{l}.$$

Here q_0 is the initial rise parameter, p_0 represents the static loading, and ρ and ν represent the amplitude and frequency of the harmonic excitation. In equation

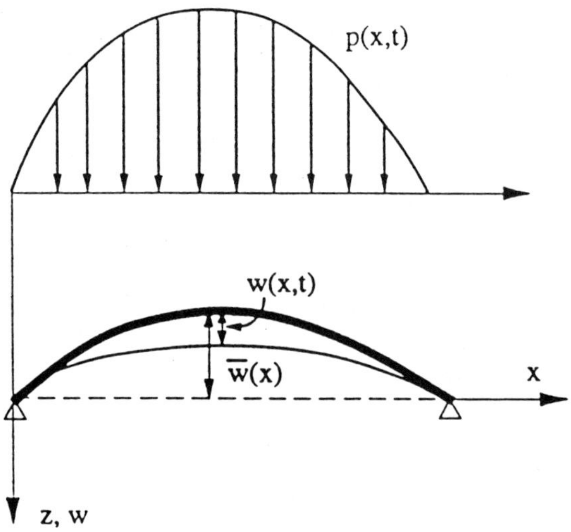

Figure 3 Geometry of the shallow arch structure subjected to lateral loading $p(x,t)$.

(3.1), $\bar{w}(x)$ represents the initial deflection of the unloaded arch. The arch is subjected to the following boundary conditions:

$$w = 0, \qquad w_{xx} = 0 \qquad \text{at } x = 0, l.$$

Now we use the Galerkin method to reduce the equation of motion to a set of ordinary differential equations by selecting appropriate shape functions. The transverse motion $w(x,t)$ of the arch is approximated by the following expression:

$$w(x,t) = q_1(t)\sin\frac{\pi x}{l} + q_2(t)\sin\frac{2\pi x}{l}. \tag{3.2}$$

The non-dimensional equations of motion describing the evolution of the amplitudes of two fundamental modes can be written in $(Q_1 = q_1 - q_0, Q_2 = q_2)$ coordinates as:

$$\begin{aligned} \ddot{Q}_1 + \beta_1\dot{Q}_1 + Q_1 + Q_1(Q_1^2 - q_0^2 + 4Q_2^2) + (q_0 - \lambda_0) &= \rho\sin\nu t \\ \ddot{Q}_2 + \beta_2\dot{Q}_2 + 16Q_2 + 4Q_2(Q_1^2 - q_0^2 + 4Q_2^2) &= 0 \end{aligned} \tag{3.3}$$

where β_1 and β_2 are the viscous damping parameters for the first and second mode respectively, and λ_0 represents the non-dimensional loading parameter. The details of this derivation and the definitions of non-dimensional variables and parameters can be found in Tien [1992]. Now we examine this system in the presence of 1:1 internal resonance and sub-harmonic parametric resonance.

Here we use the results of Kovačič and Wiggins [1992] to study the global dynamics of the two degree of freedom arch system described by (3.3). The objective is to find homoclinic or heteroclinic orbits to fixed points that are created in resonance as a result of the perturbation. The method developed by Kovačič and Wiggins [1992], combines the higher dimensional Melnikov Theory, geometrical singular perturbation theory and the theory of foliations of invariant manifolds to obtain necessary conditions for the occurrence of Silnikov type homoclinic orbits.

In order to investigate the dynamics near the equilibrium state we introduce the following scaling in the system parameters:

$$\mu = \epsilon\ \gamma,\ \beta_1 = \epsilon\ \delta_1,\ \beta_2 = \epsilon\ \delta_2 \tag{3.4}$$

where ϵ is a small positive parameter i.e., $0 < \epsilon << 1$. We observe that the system (3.3) permits $(Q_1(t) = \eta(t), Q_2(t) = 0)$ as one of the solutions. We assume the following asymptotic expansion of $\eta(t)$ in the powers of ϵ,

$$\eta(t) = \eta_0 + \sqrt{\epsilon}\ \eta_1 + \epsilon\ \eta_2 + O(\epsilon^{3/2}), \tag{3.5}$$

where η_0 is governed by a third order algebraic equation and η_1, η_2 etc. can be explicitly obtained in terms of known parameters and time:

$$\begin{aligned} &\eta_0^3 + (1 - q_0^2)\eta_0 + (q_0 - \lambda_0) = 0 \\ &\eta_1 = 0 \\ &\eta_2 = b_1 \cos(\nu t) \end{aligned} \tag{3.6}$$

where

$$b_1 = \frac{\gamma}{\omega_1^2 - \nu^2}. \tag{3.7}$$

We now consider the perturbations x_1 and x_2 in the neighborhood of the state $(Q_1(t) = \eta(t), Q_2(t) = 0)$ of (3.3) in the following fashion:

$$Q_1 = \eta(t) + \sqrt{\epsilon}\ x_1, \qquad Q_2 = 0 + \sqrt{\epsilon}\ x_2\ . \tag{3.8}$$

In this case, the nonlinear perturbation equations governing x_1 and x_2 are given as:

$$\begin{aligned} &\ddot{x}_1 + \omega_1^2 x_1 + \sqrt{\epsilon}\ \left[3\eta_0 x_1^2 + 4\eta_0 x_2^2\right] + \epsilon\left[\delta_1\dot{x}_1 + 6\eta_0\eta_2 x_1 + x_1^3 + 4x_1x_2^2\right] + O(\epsilon^{3/2}) = 0 \\ &\ddot{x}_2 + \omega_2^2 x_2 + \sqrt{\epsilon}\ \left[8\eta_0 x_1 x_2\right] + \epsilon\left[\delta_2\dot{x}_2 + 8\eta_0\eta_2 x_2 + 4x_1^2x_2 + 16x_2^3\right] + O(\epsilon^{3/2}) = 0 \end{aligned} \tag{3.9}$$

where $\omega_1^2 = 3\eta_0^2 + (1 - q_0^2)$ and $\omega_2^2 = 4[\eta_0^2 + (4 - q_0^2)]$. For the corresponding undamped system, the Hamiltonian function can be expressed as:

$$\begin{aligned} H(\mathbf{x}_p, \mathbf{x}_q) &= H_0 + \sqrt{\epsilon}H_1 + \epsilon H_2 + O(\epsilon^{3/2}) \\ &= \frac{1}{2}\ [x_{p_1}^2 +\ x_{p_2}^2 + \omega_1^2\ x_{q_1}^2 + \omega_2^2\ x_{q_2}^2\ ] \\ &\quad + \sqrt{\epsilon}\ [\eta_0\ x_{q_1}^3 + 4\eta_0\ x_{q_1}x_{q_2}^2\ ] \\ &\quad + \epsilon\ [\frac{1}{4}x_{q_1}^4 + 4x_{q_2}^4 + 2x_{q_1}^2\ x_{q_2}^2 + \eta_0\ \eta_2(3x_{q_1}^2 + 4x_{q_2}^2)\ ] + O(\epsilon^{3/2}) \end{aligned} \tag{3.10}$$

where $x_{q_1} = x_1$, $x_{q_2} = x_2$, $x_{p_1} = \dot{x}_1$ and $x_{p_2} = \dot{x}_2$. The steady state behavior and bifurcation analysis of the system is more easily described by introducing the following canonical change of variables:

$$x_{q_i} = \sqrt{2a_i}\ \cos(\omega_i t + \frac{\phi_i}{\omega_i}), \qquad x_{p_i} = -\omega_i\ \sqrt{2a_i}\ \sin(\omega_i t + \frac{\phi_i}{\omega_i}), \qquad \forall\, i = 1, 2. \tag{3.11}$$

One can check that (3.10) satisfies the sufficient conditions for the canonical transformations. We also introduce the following detuning parameters in order to examine the dynamics of the system in the vicinity of external sub-harmonic resonance and 1:1 internal resonance.

$$\omega_1^2 = \frac{1}{4}(\nu^2 - \epsilon\sigma_1), \qquad \omega_2^2 = \frac{1}{4}(\nu^2 - \epsilon\sigma_2)\ , \tag{3.12}$$

where σ_1 and σ_2 are the deviations of the excitation frequencies from twice the natural frequency of the first and second modes, respectively. In other words, σ_1 represents the measure of deviation from subharmonic resonance with respect to the first natural frequency, while $(\sigma_2 - \sigma_1)$ represents the deviation from 1:1 internal resonance. At this stage, we apply second order averaging (Bogoliubov and Mitropolsky [1961]) to obtain the following system of equations in canonical variables (a_i, ϕ_i):

$$\begin{aligned}
\dot{a}_1 &= \epsilon\big[-\delta_1 a_1 - 2k_3 a_1 a_2 \sin 2(\theta_1 - \theta_2) + \frac{6}{\nu} a_1 b_1 \eta_0 \sin 2\theta_1\ \big] + O(\epsilon^2) \\
a_1\dot{\theta}_1 &= \epsilon\big[-\frac{\sigma_1}{4\nu} a_1 - 2k_1 a_1^2 - (k_2 + k_4) a_1 a_2 - k_3 a_1 a_2 \cos 2(\theta_1 - \theta_2) \\
&\quad + \frac{3}{\nu} a_1 b_1 \eta_0 \cos 2\theta_1\ \big] + O(\epsilon^2) \\
\dot{a}_2 &= \epsilon\big[-\delta_2 a_2 + 2k_3 a_1 a_2 \sin 2(\theta_1 - \theta_2) + \frac{8}{\nu} a_2 b_1 \eta_0 \sin 2\theta_2\ \big] + O(\epsilon^2) \\
a_2\dot{\theta}_2 &= \epsilon\big[-\frac{\sigma_2}{4\nu} a_2 - 2k_5 a_2^2 - (k_2 + k_4) a_1 a_2 - k_3 a_1 a_2 \cos 2(\theta_1 - \theta_2) \\
&\quad + \frac{4}{\nu} a_2 b_1 \eta_0 \cos 2\theta_2\ \big] + O(\epsilon^2)
\end{aligned} \tag{3.13}$$

where

$$k_1 = \frac{15}{4}\frac{\eta_0^2}{\omega_1^3} - \frac{3}{4\nu}, \qquad k_2 = \frac{13}{3}\frac{\eta_0^2}{\omega_1^3} - \frac{1}{\nu}, \qquad k_3 = \frac{14\ \eta_0^2}{\omega_1^3} - \frac{2}{\nu},$$

$$k_4 = \frac{55}{3}\frac{\eta_0^2}{\omega_1^3} - \frac{3}{\nu}, \qquad k_5 = \frac{20}{3}\frac{\eta_0^2}{\omega_1^3} - \frac{12}{\nu}, \qquad \theta_i = \frac{\phi_i}{\omega_i} \quad \forall\ i = 1, 2.$$

In order to analyze the trivial equilibrium solutions, we transform the averaged equations to rectangular coordinates. We find that the trivial solution $(a_1 = 0, a_2 = 0)$ may lose its stability only through a simple bifurcation and may give rise to one or two coupled mode nontrivial equilibrium branches. The single mode steady state solutions do not exist for this system. For a detailed account of the stability and bifurcation analysis, one is referred to Malhotra and Namachchivaya [1994].

Consider the following canonical transformation to bring the averaged system to the appropriate form in order to apply the perturbation method given by Kovačič and Wiggins [1992]:

$$p_1 = a_1, \qquad p_2 = (a_1 + a_2), \qquad q_1 = (\theta_1 - \theta_2), \qquad q_2 = \theta_2. \tag{3.14}$$

This transformation leads to the following set of equations in (p_1, q_1, p_2, q_2) coordinates:

$$
\begin{aligned}
\dot{p}_1 &= \epsilon\big[-\delta_1 p_1 - 2k_3 p_1(p_2 - p_1)\sin 2q_1 + \frac{6}{\nu} b_1 p_1 \eta_0 \sin 2(q_1 + q_2)\ \big] + O(\epsilon^2) \\
\dot{q}_1 &= \epsilon\big[\frac{(\sigma_2 - \sigma_1)}{4\nu} + (p_2 - p_1)(-k_2 - k_4 + 2k_5 - k_3\cos 2q_1) \\
&\quad + p_1(-2k_1 + k_2 + k_4 + k_3\cos 2q_1) - \frac{4}{\nu} b_1 \eta_0 \cos 2q_2 \\
&\quad + \frac{3}{\nu} b_1 \eta_0 \cos 2(q_1 + q_2)\ \big] + O(\epsilon^2) \\
\dot{p}_2 &= \epsilon\big[-\delta_2 p_2 + (\delta_2 - \delta_1)p_1 + \frac{8}{\nu} b_1 \eta_0 (p_2 - p_1)\sin 2q_2 \\
&\quad + \frac{6}{\nu} b_1 p_1 \eta_0 \sin 2(q_1 + q_2)\ \big] + O(\epsilon^2) \\
\dot{q}_2 &= \epsilon\big[-\frac{\sigma_2}{4\nu} - 2k_5(p_2 - p_1) - p_1\{(k_2 + k_4) + k_3\cos 2q_1 \\
&\quad + \frac{4}{\nu} b_1 \eta_0 \cos 2q_2\ \big] + O(\epsilon^2).
\end{aligned}
\tag{3.15}
$$

We want to study the global dynamics associated with this system. This system attains the desired form if we assume the damping and forcing terms to be small as compared with other terms. For this purpose, we re-scale the variables and parameters in the following manner:

$$
\begin{aligned}
&\delta_{1,2} = \epsilon^2 \delta'_{1,2}\ , \qquad \sigma_{1,2} = \epsilon 4\nu\sigma'_{1,2}\ , \qquad b_1 = \epsilon^2 b'_1\ , \\
&p_{1,2} = \epsilon \frac{p'_{1,2}}{|k_3|}\ , \qquad q_{1,2} = (1 + \operatorname{sgn}(k_3))\frac{\pi}{4} + q'_{1,2}\ , \qquad t = \frac{t'}{\epsilon^2}.
\end{aligned}
$$

As we recall, b_1 is related to the amplitude of the forcing. On dropping the prime ($'$) from the rescaled variables and parameters, the rescaled equations, up to $O(\epsilon)$, can be expressed as:

$$
\begin{aligned}
\dot{p}_1 &= 2\ p_1\ (p_2 - p_1)\sin 2q_1 + \epsilon\big[-\delta_1 p_1 + 6 f_1 p_1 \sin 2(q_1 + q_2)\ \big] \\
\dot{q}_1 &= (\sigma_2 - \sigma_1) + (2p_1 - p_2)(\alpha - \cos 2q_1) - \gamma_1 p_2 \\
&\quad + \epsilon\big[3 f_1\ \cos 2(q_1 + q_2) - 4 f_2 \cos 2q_2\ \big] \\
\dot{p}_2 &= \epsilon\big[-\delta_2 p_2 + (\delta_2 - \delta_1)p_1 + 6 f_1 p_1 \sin 2(q_1 + q_2) + 8 f_2 (p_2 - p_1)\sin 2q_2\ \big] \\
\dot{q}_2 &= -\sigma_2 - (\gamma_1 + \alpha - \cos 2q_1)p_1 - \gamma_2 p_2 + \epsilon\big[4 f_2 \cos 2q_2\ \big]
\end{aligned}
\tag{3.16}
$$

where

$$
\alpha = \frac{k_2 + k_4 - k_1 - k_5}{|k_3|}, \qquad \gamma_1 = \frac{k_1 - k_5}{|k_3|}, \qquad \gamma_2 = \frac{2k_5}{|k_3|},
$$

$$
f_1 = \frac{b_1 \eta_0}{\nu}, \qquad f_2 = -\frac{b_1 \eta_0}{\nu}\operatorname{sgn}(k_3).
$$

Systems similar to (3.16) have been studied by Feng and Sethna [1990] for the nonresonant case and by Feng and Wiggins [1992] for the resonant case. As a first step, we would like to examine the global structure of the unperturbed system, letting $\epsilon = 0$:

$$\begin{aligned}
\dot{p}_1 &= 2\, p_1\, (p_2 - p_1) \sin 2q_1 \\
\dot{q}_1 &= (\sigma_2 - \sigma_1) + (2p_1 - p_2)(\alpha - \cos 2q_1) - \gamma_1 p_2 \\
\dot{p}_2 &= 0 \\
\dot{q}_2 &= -\sigma_2 - (\gamma_1 + \alpha - \cos 2q_1)p_1 - \gamma_2 p_2.
\end{aligned} \tag{3.17}$$

The Hamiltonian corresponding to the four-dimensional unperturbed system (3.17) is given as:

$$H_0(p_1, q_1, p_2, q_2) = (\sigma_2 - \sigma_1)\, p_1 - \sigma_2 p_2 - \gamma_1 p_1 p_2 - \frac{1}{2}\gamma_2 p_2^2 - p_1\, (p_2 - p_1)(\alpha - \cos 2q_1). \tag{3.18}$$

We notice that the first two equations are completely independent of the angle variable q_2, and $p_2 =$ constant, and thus we need to study the phase flow in (p_1, q_1) coordinates only. The canonical transformations and the periodicity of q_1 suggests that the range of interest in (p_1, q_1) coordinates is: $0 < p_1 \le p_2$ and $0 \le q_1 \le \pi$. We notice that the coordinate system (p_1, q_1, p_2, q_2) is singular at $p_1 = 0$ and $p_2 = 0$. The fixed points of the two-dimensional vector field in our range of interest are:

$$\begin{array}{lll}
a. & p_1 = p_2, & q_1 = \hat{q}_1 \\
b. & p_1 = p_2, & q_1 = \pi - \hat{q}_1 \\
c. & p_1 = p_1^+, & q_1 = \dfrac{\pi}{2} \\
d. & p_1 = p_1^-, & q_1 = 0 \\
e. & p_1 = p_1^-, & q_1 = \pi
\end{array}$$

where

$$\begin{aligned}
\hat{q}_1 &= \frac{1}{2}\cos^{-1}\left[\alpha - \gamma_1 + \frac{\sigma}{p_2}\right] \\
p_1^+ &= \frac{1}{2}\left[p_2 + \frac{\gamma_1 p_2 - \sigma}{\alpha + 1}\right] \\
p_1^- &= \frac{1}{2}\left[p_2 + \frac{\gamma_1 p_2 - \sigma}{\alpha - 1}\right] \\
\sigma &= (\sigma_2 - \sigma_1).
\end{aligned}$$

The linear stability analysis shows that the two-dimensional phase flow (p_1, q_1) has the desired features in the following range of p_2:

$$p_2^1 = \frac{\sigma}{\gamma_1 - \alpha + 1} \le p_2 \le \frac{\sigma}{\gamma_1 - \alpha - 1} = p_2^2. \tag{3.19}$$

The phase flow of (p_1, q_1) is shown in Figure 4 for the above mentioned range of p_2. This kind of phase flow occurs when $\alpha > 1$, which corresponds to $q_0^2 > 5.57$, where q_0 is the initial rise parameter of the arch. In this range of p_2, $(p_2, \hat{q}_1)$ and $(p_2, \pi - \hat{q}_1)$ exist as *saddle points*, while $(p_1^+, \frac{\pi}{2})$, $(p_1^-, 0)$ and (p_1^-, π) exist as the *center points*. The two saddle points are connected by the heteroclinic orbits A and A′ as shown in Figure 4. The system of equations describing this phase flow is integrable, and thus we can obtain the time parameterized expressions for these heteroclinic orbits. Orbit A is described by:

$$p_1 = \frac{\gamma_1 p_2 - \sigma}{\alpha - \cos 2q_1}. \tag{3.20}$$

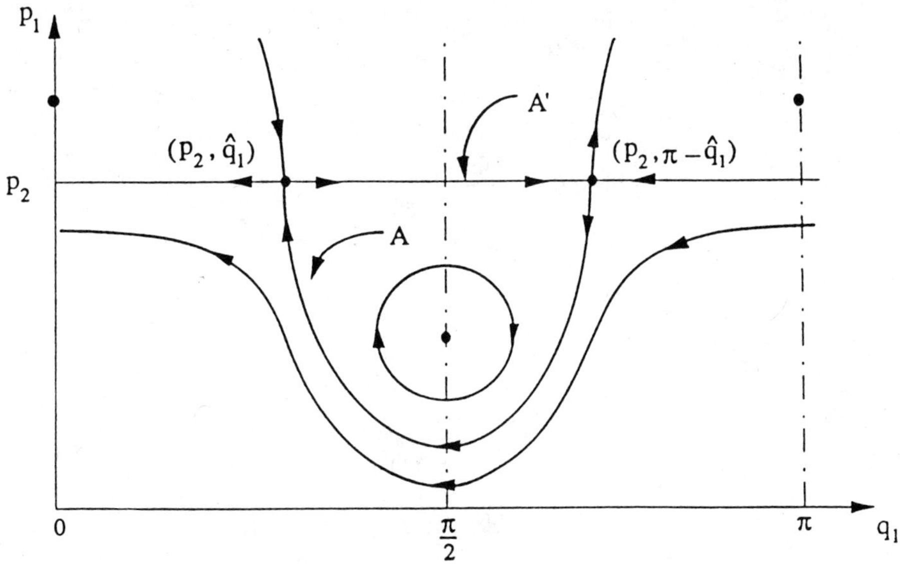

Figure 4 Unperturbed phase flow in (p_1, q_1) phase space for 1:1 resonance case.

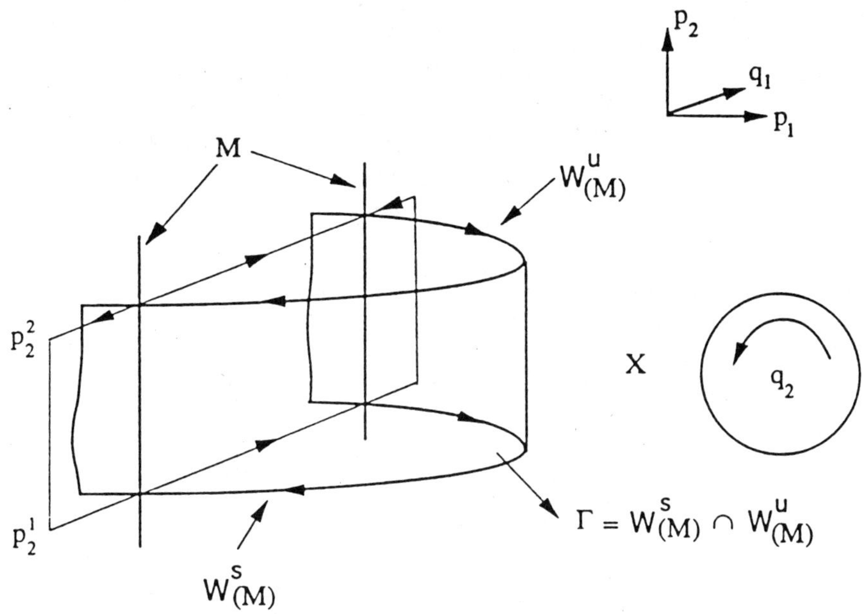

Figure 5 Geometry of the invariant manifold M and its stable and unstable manifolds.

Substituting this in the unperturbed system of equations (3.17) and integrating, we obtain the following expressions:

$$\begin{aligned}
p_1(t) &= p_2 \frac{\cosh e_A t - \cos 2\hat{q}_1}{\cosh e_A t + \cos f_A} \\
q_1(t) &= \frac{1}{2}\ \tan^{-1}\left[\frac{\sin 2\hat{q}_1 \sinh e_A t}{\cos 2\hat{q}_1\ \cosh e_A t - 1}\right] \\
p_2(t) &= p_2 \\
q_2(t) &= \omega_A t - v_A\ \tan^{-1}\left[\tan \frac{f_A}{2}\ \tanh \frac{e_A t}{2}\right] + q_{20}
\end{aligned} \tag{3.21}$$

where

$$\begin{aligned}
e_A &= 2p_2 \sin 2\hat{q}_1 \\
\cos f_A &= \frac{1 - \alpha \cos 2\hat{q}_1}{\alpha - \cos 2\hat{q}_1} < 1 \\
v_A &= 2\gamma_1 \frac{\cos f_A + \cos 2\hat{q}_1}{\sin 2\hat{q}_1 \sin f_A} \\
\omega_A &= -(\gamma_1 + \gamma_2)p_2 - \sigma_1.
\end{aligned} \tag{3.22}$$

Here, we have used $q_1(0) = \frac{\pi}{2}$ and $q_2(0) = q_{20}$. The heteroclinic orbit A$'$ is described by the following expressions:

$$\begin{aligned}
p_1(t) &= p_2 \\
q_1(t) + q_2(t) &= \omega_A t + \frac{\pi}{2} + q_{20}.
\end{aligned} \tag{3.23}$$

As we shall see later, we do not require the explicit expressions for $q_1(t)$ and $q_2(t)$ for orbit A$'$.

We now turn our attention to the existence of the normally hyperbolic invariant manifold M and its associated stable and unstable manifolds. In full four-dimensional phase space, M is a two-dimensional invariant manifold which is given as:

$$M = \{(p_1, q_1, p_2, q_2) \mid p_1 = p_2,\ q_1 = \hat{q}_1,\ p_2^1 < p_2 < p_2^2,\ 0 \le q_2 \le \pi\}. \tag{3.24}$$

The invariant manifold M has a three-dimensional stable $W^s(M)$ and a three-dimensional unstable manifold $W^u(M)$. These manifolds intersect non-transversely along a three-dimensional heteroclinic manifold Γ, which is given as:

$$\begin{aligned}
\Gamma = \{(p_1, q_1, p_2, q_2) \mid p_1 = p_1^h(t, p_2),\ q_1 = q_1^h(t, p_2),\ p_2^1 < p_2 < p_2^2, \\
q_2 = \int_0^t D_{p_2} H(p_1^h(t, p_2), q_1^h(t, p_2), p_2)\ ds + q_{20}\}
\end{aligned} \tag{3.25}$$

where $(p_1^h(t, p_2), q_1^h(t, p_2))$ is any point on the heteroclinic cycle. The dynamics restricted to the invariant manifold M is given by:

$$\begin{aligned}
\dot{p}_2 &= 0 \\
\dot{q}_2 &= -\sigma_1 - p_2(2\gamma_1 + \gamma_2).
\end{aligned} \tag{3.26}$$

We can see that $\dot{q}_2 = 0$ for $p_2 = p_2^r = \frac{-\sigma_1}{(2\gamma_1+\gamma_2)}$. This is called the resonant value of p_2, and it corresponds to the circle of fixed points in $p_2 - q_2$ plane. At $p_2 = p_2^r$, define a phase shift Δq_2 such that:

$$\Delta q_2 = q_2(+\infty, q_2^r) - q_2(-\infty, q_2^r). \tag{3.27}$$

At resonance, $\omega_A = 0$ and hence, we can show $\Delta q_2 = v_A \; f_A \text{sgn}\,(\sin 2\hat{q}_1)$. Kovačič and Wiggins [1992] show that the invariant manifold M and its stable and unstable manifolds persist under sufficiently small perturbations, but the dynamics within these manifolds generally changes dramatically. Following their analysis, we introduce the following transformation in order to study the dynamics on M_ϵ near $p_2 = p_2^r$:

$$p_2 = p_2^r + \sqrt{\epsilon}\, h, \qquad q_2 = q_2, \qquad t = \frac{\tau}{\sqrt{\epsilon}}$$

where τ is the slow time. We obtain:

$$\begin{aligned} h' &= -\delta_1 p_2^r + 6 f_1 p_2^r \sin 2(\hat{q}_1 + q_2) + \sqrt{\epsilon}\, [-\delta_1 h + 6 f_1 h \sin 2(\hat{q}_1 + q_2)\,] \\ q_2' &= -h(2\gamma_1 + \gamma_2) + \sqrt{\epsilon}\, [4 f_2 \cos 2 q_2\,]. \end{aligned} \tag{3.28}$$

The prime indicates the differentiation with respect to τ. We notice that the unperturbed system in $h - q_2$ phase space is integrable, with the following Hamiltonian:

$$\hat{H}(h, q_2) = -(2\gamma_1 + \gamma_2)\frac{h^2}{2} - \int_{q_{20}}^{q_2} [-\delta_1 + 6 f_1 \sin 2(\hat{q}_1 + s)] p_2^r \, ds. \tag{3.29}$$

The fixed points of the unperturbed vector field are:

$$\begin{aligned} p_0 &= (0, q_{2c}) = (0, \frac{1}{2}\sin^{-1}\frac{\delta_1}{6 f_1} - \hat{q}_1) \\ q_0 &= (0, q_{2s}) = (0, \frac{\pi}{2} - \frac{1}{2}\sin^{-1}\frac{\delta_1}{6 f_1} - \hat{q}_1). \end{aligned} \tag{3.30}$$

The linear stability analysis confirms that p_0 is a center type fixed point while q_0 is a saddle connected to itself by a homoclinic orbit as shown in Figure 6a. Under small perturbations, q_0 remains a hyperbolic fixed point of the same stability type, while p_0 becomes a hyperbolic sink because one can show that the leading order term of the trace of the linearization of the perturbed system (3.28) is less than zero inside the homoclinic orbit connecting q_0. In this case, the homoclinic orbit breaks and the unstable branch of the perturbed saddle point is attracted to the hyperbolic sink (Figure 6b). Let us denote these points as p_ϵ and q_ϵ, as indicated in the Figure 6b. In the perturbed system, these dynamical features are confined in an annulus A_ϵ, a region of width $O(\epsilon)$ near $p_2 = p_2^r$. We are interested in the existence of a simple homoclinic connection to p_ϵ. In other words, we would like to find out whether a trajectory in $W^u_{p_\epsilon}$ comes back into the neighborhood of A_ϵ and approaches p_ϵ as $t \to \infty$. This is accomplished in two steps; in the first step the higher dimensional Melnikov theory (Wiggins [1988]) is used to show that $W^u_{p_\epsilon} \subset W^s_{A_\epsilon}$. The second step involves the use of singular perturbation theory to show that $W^u_{p_\epsilon}$ actually asymptotes to p_ϵ. The Melnikov theory is used to find whether $W^u_{p_\epsilon}$ intersects $W^s_{A_\epsilon}$. The Melnikov Function $M(\sigma_1, \sigma_2, \delta_1, \delta_2,{}^r_2, q_{20})$ gives a perturbative measure of the distance between $W^u_{p_\epsilon}$ and $W^s_{A_\epsilon}$ and it is computed in the following manner:

$$M(\sigma_1, \sigma_2, \delta_1, \delta_2, p_2^r, q_{20}) = \int_{-\infty}^{\infty} [\frac{\partial H}{\partial p_1}\, g^{p_1} + \frac{\partial H}{\partial q_1}\, g^{q_1} + \frac{\partial H}{\partial p_2}\, g^{p_2}\,] dt \tag{3.31}$$

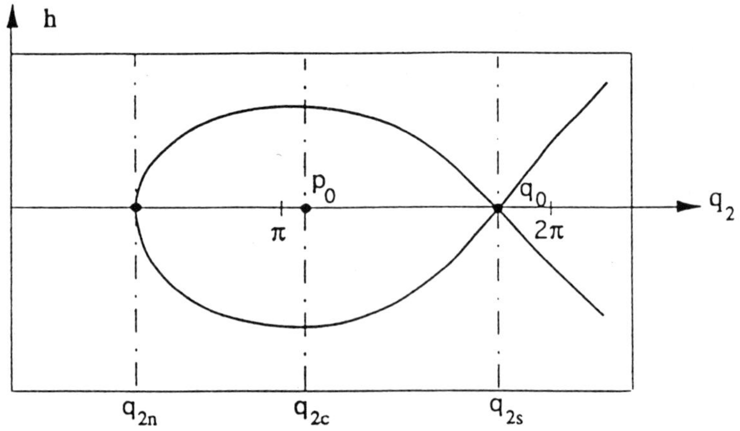

Figure 6a The dynamics on M in the neighbourhood of p_2^r (i.e., $h = 0$).

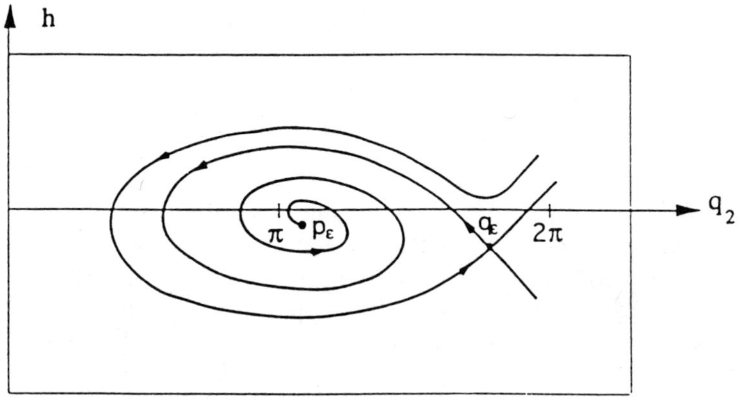

Figure 6b The dynamics on M_ε in the neighbourhood of p_2^r (i.e., $h = 0$).

where the integrand has been evaluated at any arbitrary point on the heteroclinic cycle and $p_2 = p_2^r$ and $q_2 = q_{2c}$ and $g^{p_1}, g^{q_1}, g^{p_2}$ and g^{q_2} are the $O(\epsilon)$ terms in the four-dimensional perturbed system of equations (3.16). The Hamiltonian H, corresponding to the four-dimensional perturbed system (3.16), is given as:

$$\begin{aligned} H(p_1, q_1, p_2, q_2) &= H_0 + \epsilon H_1 \\ &= (\sigma_2 - \sigma_1)\, p_1 - \sigma_2 p_2 - \gamma_1 p_1 p_2 - \frac{1}{2}\gamma_2 p_2^2 \\ &\quad - p_1\, (p_2 - p_1)(\alpha - \cos 2q_1) \\ &\quad +\ \epsilon\, [3 f_1 p_1 \cos 2(q_1 + q_2) + 4 f_2 (p_2 - p_1) \cos 2q_2\,]. \end{aligned} \tag{3.32}$$

Following a procedure described in Malhotra and Namachchivaya [1994] the Melnikov Function is obtained as:

$$M(\sigma_1, \sigma_2, \delta_1, \delta_2, p_2^r, q_{20}) = N_1 + N_2\ \sin(2q_{20}), \tag{3.33}$$

where:

$$\begin{aligned} q_{20} &= q_{2c} + \frac{1}{2} v_A \cdot f_A \cdot \operatorname{sgn}(\sin 2\hat{q}_1) \\ N_1(\sigma_1, \sigma_2, \delta_1, \delta_2, p_2^r) &= -3 f_1 p_2^r\, [\cos 2(q_{2c} + \Delta q_2 + \operatorname{sgn}(e_A)\hat{q}_1) \\ &\qquad - \cos 2(q_{2c} - \operatorname{sgn}(e_A)\hat{q}_1)\,] \\ &\quad + \frac{2\delta_1}{\sqrt{1-\alpha^2}}(\gamma_1 p_2^r + \sigma_1 - \sigma_2) \tan^{-1} \sqrt{1-\alpha^2}\ \tan \hat{q}_1 \operatorname{sgn}(e_A) \\ &\quad - \delta_2 p_2^r \Delta q_2 + (\delta_2 - \delta_1)(\gamma_1 p_2^r + \sigma_1 - \sigma_2) \int_{-\infty}^{\infty} \frac{\gamma_1 (p_2^r - p_1)}{\alpha - \cos 2q_1(t)}\, dt \\ N_2(\sigma_1, \sigma_2, \delta_1, \delta_2, p_2^r) &= 8 f_2 \int_{-\infty}^{\infty} \frac{\gamma_1 (p_2^r - p_1)^2 \cos 2q_2}{\alpha - \cos 2q_1(t)}\, dt \\ &\quad + 6 f_1 (\gamma_1 p_2^r + \sigma_1 - \sigma_2) \int_{-\infty}^{\infty} \frac{\gamma_1 (p_2^r - p_1) \cos 2(q_1 + q_2)}{\alpha - \cos 2q_1(t)} dt. \end{aligned} \tag{3.34}$$

Thus, the Melnikov Function will have a simple zero (Malhotra and Namachchivaya [1994]) if:

$$N_1 + N_2 \cdot\ \sin\ 2q_{20} = 0 \tag{3.35}$$

This condition implies that $W^u(p_\epsilon) \subset W^s(A_\epsilon)$ and hence it can be shown that $W^u(p_\epsilon)$ comes back in the neighborhood of the annulus A_ϵ. We have yet to show that $W^u(p_\epsilon)$ indeed returns to the annulus and falls in the domain of attraction of p_ϵ, which can be approximated by the area enclosed by the homoclinic orbit connecting q_0. This can be shown if:

$$q_{2n} < (q_{2c} + \Delta q_2) < q_{2s} \tag{3.36}$$

where q_{2s} and q_{2c} are given by (3.30) and q_{2n} is obtained by solving the following equation:

$$\hat{H}(0, q_{2n}) - \hat{H}(0, q_{2s}) = 0$$

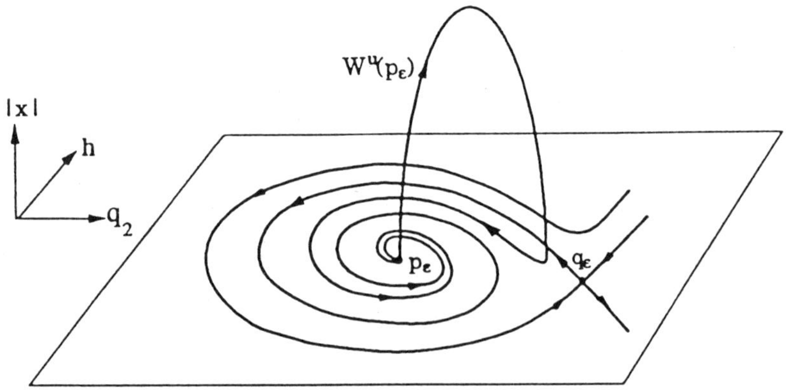

Figure 7 Orbit homoclinic to p_ε.

which leads to the following transcendental equation:

$$\delta_1(q_{2n} + \hat{q}_1 - \frac{\pi}{2} + \frac{1}{2}\sin^{-1}\frac{\delta_1}{6f_1}) \; + \; 3f_1 \; [\cos 2(\hat{q}_1 + q_{2n}) \\ - \cos 2(\frac{\pi}{2} - \frac{1}{2}\sin^{-1}\frac{\delta_1}{6f_1}) \;] = 0. \tag{3.37}$$

For the system parameters $(\sigma_1, \sigma_2, \delta_1, \delta_2, p_2^r, q_{20})$ satisfying (3.35) and (3.36) simultaneously, there may exist a simple homoclinic orbit connecting p_ϵ to itself (Figure 7). From a theorem by Silnikov, this mechanism may give rise to chaotic behavior in the sense of the Smale Horseshoe provided certain conditions on the stability properties of the fixed point are satisfied. In our case, this condition implies that the eigenvalues in $p_1 - q_1$ plane should not be identical, as the other two eigenvalues are of $O(\epsilon)$. The Melnikov Function corresponding to the other segment of the heteroclinic cycle (i.e., A′) is found to be identically zero for all parameter values, and hence no Silnikov type orbits can occur for A′.

References

Bogoliubov, N. and Mitropolsky, Y.A. [1961] *Asymptotical Methods in the Theory of Nonlinear Oscillations*, Gordon and Breach, New York.

Feng, Z. and Sethna, P.R. [1990] *Global bifurcation and chaos in parametrically forced systems with one-one resonance* Dynamics and Stability of Systems **5**, 201–225.

_____ *Global bifurcations in the Motion of Parametrically Excited Thin Plates*, To appear in Nonlinear Dynamics.

Feng, Z. and Wiggins, S. [1993] *On the existence of chaos in a class of two-degree-of-freedom, damped, strongly parametrically forced mechanical systems with broken $O(2)$ symmetry* Jour. Appl. Math. Phys. (ZAMP) **44**, 1–48.

Fu, F.C.L. and Nemat-Nasser, S. [1972] *On the Stability of Steady-State Response of Certain Nonlinear Dynamic Systems Subjected to Harmonic Excitations*

Ingeneur-Archiv **41**, 407-420.

Guckenheimer, J. and Holmes, P. [1983] *Dynamical Systems and Bifurcations of Vector Fields*, Springer-Verlag, New York.

Hsu, C.S. [1969] *Stability of shallow arches against snap-through under time wise step loads* AIAA J. **35(1)**, 31–39.

Humphrey, J.S. [1966] *On dynamic snap buckling of shallow arches* AIAA J. **4(5)**, 878–886.

Lock, M.H. [1965] *The snapping of a shallow sinusoidal arch under a step pressure load*, Air force Systems Command, Los Angeles (CA) Report No. TRD-469 (5240-10-16).

Kovačič, G. and Wiggins, S. [1992] *Orbits homoclinic to resonances, with an application to chaos in a model of the forced and damped sine-Gordon equation* Physica D **57**, 185–225.

Malhotra, N. and Namachchivaya, N. Sri [1994] *Global Dynamics of Shallow Arch Structures*, in preparation.

Melnikov, M.H. [1963] *On the stability of the center for time periodic perturbations* Trans. Mosc. Math. Soc. **12(1)**, 1–57.

Mettler, E. [1962] *Dynamic Buckling*, Handbook of Engineering Mechanics (W. Flugge, ed.), McGraw Hill, New York.

Namachchivaya, N. Sri and Doyle, M.M. [1988] *Chaotic Motion of a shallow arch*, Proceedings of 29th Structures, Structural Dynamics and Materials Conference, pp. 198–209.

Namachchivaya, N. Sri, Doyle, M.M., Langford, W.F. and Evans, N. [1994] *Normal Form for generalized Hopf Bifurcation with non-semisimple 1:1 resonance*, To appear in Jour. Appl. Math. Phys. (ZAMP), **45**, 312-335.

Namachchivaya, N. Sri and Malhotra, N. *Parametrically Excited Hopf Bifurcation with Nonsemisimple 1:1 Resonance*, Proceedings of the ASME meeting on Nonlinear Vibrations, Anaheim, California, November 1992.

O'Reilly, O.M., Malhotra, N. and Namachchivaya, N. Sri [1994] *Destabilization of Equilibria of Reversible Dynamical Systems*, Submitted to Nonlinear Dynamics.

Tien, W.M. [1992] *Chaotic and Stochastic dynamics of structural systems*, Ph.D. Thesis, Univ. of Illinois, Urbana-Champaign.

Tien, W.M., Namachchivaya, N. Sri and Malhotra, N. [1994] *Nonlinear dynamics of a shallow arch under periodic excitation, part II: 1:1 internal resonance*, To appear in Int. Jour. of Nonlin. Mech.

van Gils, S.A. Krupa, M. and Langford, W.F. [1990] *Hopf Bifurcation with non-semisimple 1:1 resonance* Nonlinearity **3**, 825–850.

Wiggins, S. [1988] *Global Bifurcations and Chaos*, Springer-Verlag, New York.

——— [1990] *Introduction to Applied Nonlinear Dynamical Systems and Chaos*, Springer-Verlag, New York.

Fields Institute Communications
Volume **4**, 1995

A General Reduction Method for Periodic Solutions Near Equilibria in Hamiltonian Systems

André Vanderbauwhede
Department of Pure Mathematics and Computer Algebra
University of Gent
Krijgslaan 281, B-9000 Gent, Belgium

Jan-Cees van der Meer
Department of Mathematics and Informatics
Technical University of Eindhoven
P.O. Box 513, 5600 MB, Eindhoven, The Netherlands

Abstract. We introduce a new and general reduction method for the study of periodic solutions near equilibria in systems of ordinary differential equations, with a period near a given period T_0. The method combines elements from the normal form approach with the Liapunov-Schmidt method, and reduces the existence problem for periodic solutions of the original system to the same problem for a reduced equation, whose phase space is spanned by the generalized eigenspaces corresponding to the purely imaginary eigenvalues of the linearization at the equilibrium which are in resonance with T_0. The reduced system inherits such properties as equivariance, reversibility, or being Hamiltonian from the original system, but also has an additional S^1-symmetry. Moreover: (1) the periodic solutions of the reduced system have a particular form which immediately leads to the bifurcation equations, and (2) the reduced system can be calculated up to any order by a normal form reduction of the original system. The details of the reduction and the proofs are worked out for the case of Hamiltonian systems, but can easily be adapted to other types of systems.

1 Introduction

Consider a parameter-dependent autonomous system

$$\dot{x} = f(x, \lambda) \tag{1.1}$$

1991 *Mathematics Subject Classification.* Primary 58F22; Secondary 58F05, 34C20, 34C23.

The research in this paper was supported by the E.E.C. Science Project on "Bifurcation Theory and its Applications".

with $x \in \mathbb{R}^n, \lambda \in \mathbb{R}^m$, and $f : \mathbb{R}^n \times \mathbb{R}^m \to \mathbb{R}^n$ a C^∞-smooth mapping such that $f(0,\lambda) = 0$ for all $\lambda \in \mathbb{R}^m$. In this paper we will obtain a general reduction result for the following classical Hopf bifurcation problem:

(P) Given $\lambda_0 \in \mathbb{R}^m$ and $T_0 > 0$, describe all small T-periodic solutions of $(1.1)_\lambda$, for all $(\lambda, T) \in \mathbb{R}^m \times \mathbb{R}_+$ near (λ_0, T_0).

In contrast with most treatments of Hopf bifurcation we will not impose any restrictions on the spectrum of the linearization

$$A_0 := D_x f(0, \lambda_0) \in \mathcal{L}(\mathbb{R}^n), \tag{1.2}$$

except for the hypothesis that zero should not be an eigenvalue of A_0. This hypothesis excludes steady state bifurcations at $\lambda = \lambda_0$, and also implies that the assumption $f(0,\lambda) = 0$ can locally (near $\lambda = \lambda_0$) be satisfied, by only assuming that $f(0, \lambda_0) = 0$.

There are two more or less standard approaches to the problem (P), namely (a) the Liapunov-Schmidt method, and (b) a reduction based on normal form theory. Since it is our aim to present an approach which in a sense combines both methods let us briefly review them here; technical details will be given in the later sections (see also Chapter XVI of Golubitsky, Stewart and Schaeffer [1988] for a similar discussion).

The basic idea of the Liapunov-Schmidt method is to reformulate (P) as an equation in a Banach space of T_0-periodic mappings $x : \mathbb{R} \to \mathbb{R}^n$; in order to do so one first has to fix the period at T_0, by making a time rescale with an unknown scaling parameter σ, related to the actual period T. Appropriate splittings and an application of the implicit function theorem then gives the so-called determining or bifurcation equations, from which one can obtain the bifurcation diagrams (at least in principle — in practice this has been worked out only for the simpler cases). A major property of this approach is the fact that it leads to equations which have an explicit S^1(circle)-symmetry, related to the fact that if $\tilde{x}(t)$ is a T-periodic solution of (1.1), then so is $(S_\varphi \tilde{x})(t) := \tilde{x}(t+\varphi)$ for each $\varphi \in \mathbb{R}/T\mathbb{Z} \cong S^1$ (observe that $S_T \tilde{x} = \tilde{x}$); in a sense the Liapunov-Schmidt approach makes this "implicit symmetry" of (1.1) explicit. The S^1-equivariance of the bifurcation equations plays an important role in many treatments of Hopf bifurcation and forms a crucial element in explaining a number of features of such bifurcations (see e.g. Golubitsky, Stewart and Schaeffer [1988] and Vanderbauwhede [1982]). The method has also its disadvantages, the most prominent being that it concentrates exclusively on periodic solutions and ignores the other dynamic behavior of (1.1). As a consequence it is for example not clear how (in general) one can obtain the stability of the periodic solutions from the bifurcation equations which determine the existence of these solutions.

In contrast, the normal form approach keeps full track of the dynamics, since it essentially consists in making coordinate transformations which put (1.1) into an appropriate "simpler" form. As a preliminary (although not necessary) step one can use center manifold theory to conclude that the periodic orbits we are looking for all belong to a locally invariant submanifold associated to the purely imaginary eigenvalues of A_0; restricting the flow to this center manifold lowers the dimension n of the system under consideration, and also allows to assume that all eigenvalues of A_0 are purely imaginary. The normal form reduction then transforms (1.1) in an equation of the form

$$\dot{x} = f_{NF}(x, \lambda) + R(x, \lambda), \tag{1.3}$$

where $f_{NF} : \mathbb{R}^n \times \mathbb{R}^m \to \mathbb{R}^n$ commutes with the flow generated by the semisimple part of A_0 (see Section 2 for the definitions), while $R(x, \lambda) = O(\|x\|^{k+1})$ as $x \to 0$, for some arbitrarily chosen $k \geq 2$. The idea is then to approximate (1.3) by the truncated "normal form equation"

$$\dot{x} = f_{NF}(x, \lambda). \tag{1.4}$$

The symmetry of this equation allows a simple Ansatz to obtain periodic solutions of (1.4); the determining equations resulting from this Ansatz have an S^1-symmetry. In order to get periodic solutions for the actual equation (1.3) one has then to show that the periodic solutions of (1.4) persist in some way when one perturbs (1.4) with the higher order term $R(x, \lambda)$. This procedure raises the question whether it gives us all the periodic solutions we were looking for; the answer is positive, but to prove this one has to resort to the Liapunov-Schmidt method. The main disadvantage of the normal form approach however is that it centers around an approximate equation, and requires a persistence proof to get results for the full equation; in particular the higher order perturbation $R(x, \lambda)$ breaks the S^1-symmetry of the normal form set-up. From the other side it is in many cases possible to obtain from (1.4) stability information for the periodic solutions of (1.3) (again see Golubitsky, Stewart and Schaeffer [1988]). Moreover, (1.4) also contains information about other types of solutions of (1.3) which we do not consider here, such as quasi-periodic, homoclinic and heteroclinic solutions. For an introduction to the normal form approach, including a lot of simple examples, we refer to the recent book of Iooss and Adelmeyer [1992].

In this paper we propose an alternative approach which combines both these methods, taking advantage of their strong points and trying to avoid as much as possible their weaknesses. The basic idea is to first bring system (1.1) into normal form (1.3), and then apply a Liapunov-Schmidt type reduction on this equation. (The difference is that we reduce to the generalized nullspace, and not to the nullspace as in the usual Liapunov-Schmidt reduction). As a result the problem (P) for (1) reduces to a similar problem for an S^1-equivariant system on this generalized nullspace; the S^1-equivariance implies that the periodic solutions of the reduced equation have a particular form, which then immediately gives the determining equations. Moreover, the reduced vectorfield is up to higher order terms given by a restriction of the normal form vectorfield f_{NF}, and therefore it carries the dynamical information of this normal form.

Similarly as for both other methods also our approach can be worked out such that additional structures of (1.1) are preserved. For example, if (1.1) is Γ-equivariant then the reduced system will be $\Gamma \times S^1$-equivariant. And if the original system is Hamiltonian or reversible, then the reduced system will be S^1-equivariant and respectively Hamiltonian or reversible. The Hamiltonian case is the most difficult one to prove, since we have to take special care of the variational structure of the problem. Therefore we will, in the subsequent sections, describe the reduction for Hamiltonian systems, hoping that the reader will be able to extract the corresponding results for general, equivariant or reversible vectorfields (see also Knobloch [1992] for the equivariant case). Our results for Hamiltonian systems can already be found (in a somewhat hidden form) in earlier work of Duistermaat and van der Meer (see Duistermaat [1984] and van der Meer [1986]), but the proofs in both these references are incomplete. Also, in the Hamiltonian context the Weinstein-Moser reduction, and certain aspects of our treatment show similarities with approaches described in earlier work of Weinstein [1973], Moser [1976] and Rabinowitz [1986]. Results similar to our main Theorem 3.2 below can also be formulated for the bifurcation of periodic points from a fixed point of a mapping; in that case the additional S^1-equivariance is replaced by a $\mathbb{Z}_q$-equivariance, where q is the period. When applied to the Poincaré map associated to a periodic orbit of (1.1) this can be used to study subharmonic bifurcations from such periodic orbit.

Before stating and proving our main result we review in the next section some mathematical preliminaries on Hamiltonian systems and on Jordan-Chevalley decompositions.

2 Preliminaries

The natural setting for a study of Hamiltonian systems is that of a symplectic manifold; however, if one is only interested in local problems (such as in our case) one can use the classical Darboux theorem (see for example Abraham and Marsden [1978]) to assume that the symplectic form is constant, i.e. it is sufficient to work in a *symplectic vectorspace.* So let (V, ω) be a symplectic vectorspace, i.e. V is an even-dimensional real vectorspace and $\omega : V \times V \to \mathbb{R}$ is an anti-symmetric and nondegenerate bilinear form over V:

(a) $\omega(x,y) = -\omega(y,x), \quad \forall x, y \in V;$

(b) $\omega(x,y) = 0, \ \forall y \in V \quad \Rightarrow x = 0.$

Such bilinear form ω is called a *symplectic form* over V. The nondegeneracy of ω implies that to each C^∞-function $H : V \to \mathbb{R}$ there corresponds a C^∞-vectorfield $X_H : V \to V$, defined by the relation

$$\omega(X_H(x), y) = DH(x) \cdot y, \quad \forall x, y \in V. \tag{2.1}$$

We call H a *Hamilton function* (or *Hamiltonian*), and X_H the corresponding *Hamiltonian vectorfield.* We will denote the flow of the Hamilton equation

$$\dot{x} = X_H(x) \tag{2.2}$$

by $\varphi_H(t, x)$ or $\varphi_H^t(x)$.

A C^∞-mapping $\psi : V \to V$ is called *symplectic* (or *canonical*) if

$$\omega(D\psi(x) \cdot y, D\psi(x) \cdot z) = \omega(y, z), \quad \forall x, y, z \in V. \tag{2.3}$$

The transformation $x = \psi(y)$, where $\psi \in C^\infty(V)$ is a symplectic *diffeomorphism*, transforms (2.2) in the equation

$$\dot{y} = D\psi(y)^{-1} X_H(\psi(y)) = X_{H \circ \psi}(y); \tag{2.4}$$

this last equality follows immediately from (2.3) and (2.1):

$$\begin{aligned} \omega(D\psi(y)^{-1} X_H(\psi(y)), z) &= \omega(X_H(\psi(y)), D\psi(y) \cdot z) \\ &= DH(\psi(y)) \cdot D\psi(y) \cdot z = D(H \circ \psi)(y) \cdot z. \end{aligned}$$

So the transformed equation is again a Hamilton equation, with Hamiltonian $H \circ \psi$. An important class of symplectic diffeomorphisms is given by the time maps φ_H^t of

Hamiltonian vectorfields (with $t \in \mathbb{R}$ arbitrary but fixed). Indeed, differentiating (2.1) shows that

$$\omega(DX_H(x) \cdot y, z) = \omega(DX_H(x) \cdot z, y); \quad \forall x, y, z \in V; \tag{2.5}$$

it follows that

$$\begin{aligned} \frac{d}{dt}\omega(D\varphi_H^t(x) \cdot y, D\varphi_H^t(x) \cdot z) &= \omega(DX_H(\varphi_H^t(x)) \cdot D\varphi_H^t(x) \cdot y, D\varphi_H^t(x) \cdot z) \\ &+ \omega(D\varphi_H^t(x) \cdot y, DX_H(\varphi_H^t(x)) \cdot D\varphi_H^t(x) \cdot z) = 0, \end{aligned}$$

and hence

$$\omega(D\varphi_H^t(x) \cdot y, D\varphi_H^t(x) \cdot z) = \omega(y, z), \quad \forall t \in \mathbb{R}. \tag{2.6}$$

In particular, using the symplectic transformation $x = \varphi_F^t(y)$ on (2.2), we conclude that for each $H, F \in C^\infty(V; \mathbb{R})$ we have

$$X_{H \circ \varphi_F^t}(x) = D\varphi_F^t(x)^{-1} X_H(\varphi_F^t(x)), \quad \forall t \in \mathbb{R}. \tag{2.7}$$

Next we introduce the *Poisson bracket* of two functions $F, G \in C^\infty(V; \mathbb{R})$; this is the function $\{F, G\} \in C^\infty(V; \mathbb{R})$ given by

$$\{F, G\}(x) := \omega(X_G(x), X_F(x)) = DG(x) \cdot X_F(x). \tag{2.8}$$

This Poisson bracket is clearly antisymmetric (i.e. $\{F, G\} = -\{G, F\}$), and from (2.8) and (2.5) we see that

$$\begin{aligned} D\{F, G\}(x) \cdot y &= \omega(DX_G(x) \cdot y, X_F(x)) + \omega(X_G(x), DX_F(x) \cdot y) \\ &= \omega(DX_G(x) \cdot X_F(x) - DX_F(x) \cdot X_G(x), y); \end{aligned} \tag{2.9}$$

it follows that

$$X_{\{F,G\}} = [X_F, X_G], \tag{2.10}$$

where the *Lie bracket* $[X, Y] \in C^\infty(V)$ of two vectorfields $X, Y \in C^\infty(V)$ is defined by

$$[X, Y](x) := DY(x) \cdot X(x) - DX(x) \cdot Y(x). \tag{2.11}$$

From (2.8), (2.9) and (2.5) one also easily shows that the Poisson bracket satisfies the *Jacobi identity*

$$\{F, \{G, H\}\} + \{G, \{H, F\}\} + \{H, \{F, G\}\} = 0, \tag{2.12}$$

valid for all $F, G, H \in C^\infty(V; \mathbb{R})$. We conclude that the Poisson bracket gives $C^\infty(V; \mathbb{R})$ the structure of an infinite dimensional Lie algebra.

The Poisson bracket can be used to define for each $F \in C^\infty(V; \mathbb{R})$ a linear operator $\mathrm{ad}(F)$ on $C^\infty(V; \mathbb{R})$, given by

$$\mathrm{ad}(F)(G) := \{F, G\}, \quad \forall F, G \in C^\infty(V; \mathbb{R}). \tag{2.13}$$

It follows from the Jacobi identity (2.12) that

$$\mathrm{ad}(\{F, G\}) = [\mathrm{ad}(F), \mathrm{ad}(G)], \tag{2.14}$$

where the bracket at the right hand side denotes the commutator of two linear operators ($[A,B] := AB - BA$). The mapping $F \mapsto \mathrm{ad}(F)$ is called the *adjoint representation* of the Lie algebra $(C^\infty(V;\mathbb{R}), \{.,.\})$.

Using these definitions we observe the following. If $F, H \in C^\infty(V;\mathbb{R})$ and $H_t := H \circ \varphi_F^t$ (for $t \in \mathbb{R}$) then

$$\frac{d}{dt} H_t(x) = DH(\varphi_F^t(x)) \cdot X_F(\varphi_F^t(x)) = \{F,H\}(\varphi_F^t(x)). \tag{2.15}$$

So H is a first integral for the vectorfield X_F if and only if $\{F,H\} = 0$; more precisely, we have the following:

Lemma 2.1 *Let $F, H \in C^\infty(V;\mathbb{R})$; then the following are equivalent:*
(i) $\{F,H\} = 0$;
(ii) $H \circ \varphi_F^t = H, \quad \forall t \in \mathbb{R}$.
Moreover, (i) *and* (ii) *imply*
(iii) $X_H(\varphi_F^t(x)) = D\varphi_F^t(x) \cdot X_H(x), \quad \forall t \in \mathbb{R}$,
while (iii) *is equivalent to* (i) *and* (ii) *if F has at least one critical point.*

Proof The equivalence of (i) and (ii) follows immediately from (2.15) and $\varphi_F^0(x) = x$. Also, (2.7) shows that (iii) is equivalent to $X_{H \circ \varphi_F^t} = X_H$; hence (i) and (ii) imply (iii), while it is easily seen that (iii) implies $H(\varphi_F^t(x)) - H(\varphi_F^t(x_0)) = H(x) - H(x_0)$, for each $x, x_0 \in V$. Therefore, if F has a critical point $x_0 \in V$ (i.e. $DF(x_0) = 0$) then $X_F(x_0) = 0$, $\varphi_F^t(x_0) = x_0$ and $H(\varphi_F^t(x)) = H(x)$. □

Corollary 2.2 *We have for each $H \in C^\infty(V;\mathbb{R})$ that*

$$H(\varphi_H^t(x)) = H(x) \text{ and } X_H(\varphi_H^t(x)) = D\varphi_H^t(x) \cdot X_H(x), \forall t \in \mathbb{R}. \tag{2.16}$$

Replacing H by F in (2.16) we can rewrite (2.15) as

$$\frac{d}{dt} H_t(x) = DH(\varphi_F^t(x)) \cdot D\varphi_F^t(x) \cdot X_F(x) = DH_t(x) \cdot X_F(x),$$

i.e.

$$\frac{d}{dt} H_t = \{F, H_t\} = \mathrm{ad}(F) \cdot (H_t). \tag{2.17}$$

This relation will be important later for our normal form reduction.

For the remainder of this section we turn to *linear* Hamiltonian systems; those correspond to quadratic Hamiltonians. We start with the following.

Lemma 2.3 *A linear operator $A \in \mathcal{L}(V)$ defines a Hamiltonian vectorfield on (V, ω) if and only if*

$$\omega(Ax, y) = \omega(Ay, x), \quad \forall x, y \in V; \tag{2.18}$$

a corresponding Hamiltonian is then given by

$$Q_A(x) := \tfrac{1}{2}\omega(Ax, x), \quad \forall x \in V. \tag{2.19}$$

Proof If $Ax = X_H(x)$ for an appropriate $H \in C^\infty(V;\mathbb{R})$ then

$$DH(z) \cdot y = \omega(Az, y) \text{ and } D^2H(z) \cdot (x,y) = \omega(Ax, y), \quad \forall x, y, z \in V,$$

which implies (2.18). Conversely, if $A \in \mathcal{L}(V)$ satisfies (2.18) and Q_A is defined by (2.19) then

$$DQ_A(x) \cdot y = \tfrac{1}{2}\omega(Ay, x) + \tfrac{1}{2}\omega(Ax, y) = \omega(Ax, y),$$

i.e. $A = X_{Q_A}$. This proves the lemma. □

Linear operators $A \in \mathcal{L}(V)$ satisfying (2.18) are called *infinitesimally symplectic*, and the space of such operators is denoted by $\mathfrak{sp}(V)$. Since for $A, B \in \mathfrak{sp}(V)$ we have

$$\begin{aligned}\omega([A,B]x, y) &= \omega(ABx, y) - \omega(BAx, y) = \omega(Ay, Bx) - \omega(By, Ax) \\ &= \omega(x, BAy) - \omega(x, ABy) = \omega([A,B]y, x);\end{aligned}$$

we conclude that $\mathfrak{sp}(V)$, equipped with the commutator bracket, forms a Lie algebra. In fact, $\mathfrak{sp}(V)$ is the Lie algebra of the Lie group $\mathfrak{Sp}(V)$ consisting of symplectic linear operators (the condition for $A \in \mathcal{L}(V)$ to be symplectic is that $\omega(Ax, Ay) = \omega(x, y), \forall x, y \in V$).

Let $\mathcal{E}_0 = \mathcal{E}_0(V)$ be the space of quadratic functions $Q : V \to \mathbb{R}$ (i.e. such that $Q(x) = \frac{1}{2}D^2Q(0) \cdot (x, x) = \frac{1}{2}DQ(x) \cdot x$). By Lemma 2.3 we have for each $Q \in \mathcal{E}_0$ that $X_Q \in \mathfrak{sp}(V)$, and

$$Q(x) = \tfrac{1}{2}DQ(x) \cdot x = \tfrac{1}{2}\omega(X_Q(x), x) = Q_{X_Q}(x);$$

together with the relation $A = X_{Q_A}$, which holds for all $A \in \mathfrak{sp}(V)$, this shows that the mapping $A \in \mathfrak{sp}(V) \mapsto Q_A \in \mathcal{E}_0(V)$ is a linear isomorphism. Moreover we have for all $A, B \in \mathfrak{sp}(V)$ and for all $x \in V$ that

$$\begin{aligned}\{Q_A, Q_B\}(x) &= \omega(Bx, Ax) = \tfrac{1}{2}\omega(BAx, x) + \tfrac{1}{2}\omega(x, ABx) \\ &= -\tfrac{1}{2}\omega([A,B]x, x) = -Q_{[A,B]}(x);\end{aligned}$$

this proves that $\mathcal{E}_0(V)$, equipped with the Poisson bracket, forms a Lie algebra isomorphic to $\mathfrak{sp}(V)$. More precisely, the mapping $A \mapsto -Q_A = Q_{-A}$ forms a Lie algebra isomorphism from $(\mathfrak{sp}(V), [.,.])$ (commutator bracket) onto $(\mathcal{E}_0(V), \{.,.\})$.

Taking $F = Q_A$ (with $A \in \mathfrak{sp}(V)$) in Lemma 2.1 gives the following.

Lemma 2.4 *For each $H \in C^\infty(V;\mathbb{R})$ and each $A \in \mathfrak{sp}(V)$ the following statements are equivalent:*

(i) $\{H, Q_A\} = 0$;

(ii) $H(e^{At}x) = H(x), \quad \forall t \in \mathbb{R}$;

(iii) $X_H(e^{At}x) = e^{At}X_H(x), \quad \forall t \in \mathbb{R}$.

Considering $A \in \mathfrak{sp}(V)$ as an element of $\mathcal{L}(V)$ we know from the classical Jordan-Chevalley theorem that there is a unique decomposition $A = A_s + A_n$, where $A_s, A_n \in \mathcal{L}(V)$ are such that (i) A_s is semisimple (i.e. complex diagonalizable), (ii) A_n is nilpotent, and (iii) A_s and A_n commute. Using the fact that $\mathfrak{sp}(V)$ is a *semisimple* Lie algebra one can actually prove more: not only will A_s and A_n themselves belong to $\mathfrak{sp}(V)$, but also their images under any representation of the Lie algebra $\mathfrak{sp}(V)$ will be semisimple, respectively nilpotent (see for example, Humphreys [1972]; full details will also be given in Sanders, Vanderbauwhede and van der Meer (in preparation)). Moreover, we can use the isomorphism between $\mathfrak{sp}(V)$ and $\mathcal{E}_0(V)$ to obtain a similar decomposition of the elements of $\mathcal{E}_0(V)$. We can summarize these results as follows.

Let W be a (finite-dimensional) vectorspace; the space $\mathcal{L}(W)$ of linear operators on W, equipped with the commutator bracket $[A, B] := AB - BA$, forms a Lie algebra, usually denoted by $\mathfrak{gl}(W)$. A representation of the Lie algebra $(\mathcal{E}_0(V), \{.,.\})$ over W is a Lie algebra homomorphism from $\mathcal{E}_0(V)$ into $\mathfrak{gl}(W)$, i.e. a linear mapping $\varphi : \mathcal{E}_0(V) \to \mathcal{L}(W)$ such that

$$\varphi(\{Q, P\}) = [\varphi(Q), \varphi(P)], \quad \forall Q, P \in \mathcal{E}_0(V).$$

In particular, the mapping $Q \in \mathcal{E}_0(V) \mapsto -X_Q = X_{-Q} \in \mathfrak{sp}(V)$ defines a representation of $\mathcal{E}_0(V)$ over V.

Theorem 2.5 (Jordan-Chevalley decomposition) *For each $Q \in \mathcal{E}_0(V)$ there exists a unique decomposition $Q = S + N$, with $S, N \in \mathcal{E}_0(V)$ such that $\{S, N\} = 0$ and such that for each representation φ of the Lie algebra $(\mathcal{E}_0(V), \{.,.\})$ over a vectorspace W we have that $\varphi(S) \in \mathcal{L}(W)$ is semisimple and $\varphi(N) \in \mathcal{L}(W)$ is nilpotent.* □

We call $Q = S + N$ the *Jordan-Chevalley decomposition* of $Q \in \mathcal{E}_0(V)$. Combining Theorem 2.5 with some elementary algebra gives the following (see van der Meer [1982]).

Lemma 2.6 *Let $Q = S + N$ be the Jordan-Chevalley decomposition of $Q \in \mathcal{E}_0(V)$, and $\varphi : \mathcal{E}_0(V) \to \mathcal{L}(W)$ a representation of $\mathcal{E}_0(V)$ over a vectorspace W. Then:*

(i) $W = Im\ \varphi(S) \oplus ker\ \varphi(S)$;

(ii) $ker\ \varphi(Q) = ker\ \varphi(S) \cap ker\ \varphi(N)$;

(iii) $Im\ \varphi(Q) = Im\ \varphi(S) \oplus (ker\ \varphi(S) \cap Im\ \varphi(N))$.

3 The main result

Let (V, ω) be a (finite-dimensional) symplectic vectorspace, Λ a Banach space, and $H \in C^\infty(V \times \Lambda; \mathbb{R})$. Consider then the parameter dependent Hamiltonian system

$$\dot{x} = X_H(x, \lambda), \tag{3.1}$$

where, for each $\lambda \in \Lambda$, $X_H(., \lambda)$ is the Hamiltonian vectorfield corresponding to the Hamiltonian $H_\lambda = H(., \lambda)$. We make the following assumptions:

(H) (i) $H(0, \lambda) = 0$ and $D_x H(0, \lambda) = 0$ for all $\lambda \in \Lambda$;

(ii) $\lambda_0 \in \Lambda$ and $T_0 > 0$ are such that if we put $A_0 := D_x X_H(0, \lambda_0) \in \mathcal{L}(V)$ then the linear equation

$$\dot{x} = A_0 x \tag{3.2}$$

has some nontrivial T_0-periodic solutions;

(iii) A_0 is nonsingular.

As described in the introduction we want to study, for all $(\lambda, T) \in \Lambda \times \mathbb{R}_+$ near (λ_0, T_0), all sufficiently small T-periodic solutions of $(3.1)_\lambda$.

Defining $Q_0 \in \mathcal{E}_0(V)$ as the quadratic part of $H(\cdot, \lambda_0)$, i.e.

$$Q_0(x) := \tfrac{1}{2} D_x^2 H(0, \lambda_0) \cdot (x, x), \tag{3.3}$$

it follows directly from the definitions that $\omega(A_0x, y) = D_x^2 H(0, \lambda_0) \cdot (x, y) = DQ_0(x) \cdot y$; this shows that $A_0 \in \mathfrak{sp}(V)$ and $A_0 = X_{Q_0}$. Let $Q_0 = S + N$ be the Jordan-Chevalley decomposition of Q_0 (see Theorem 2.5); the corresponding decomposition of A_0 is then given by $A_0 = A_s + A_n$, with $A_s := X_S \in \mathfrak{sp}(V)$ semisimple and $A_n := X_N \in \mathfrak{sp}(V)$ nilpotent. Finally we set

$$U := \ker\,(e^{A_s T_0} - I); \tag{3.4}$$

U is equal to the direct sum of the (real) generalized eigenspaces of A_0 corresponding to pairs of eigenvalues of the form $\pm ik\omega_0$, with $k \in \mathbb{N}$ and $\omega_0 := \frac{2\pi}{T_0}$. We denote by $\widetilde{S}$ and $\widetilde{N}$ the restrictions of respectively S and N to the subspace U of V. Before stating our main result we prove the following.

Lemma 3.1 *U forms a symplectic subspace of V, i.e. the restriction $\widetilde{\omega}$ of the symplectic form ω to $U \times U$ is a symplectic form over U.*

Proof Since A_s is semisimple we have $V = U \oplus \mathrm{Im}\,(e^{A_s T_0} - I)$, while the fact that A_s is infinitesimally symplectic implies that $e^{A_s T_0}$ is symplectic. It follows that for each $u \in U$ and each $w = (e^{A_s T_0} - I)v \in \mathrm{Im}\,(e^{A_s T_0} - I)$ we have

$$\omega(u, w) = \omega(u, (e^{A_s T_0} - I)v) = \omega((e^{-A_s T_0} - I)u, v) = 0,$$

by the definition (3.4) of U. Hence $\widetilde{\omega} := \omega|_{U \times U}$ is a nondegenerate bilinear form on U; since it is obviously also anti-symmetric it forms a symplectic form on U. □

Using Lemma 3.1 we can associate to each $\widetilde{H} \in C^{k+1}(U \times \Lambda; \mathbb{R})$ (with $k \geq 2$) a Hamiltonian vectorfield $X_{\widetilde{H}} \in C^k(U \times \Lambda; U)$, defined by $\widetilde{\omega}(X_{\widetilde{H}}(u, \lambda), \bar{u}) = D_u \widetilde{H}(u, \lambda) \cdot \bar{u}$, $\forall u, \bar{u} \in U$. We denote the associated Poisson bracket on $C^\infty(U; \mathbb{R})$ (or $C^{k+1}(U; \mathbb{R})$) by $\{.,.\}_U$. Observe also that

$$A_s u = X_{\widetilde{S}}(u), \quad \forall u \in U; \tag{3.5}$$

it follows then from the definition (3.4) of U that $\widetilde{S}$ generates a symplectic S^1-action on U, given by

$$\xi \in \mathbb{R}/T_0\mathbb{Z} \cong S^1 \;\mapsto\; \varphi_{\widetilde{S}}^{\xi} = e^{A_s \xi}|_U. \tag{3.6}$$

Our main result now takes the following form.

Theorem 3.2 *Under the hypotheses (H), and with the notations and conventions described above, there exists for each $k \geq 2$ a number $\varepsilon_k > 0$, a C^{k+1}-mapping $\widetilde{H} : U \times \Lambda \to \mathbb{R}$, and a C^k-mapping $\Psi : U \times \Lambda \to V$, with the following properties:*

(i) $\widetilde{H}(0, \lambda) = 0$ *and* $D_u \widetilde{H}(0, \lambda) = 0$ *for all* $\lambda \in \Lambda$, *and* $\widetilde{H}(u, \lambda_0) = Q_0(u) + O(\|u\|^3)$ *as* $u \to 0$;

(ii) $\Psi(0, \lambda) = 0$ *for all* $\lambda \in \Lambda$, *and* $D_u \Psi(0, \lambda_0) \cdot u = u, \forall u \in U$;

(iii) $\widetilde{H}$ *and* Ψ *are* C^∞ *on* $(U - \{0\}) \times \Lambda$;

(iv) $\{\widetilde{H}_\lambda, \widetilde{S}\}_U = 0, \quad \forall \lambda \in \Lambda$;

(v) $\tilde{x} : \mathbb{R} \to V$ *is a* T*-periodic solution of* $(3.1)_\lambda$, *with*

$$\sup_{t \in \mathbb{R}} \|\tilde{x}(t)\| < \varepsilon_k, \; |T - T_0| < \varepsilon_k \;\; \text{and} \;\; \|\lambda - \lambda_0\| < \varepsilon_k,$$

if and only if

$$\tilde{x}(t) = \Psi(\tilde{u}(t), \lambda), \quad \forall t \in \mathbb{R}, \tag{3.7}$$

where $\tilde{u} : \mathbb{R} \to U$ is a sufficiently small T-periodic solution of the reduced Hamiltonian system

$$\dot{u} = X_{\widetilde{H}}(u, \lambda). \tag{3.8}$$

Moreover:

(vi) *all such periodic solutions $\tilde{u}(t)$ of (3.8) have the form*

$$\tilde{u}(t) = e^{(1+\sigma)A_s t}u, \tag{3.9}$$

with $u \in U$ and $\sigma = \frac{T_0}{T} - 1 \in \mathbb{R}$ sufficiently small;

(vii) *if the original Hamiltonian $H(x, \lambda)$ is in normal form up to order $(k+1)$, that is, if*

$$H(x, \lambda) = H_{NF}(x, \lambda) + R(x, \lambda) \tag{3.10}$$

with $\{H_{NF,\lambda}, S\} = 0$ and $R(x, \lambda) = O(\|x\|^{k+2})$ as $x \to 0$, then the reduction can be done in such a way that

$$\widetilde{H}(u, \lambda) = H_{NF}(u, \lambda) + O(\|u\|^{k+2}) \tag{3.11}$$

and

$$\Psi(u, \lambda) = u + O(\|u\|^{k+1}) \tag{3.12}$$

as $u \to 0$. □

As an immediate consequence of this theorem we see that the *determining equations* for our periodic solutions can be obtained by bringing (3.9) into (3.8); using (3.6), (iv) and Lemma 2.4 these equations take the form

$$(1+\sigma)A_s(u) = X_{\widetilde{H}}(u, \lambda), \tag{3.13}$$

or, equivalently,

$$(1+\sigma)D_u\widetilde{S}(u) = D_u\widetilde{H}(u, \lambda). \tag{3.14}$$

We conclude that the periodic solutions we are looking for correspond to the *relative equilibria* of the S^1-symmetric Hamiltonian vectorfield $X_{\widetilde{H}}$, or to the critical points of the *energy-momentum map*

$$u \mapsto (\widetilde{H}(u, \lambda), \widetilde{S}(u)). \tag{3.15}$$

(see for example, Marsden [1992] for more details on these topics). This suggests the use of variational techniques for the study of (3.13) or (3.14), such as in the already cited work of Moser and Weinstein, or the use of S^1-equivariant singularity theory to further simplify these equations; in the lecture notes of van der Meer [1986] one can see how this last approach works for the study of the generic Hamiltonian Hopf bifurcation; we refer to van der Meer (to appear) and Sanders, Vanderbauwhede, and van der Meer (in preparation) for an application to the degenerate Hamiltonian Hopf bifurcation. Further applications to reversible systems will be given in forthcoming work of Knobloch and Vanderbauwhede (in preparation).

In the next sections we give the proof of Theorem 3.2; in Section 4 we perform a normal form reduction which brings the Hamiltonian in the form (3.11), and in Section 5 we carry out a Liapunov-Schmidt type reduction which gives us the main result. In the appendix we give a short proof of the equivariant Morse lemma needed in section 5.

4 The normal form reduction

The aim of this section is to prove the following normal form result.

Theorem 4.1 *Under the hypothesis (H) there exists for each $k \geq 0$ a neighborhood ω_k of λ_0 in Λ and a smooth mapping $\psi : V \times \omega_k \to V$, with $\psi(0, \lambda) = 0$ for all $\lambda \in \omega_k$ and $D_x\psi(0, \lambda_0) = I_V$, such that for all $\lambda \in \omega_k$ the following holds:*

(i) $\psi_\lambda := \psi(\cdot, \lambda)$ *is a symplectic diffeomorphism on* V*;*

(ii) $H(\psi(x, \lambda), \lambda) = H_{NF}(x, \lambda) + R(x, \lambda)$, *with*

$$\{H_{NF,\lambda}, S\} = 0 \tag{4.1}$$

and $R(x, \lambda) = O(\|x\|^{k+3})$ *as* $x \to 0$, *uniformly for* $\lambda \in \omega_k$. □

We call H_{NF} a *normal form up to order* $(k+2)$ of the Hamiltonian H. In fact, H_{NF} as given by the theorem is a normal form with respect to the semisimple part S of Q_0; taking also the nilpotent part N into account one can actually impose more conditions on the normal form than just (4.1) (see Cushman and Sanders [1986] and Sanders, Vanderbauwhede, and van der Meer (in preparation) for more details), but the formulation used here is sufficient to prove our main Theorem 3.2. Another important point to observe is that, as will be seen from the proof, the neighborhood ω_k may shrink down to just $\{\lambda_0\}$ as $k \to \infty$. This normal form result is certainly not new, but we include it here for completeness and because we want to emphasize the property (4.1) of the normal form Hamiltonian. For more references on normal form theory see the extensive bibliography of van der Meer [1986]; see also Vanderbauwhede [1989] for the parameter-dependent case, and Iooss and Adelmeyer [1992] for a somewhat different approach.

The idea of the proof is to construct $\psi(x, \lambda)$ as the (*time* = 1)-map of an appropriate Hamiltonian system; that is, we show how to construct a (parameter dependent) Hamiltonian $F(x, \lambda)$ with corresponding Hamiltonian flow $\varphi_F^t(x, \lambda)$, such that $\psi(x, \lambda) = \varphi_F^{t=1}(x, \lambda)$ satisfies the condition (ii) of the theorem; the condition (i) is then automatically satisfied.

We start with some notation. Let

$$\mathcal{M}_0 = \mathcal{M}_0(V) := \{F \in C^\infty(V, \mathbb{R}) \mid F(0) = 0, DF(0) = 0\}. \tag{4.2}$$

For each $k \geq 0$ we define finite-dimensional subspaces $\mathcal{E}_k = \mathcal{E}_k(V)$ and $\mathcal{F}_k = \mathcal{F}_k(V)$ of $\mathcal{M}_0$ as follows: $\mathcal{E}_k(V)$ is the space of all $F \in \mathcal{M}_0$ which are homogeneous polynomials of degree $(k+2)$, while $\mathcal{F}_k(V) := \mathcal{E}_0(V) \oplus \mathcal{E}_1(V) \oplus \cdots \oplus \mathcal{E}_k(V)$. We also define projections $\widetilde{J}_k$ and J_k $(k \geq 0)$ in $\mathcal{M}_0$ onto respectively $\mathcal{E}_k$ and $\mathcal{F}_k$ by

$$(\widetilde{J}_k F)(x) := \tfrac{1}{(k+2)!} D^{(k+2)} F(0) \cdot x^{(k+2)} \text{ and } J_k := \sum_{i=0}^{k} \widetilde{J}_i\,, \tag{4.3}$$

where $F \in \mathcal{M}_0$ and $x^{(l)} = (x, \cdots, x) \in V^l$ for $x \in V$ and $l \in \mathbb{N}$. The use of a shifted index (i.e. $(k+2)$ instead of k) is motivated by the fact that with our definitions we have

$$\{\mathcal{E}_k(V), \mathcal{E}_l(V)\} \subset \mathcal{E}_{k+l}(V), \quad \forall k, l \geq 0. \tag{4.4}$$

In particular we see from (4.4) that $(\mathrm{ad}\, Q)(\mathcal{E}_k) \subset \mathcal{E}_k$ for each $Q \in \mathcal{E}_0(V)$ and $k \geq 0$; denoting the restriction of $\mathrm{ad}\, Q$ to $\mathcal{E}_k(V)$ by $\mathrm{ad}_k(Q)$ it follows that the mapping

$$Q \in \mathcal{E}_0(V) \mapsto \mathrm{ad}_k(Q) \in \mathcal{L}(\mathcal{E}_k(V))$$

forms a representation of the Lie algebra $\mathcal{E}_0(V)$ over $\mathcal{E}_k(V)$. Applying Lemma 2.6 to the quadratic part Q_0 of H_{λ_0} we have then

$$\mathcal{E}_k(V) = \mathrm{Im}\ \mathrm{ad}_k(S) \oplus \ker\ \mathrm{ad}_k(S), \quad \forall k \geq 0. \tag{4.5}$$

We denote the projection in $\mathcal{E}_k(V)$ onto $\mathrm{Im}\ \mathrm{ad}_k(S)$ and parallel to $\ker\ \mathrm{ad}_k(S)$ by $\widetilde{\pi}_k \in \mathcal{L}(\mathcal{E}_k(V))$, and set

$$\pi_k := \sum_{i=1}^{k} \widetilde{\pi}_i \cdot \widetilde{J}_i \in \mathcal{L}(\mathcal{F}_k(V)); \tag{4.6}$$

π_k is the projection in $\mathcal{F}_k$ onto $\mathcal{F}_k \cap \mathrm{Im}\ (\mathrm{ad}\, S)$ and parallel to $\mathcal{F}_k \cap \ker\ (\mathrm{ad}\, S)$, such that for $F \in \mathcal{F}_k(V)$ we have $\{F, S\} = 0$ if and only if $\pi_k F = 0$. Finally we define for each $F \in \mathcal{F}_k(V)$ a linear operator $\mathrm{ad}^{(k)} F \in \mathcal{L}(\mathcal{F}_k(V))$ by

$$(\mathrm{ad}^{(k)} F) \cdot G := J_k\{F, G\}, \quad \forall G \in \mathcal{F}_k(V); \tag{4.7}$$

observe that for each $F, G \in \mathcal{M}_0$ and each $k \geq 0$ we have

$$J_k\{F, G\} = J_k\{J_k F, J_k G\} = (\mathrm{ad}^{(k)} J_k F) \cdot (J_k G). \tag{4.8}$$

Now fix some $k \in \mathbb{N}$, some $\lambda \in \Lambda$ and some $F \in \mathcal{F}_k(V)$; we first calculate $G_t := J_k(H_\lambda \circ \varphi_F^t) \in \mathcal{F}_k(V)$ for $t \in \mathbb{R}$. From (2.17) we have

$$\frac{d}{dt}(H_\lambda \circ \varphi_F^t) = \{F\, ,\, H_\lambda \circ \varphi_F^t\},$$

and hence, by (4.8):

$$\frac{d}{dt} G_t = (\mathrm{ad}^{(k)} F) \cdot G_t, \quad \forall t \in \mathbb{R}. \tag{4.9}$$

Since $G_{t=0} = J_k H_\lambda$ it follows that

$$G_t = \exp(t\, \mathrm{ad}^{(k)} F) \cdot J_k H_\lambda, \quad \forall t \in \mathbb{R}. \tag{4.10}$$

We conclude that the conditions of Theorem 4.1 will be satisfied (with $\psi = \varphi_F^1$ and $H_{NF,\lambda} = J_k(H_\lambda \circ \varphi_F^1)$) if we can determine $F \in \mathcal{F}_k(V)$ such that

$$\pi_k(\exp(\mathrm{ad}^{(k)} F) \cdot J_k H_\lambda) = 0. \tag{4.11}$$

We will use induction on k and the implicit function theorem to solve (4.11); more precisely, we will prove the following lemma, which immediately implies Theorem 4.1.

Lemma 4.2 *For each $k \geq 0$ there exists a neighborhood ω_k of λ_0 in Λ and a smooth function $\widetilde{F}_k : \omega_k \to \mathcal{F}_k(V) \cap Im$ (ad S) with $J_0 \widetilde{F}_k(\lambda_0) = 0$ and such that $F = \widetilde{F}_k(\lambda)$ satisfies (4.11), i.e. such that*

$$\pi_k(\exp(\mathrm{ad}^{(k)} \widetilde{F}_k(\lambda)) \cdot J_k H_\lambda = 0, \quad \forall \lambda \in \omega_k. \tag{4.12}$$

Proof For each $k \geq 0$ we define a smooth mapping $A_k : (\mathcal{F}_k(V) \cap \mathrm{Im}(\mathrm{ad}\, S)) \times \Lambda \to \mathcal{F}_k(V) \cap \mathrm{Im}(\mathrm{ad}\, S)$ by

$$A_k(F, \lambda) := \pi_k(\exp(\mathrm{ad}^{(k)} F) \cdot J_k H_\lambda). \tag{4.13}$$

For $k = 0$ we have $A_0(0, \lambda_0) = 0$, since $J_0 H_{\lambda_0} = Q_0$ and $\{Q_0, S\} = 0$. Moreover, we have for each $F \in \mathcal{E}_0(V) \cap \mathrm{Im}(\mathrm{ad}\, S) = \mathrm{Im}(\mathrm{ad}_0\, S)$ that

$$D_F A_0(0, \lambda_0) \cdot F = \pi_0(\mathrm{ad}^{(0)} F) \cdot Q_0 = -\pi_0(\mathrm{ad}_0\, Q_0)(F) = -(\mathrm{ad}_0\, Q_0)(F).$$

Now $\mathrm{ad}_0\, Q_0 = \mathrm{ad}_0\, S + \mathrm{ad}_0\, N$ is invertible on $\mathrm{Im}(\mathrm{ad}_0\, S)$, and hence we can use the implicit function theorem to conclude that there exists a neighborhood ω_0 of λ_0 in Λ and a smooth mapping $\widetilde{F}_0 : \omega_0 \to \mathcal{E}_0(V) \cap \mathrm{Im}(\mathrm{ad}\, S)$ with $\widetilde{F}_0(\lambda_0) = 0$ such that $A_0(\widetilde{F}_0(\lambda), \lambda) = 0$ for all $\lambda \in \omega_0$.

Next let $k \geq 1$, and suppose that the lemma has been proven for $k - 1$, i.e. there exist a neighborhood ω_{k-1} of λ_0 in Λ and a smooth function $\widetilde{F}_{k-1} : \omega_{k-1} \to \mathcal{F}_{k-1}(V) \cap \mathrm{Im}(\mathrm{ad}\, S)$ with $J_0 \widetilde{F}_{k-1}(\lambda_0) = 0$ and such that $A_{k-1}(\widetilde{F}_{k-1}(\lambda), \lambda) = 0$ for all $\lambda \in \omega_{k-1}$. We have then for each $\lambda \in \omega_{k-1}$ and for each $F \in \mathcal{E}_k(V) \cap \mathrm{Im}(\mathrm{ad}\, S) = \mathrm{Im}(\mathrm{ad}_k\, S)$ that

$$A_k(\widetilde{F}_{k-1}(\lambda) + F, \lambda) = A_{k-1}(\widetilde{F}_{k-1}(\lambda), \lambda) + B_k(F, \lambda) = B_k(F, \lambda),$$

with $B_k : (\mathrm{Im}(\mathrm{ad}_k\, S)) \times \omega_{k-1} \to \mathrm{Im}(\mathrm{ad}_k\, S)$ defined by

$$B_k(F, \lambda) := \widetilde{\pi}_k\, \widetilde{J}_k(\exp(\mathrm{ad}^{(k)}(\widetilde{F}_{k-1}(\lambda) + F)) \cdot J_k H_\lambda). \tag{4.14}$$

Hence it is sufficient to solve the equation $B_k(F, \lambda) = 0$. Now it is easy to verify that

$$\begin{aligned} B_k(F, \lambda_0) &= \widetilde{\pi}_k \widetilde{J}_k(\exp(\mathrm{ad}^{(k)} \widetilde{F}_{k-1}(\lambda_0)) \cdot J_k H_{\lambda_0}) + (\mathrm{ad}^{(k)} F) \cdot Q_0 \\ &= \widetilde{\pi}_k \widetilde{J}_k(\exp(\mathrm{ad}^{(k)} \widetilde{F}_{k-1}(\lambda_0)) \cdot J_k H_{\lambda_0}) - (\mathrm{ad}_k\, Q_0) \cdot F; \end{aligned}$$

since $\mathrm{ad}_k(Q_0) = \mathrm{ad}_k(S) + \mathrm{ad}_k(N)$ is invertible on $\mathrm{Im}(\mathrm{ad}_k\, S)$ we can again apply the implicit function theorem to conclude that there exist a neighborhood $\omega_k \subset \omega_{k-1}$ of λ_0 in Λ and a smooth mapping $\overline{F}_k : \omega_k \to \mathrm{Im}(\mathrm{ad}_k\, S)$ such that $B_k(\overline{F}_k(\lambda), \lambda) = 0$ for $\lambda \in \omega_k$. Setting $\widetilde{F}_k(\lambda) = \widetilde{F}_{k-1}(\lambda) + \overline{F}_k(\lambda)$ then proves the lemma and Theorem 4.1. □

5 Proof of the main result

In this section we carry out the Liapunov-Schmidt reduction which will prove our main Theorem 3.2. Fixing some $k \geq 2$ we can use Theorem 4.1 to assume that the Hamiltonian $H(x, \lambda)$ appearing in (3.1) has the form

$$H(x, \lambda) = H_{NF}(x, \lambda) + R(x, \lambda), \tag{5.1}$$

with $\{S, H_{NF,\lambda}\} = 0$ for all λ in a neighborhood ω_k of λ_0 in Λ and $R(x,\lambda) = O(\|x\|^{k+2})$ as $x \to 0$, uniformly in $\lambda \in \omega_k$. We write the corresponding Hamiltonian system as

$$\dot{x} = X_H(x,\lambda) = X_{NF}(x,\lambda) + X_R(x,\lambda). \tag{5.2}$$

It follows from Lemma 2.4 that

$$X_{NF}(e^{A_s\theta}x,\lambda) = e^{A_s\theta}X_{NF}(x,\lambda), \quad \forall\theta \in \mathbb{R}, \tag{5.3}$$

while also $X_R(x,\lambda) = O(\|x\|^{k+1})$ as $x \to 0$, uniformly in $\lambda \in \omega_k$.

The first step is a time rescaling in (5.2); for each $T > 0$ near T_0 we set $\sigma := \frac{T_0}{T} - 1$, and observe that $\tilde{x} : \mathbb{R} \to V$ is a T-periodic solution of (5.2) if and only if $\hat{x}(t) := \tilde{x}((1+\sigma)^{-1}t)$ is a T_0-periodic solution of the equation

$$(1+\sigma)\dot{x} = X_H(x,\lambda). \tag{5.4}$$

Therefore, our problem is equivalent to that of finding, for all (σ,λ) near $(0,\lambda_0)$, all small T_0-periodic solutions of (5.4).

Let Z be the Banach space of all continuous T_0-periodic mappings $z : \mathbb{R} \to V$ equipped with the supremum norm. We let X be the subspace of all $x \in Z$ which are of class C^1, and equip X with the C^1-supremum norm to make it a Banach space. On Z we also define a symplectic form $\omega_Z : Z \times Z \to \mathbb{R}$ by

$$\omega_Z(z_1,z_2) := \frac{1}{T_0}\int_0^{T_0} \omega(z_1(t), z_1(t))\,dt, \quad \forall z_1, z_2 \in Z. \tag{5.5}$$

Finally we define a smooth mapping $F : X \times \mathbb{R} \times \Lambda \to Z$ by

$$F(x,\sigma,\lambda)(t) := -(1+\sigma)\dot{x}(t) + X_H(x(t),\lambda), \quad \forall t \in \mathbb{R}. \tag{5.6}$$

We then have to solve the equation

$$F(x,\sigma,\lambda) = 0 \tag{5.7}$$

for (x,σ,λ) near $(0,0,\lambda_0)$ in $X \times \mathbb{R} \times \Lambda$.

First we collect the main properties of the mapping F.

1. F is C^∞-smooth, and $F(0,\sigma,\lambda) = 0$ for all (σ,λ).
2. F is (formally) Hamiltonian, as follows. Define functionals $\mathcal{S} : X \to \mathbb{R}$, $\mathcal{H} : Z \times \Lambda \to \mathbb{R}$ and $\mathcal{F} : X \times \mathbb{R} \times \Lambda \to \mathbb{R}$ by

$$\mathcal{S}(x) := \tfrac{1}{2}\omega_z(\dot{x},x), \quad \mathcal{H}(z,\lambda) := \frac{1}{T_0}\int_0^{T_0} H(z(t),\lambda)\,dt, \tag{5.8}$$

and

$$\mathcal{F}(x,\sigma,\lambda) := -(1+\sigma)\mathcal{S}(x) + \mathcal{H}(x,\lambda). \tag{5.9}$$

It is then easily seen that

$$D_x\mathcal{F}(x,\sigma,\lambda)\cdot\tilde{x} = \omega_Z(F(x,\sigma,\lambda),\tilde{x}), \quad \forall\tilde{x} \in X. \tag{5.10}$$

3. F is S^1-equivariant: if one defines a linear action $\Gamma : \mathbb{R}/T_0\mathbb{Z} \cong S^1 \to \mathcal{L}(Z)$ of S^1 over Z by $(T_\varphi z)(t) := z(t+\varphi)$ (for $\varphi \in S^1$ and $z \in Z$), then this action is symplectic:

$$\omega_Z(\Gamma_\varphi z_1, \Gamma_\varphi z_2) = \omega_Z(z_1, z_2), \quad \forall \varphi \in S^1, \tag{5.11}$$

and we have

$$\mathcal{F}(\Gamma_\varphi x, \sigma, \lambda) = \mathcal{F}(x, \sigma, \lambda) \text{ and } F(\Gamma_\varphi x, \sigma, \lambda) = \Gamma_\varphi F(x, \sigma, \lambda), \quad \forall \varphi \in S^1. \tag{5.12}$$

Next we set $L := D_x F(0, 0, \lambda_0) \in \mathcal{L}(X, Z)$; explicitly L has the form

$$(Lx)(t) = -\dot{x}(t) + A_0 x(t) = -\dot{x}(t) + A_s x(t) + A_n x(t),$$

and thus

$$L = L_s + L_n, \ (L_s x)(t) := -\dot{x}(t) + A_s x(t), \ (L_n x)(t) := A_n x(t). \tag{5.13}$$

These linear operators have the following properties:

4. $L_s L_n = L_n L_s$.
5. L_n is nilpotent.
6. $\ker L_s = \zeta(U)$, where $\zeta : U \to X$ is defined by $\zeta(u)(t) := e^{A_s t}u$, $\forall t \in \mathbb{R}$.
7. $Y := \text{Im } L_s = \{z \in Z \mid \omega_Z(\zeta(u), z) = 0, \ \forall u \in U\}$. Together with (**6**) this implies that L_s is a Fredholm operator with index 0. To prove (**7**), use the variation of constants formula in combination with Lemma 3.1, which implies that
$$\text{Im}(e^{A_s T_0} - I) = \{x \in V \mid \omega(u, x) = 0, \ \forall u \in U\}.$$
8. $Z = \zeta(U) \oplus Y$; this follows from (**6**), (**7**), Lemma 3.1 and the fact that
$$\omega_Z(\zeta(u_1), \zeta(u_2)) = \widetilde{\omega}(u_1, u_2), \ \forall u_1, u_2 \in U.$$
9. $L = L_s + L_n$ maps $Y \cap X$ into Y, and $L|_{Y \cap X} \in \mathcal{L}(X \cap Y, Y)$ is invertible.
10. $\ker L = \ker L_s \cap \ker L_n = \zeta(U \cap \ker A_n) \subset \zeta(U)$.

Let $P_0 \in \mathcal{L}(V)$ be the projection with Im $P_0 = U$ and $\ker P_0 = \text{Im}(e^{A_s T_0} - I)$, and define $P \in \mathcal{L}(Z, U)$ by

$$Pz := \frac{1}{T_0} \int_0^{T_0} e^{-A_s t} P_0 z(t)\, dt. \tag{5.14}$$

Then we have the following properties:

11. $\omega(P_0 x_1, x_2) = \omega(P_0 x_1, P_0 x_2) = \omega(x_1, P_0 x_2), \quad \forall x_1, x_2 \in V$.
12. $\ker P = Y$.
13. $\zeta \circ P \in \mathcal{L}(Z)$ is a projection, with $\text{Im}(\zeta \circ P) = \zeta(U)$ and $\ker(\zeta \circ P) = Y$.
14. $\zeta(e^{A_s \varphi} u) = \Gamma_\varphi \zeta(u)$, $\forall \varphi \in S^1$.
15. $P(\Gamma_\varphi z) = e^{A_s \varphi} Pz$, $\forall \varphi \in S^1$.
16. $(\zeta \circ P)\Gamma_\varphi = \Gamma_\varphi(\zeta \circ P)$, $\forall \varphi \in S^1$.

Using these properties we can write each $x \in X$ in the form $x = \zeta(u) + y$, with $u \in U$ and $y \in X \cap Y$; the equation (5.7) is then equivalent to the system

$$\begin{aligned} &(a) \quad (I - \zeta_\circ P)F(\zeta(u) + y, \sigma, \lambda) = 0, \\ &(b) \qquad\quad PF(\zeta(u) + y, \sigma, \lambda) = 0. \end{aligned} \tag{5.15}$$

By property (**9**) and the implicit function theorem we can solve (5.15(a)) for $y = \tilde{y}(u, \sigma, \lambda)$; bringing this solution into (5.15(b)) gives us the *bifurcation equation*

$$G(u, \sigma, \lambda) := PF(\zeta(u) + \tilde{y}(u, \sigma, \lambda), \sigma, \lambda) = 0. \tag{5.16}$$

The mappings $\tilde{y} : U \times \mathbb{R} \times \Lambda \to X \cap Y$ and $G : U \times \mathbb{R} \times \Lambda \to U$ have the following properties, which follow easily from the implicit function theorem, in particular the uniqueness part:

17. $\tilde{y}$ and G are C^∞-smooth and uniquely defined on a neighborhood of $(0, 0, \lambda_0)$ in $U \times \mathbb{R} \times \Lambda$.

18. $\tilde{y}(0, \sigma, \lambda) = 0$ and $G(0, \sigma, \lambda) = 0$.

19. $F(\zeta(u) + \tilde{y}(u, \sigma, \lambda), \sigma, \lambda) = \zeta(G(u, \sigma, \lambda))$.

20. $\tilde{y}(e^{A_s\varphi}u, \sigma, \lambda) = \Gamma_\varphi \tilde{y}(u, \sigma, \lambda), \ \forall \varphi \in S^1$.

21. $G(e^{A_s\varphi}u, \sigma, \lambda) = e^{A_s\varphi}G(u, \sigma, \lambda), \ \forall \varphi \in S^1$, i.e. the equation (5.16) is S^1-equivariant.

Moreover, using (5.2) and (5.3) we also have:

22. $X_{NF}(u, \lambda) \in U$ for $u \in U$, since $X_{NF}(u, \lambda) = X_{NF}(e^{A_s T_0}u, \lambda) = e^{A_s T_0}X_{NF}(u, \lambda)$.

23. $F(\zeta(u), \sigma, \lambda)(t) = e^{A_s t}[-(1 + \sigma)A_s u + X_{NF}(u, \lambda)] + X_R(e^{A_s t}u, \lambda)$, from which it follows that $(I - \zeta_\circ P)F(\zeta(u), \sigma, \lambda) = O(\|u\|^{k+1})$ as $u \to 0$, uniformly for (σ, λ) near $(0, \lambda_0)$.

24. $\tilde{y}(u, \sigma, \lambda) = O(\|u\|^{k+1})$ as $u \to 0$.

25. $G(u, \sigma, \lambda) = -(1 + \sigma)A_s u + X_{NF}(u, \lambda) + O(\|u\|^{k+1})$.

Before we continue the proof, we make the following observations:

a. If $\{S, H_\lambda\} = 0$ for all λ, that is if $H = H_{NF}$, then $\tilde{y}(u, \sigma, \lambda) \equiv 0$ and $G(u, \sigma, \lambda) = -(1 + \sigma)A_s u + X_H(u, \lambda)$. This implies that for (σ, λ) near $(0, \lambda_0)$ all small solutions of (5.7) have the form $x = \zeta(u)$, with (u, σ, λ) satisfying the equation

$$(1 + \sigma)A_s u = X_H(u, \lambda). \tag{5.17}$$

Hence for (T, λ) near (T_0, λ_0) all small T-periodic solutions of $(3.1)_\lambda$, have the form $\tilde{x}(t) = e^{(1+\sigma)A_s t}u$, with $\sigma = \frac{T_0}{T} - 1$ and (u, σ, λ) satisfying (5.17).

b. If $\widetilde{H} : U \times \Lambda \to \mathbb{R}$ is S^1-invariant, then applying (**a**) to (3.8) shows that for $(T, \lambda) = ((1 + \sigma)^{-1}T_0, \lambda)$ near (T_0, λ_0) all small T-periodic solutions of $(3.8)_\lambda$ have the form (3.9), with (u, σ, λ) a solution of (3.13).

c. From (**a**) and (**b**) we see that Theorem 3.2 holds when $\{S, H_\lambda\} = 0$ for all λ: just take $\widetilde{H}(u, \lambda) = H(u, \lambda)$ and $\Psi(u, \lambda) = u$. Observe that the determining equation (5.17) is equivalent to the equation

$$D_u \mathcal{F}_0(u, \sigma, t) = 0, \tag{5.18}$$

where the S^1-invariant functional $\mathcal{F}_0 : U \times \mathbb{R} \times \Lambda \to \mathbb{R}$ is given by $\mathcal{F}_0(u,\sigma,\lambda) := \mathcal{F}(\zeta(u),\sigma,\lambda) = -(1+\sigma)\widetilde{S}(u) + H(u,\lambda)$.

d. By the same argument Theorem 3.2 will be proven in the general case when we can show the existence of an S^1-invariant functional $\mathcal{F}_0 : U \times \mathbb{R} \times \Lambda \to \mathbb{R}$ of the form

$$\mathcal{F}_0(u,\sigma,\lambda) = -(1+\sigma)\widetilde{S}(u) + \widetilde{H}(u,\lambda) \tag{5.19}$$

such that the bifurcation equation (5.16) is equivalent to (5.18). Setting $\widetilde{\mathcal{F}}(u,\sigma,\lambda) := \mathcal{F}(\zeta(u) + \tilde{y}(u,\sigma,\lambda),\sigma,\lambda)$ and using property (**7**) it is easily seen that $D_u\widetilde{\mathcal{F}}(u,\sigma,\lambda)\cdot\tilde{u} = \widetilde{\omega}(G(u,\sigma,\lambda),\tilde{u})$, and hence (5.16) is equivalent to $D_u\widetilde{\mathcal{F}}(u,\sigma,\lambda) = 0$. However, $\widetilde{F}$ does not have the simple form (5.19), because of the presence of higher order terms depending on σ and λ and coming from $\tilde{y}(u,\sigma,\lambda)$. Therefore we need some further reduction.

Our first step in this further reduction is to eliminate σ from the higher order terms. Since $F(0,\sigma,\lambda) = 0$ we have $D_\sigma G(0,0,\lambda_0) = 0$, and hence we cannot directly solve (5.16) for σ. The way out is to introduce polar coordinates on U. Let $(.,.)$ be an inner product on U for which the S^1-action (3.6) is orthogonal, i.e. such that

$$(e^{A_s\varphi}u, e^{A_s\varphi}\tilde{u}) = (u,\tilde{u}), \quad \forall u,\tilde{u}\in U, \forall\varphi\in S^1. \tag{5.20}$$

Let $\Sigma := \{\theta \in U \mid (\theta,\theta) = 1\}$ be the unit sphere in U, and define $g : \mathbb{R}\times\Sigma\times\mathbb{R}\times\Lambda \to \mathbb{R}$ by

$$g(\rho,\theta,\lambda) := \begin{cases} (A_s\theta, \rho^{-1}G(\rho\theta,\sigma,\lambda)) & \text{for } \rho \neq 0, \\ (A_s\theta, D_uG(0,\sigma,\lambda).\theta) & \text{for } \rho = 0. \end{cases} \tag{5.21}$$

If $(u,\sigma,\lambda) = (\rho\theta,\sigma,\lambda)$ is a solution of (5.16) with $\rho \neq 0$ then we have necessarily

$$g(\rho,\theta,\sigma,\lambda) = 0. \tag{5.22}$$

Now the function g is smooth, with

$$g(0,\theta,\sigma,\lambda_0) = -\sigma(A_s\theta, A_s\theta) + (A_s\theta, A_n\theta); \tag{5.23}$$

it follows that $D_\sigma g(0,\theta,\sigma,\lambda_0) = -(A_s\theta,A_s\theta) \neq 0$, since we have assumed that A_0 and hence also A_s are nonsingular. Consequently (5.22) has for each $\theta\in\Sigma$ and for each (ρ,λ) near $(0,\lambda_0)$ a unique solution $\sigma = \tilde{\sigma}(\rho,\theta,\lambda)$. One easily verifies that

29. $\tilde{\sigma}(\rho, e^{A_s\varphi}\theta,\lambda) = \tilde{\sigma}(\rho,\theta,\lambda)$, $\forall\varphi\in S^1$.

30. $\tilde{\sigma}(-\rho,-\theta,\lambda) = \tilde{\sigma}(\rho,\theta,\lambda)$.

31. $\tilde{\sigma}(0,\theta,\lambda_0) = 0$ for $\theta\in\Sigma\cap\ker A_n$.

Remark The foregoing assumes that g is defined and satisfies (5.23) for all $\sigma\in\mathbb{R}$, while according to (**17**) $\tilde{y}(u,\sigma,\lambda)$ and $G(u,\sigma,\lambda)$ are only well defined for σ near zero. The solution to this difficulty is to extend $\tilde{y}(u,\sigma,\lambda)$ to all $\sigma\in\mathbb{R}$ in such a way that $\tilde{y}(u,\sigma,\lambda) = O(\|u\|^{k+1})$ as $u\to 0$, uniformly for all $\sigma\in\mathbb{R}$ and for λ near λ_0, and then to use (5.16) and (5.21) to define G and g correspondingly. However we should be aware that the results make sense only for σ near zero.

Using $\tilde{\sigma}$ we define $\sigma^* : (U - \{0\}) \times \Lambda \to \mathbb{R}$ by $\sigma^*(\rho\theta, \lambda) := \tilde{\sigma}(\rho, \theta, \lambda)$, and then substitute $\sigma^*(u, \lambda)$ for σ in the mappings $\tilde{y}$ and G. This gives us mappings $y^* : U \times \Lambda \to X \cap Y$, $x^* : U \times \Lambda \to X$ and $G^* : U \times \Lambda \to U$, defined as follows:

$$y^*(u, \lambda) := \tilde{y}(u, \sigma^*(u, \lambda), \lambda) \text{ if } u \neq 0,\ y^*(0, \lambda) := 0, \tag{5.24}$$

$$x^*(u, \lambda) := \zeta(u) + y^*(u, \lambda), \tag{5.25}$$

and

$$G^*(u, \lambda) := G(u, \sigma^*(u, \lambda), \lambda) \text{ if } u \neq 0;\ G^*(0, \lambda) := 0 \tag{5.26}$$

These mappings have the following properties:

32. σ^* is C^∞-smooth, and $\sigma^*(e^{A_s\varphi}u, \lambda) = \sigma^*(u, \lambda)$, $\forall \varphi \in S^1$.

33. $y^*(u, \lambda) = O(\|u\|^{k+1})$ as $u \to 0$.

34. y^* and x^* are of class C^k, and C^∞ for $u \neq 0$.

35. $x^*(e^{A_s\varphi}u, \lambda) = \Gamma_\varphi x^*(u, \lambda)$, $\forall \varphi \in S^1$.

36. G^* is continuous, and C^∞ for $u \neq 0$.

37. $G^*(e^{A_s\varphi}u, \lambda) = e^{A_s\varphi}G(u, \lambda)$, $\forall \varphi \in S^1$.

38. $(G^*(u, \lambda), A_s u) = 0$ for all $(u, \lambda) \in U \times \Lambda$ close to $(0, \lambda_0)$.

39. $F(x^*(u, \lambda), \sigma^*(u, \lambda), \lambda) = \zeta(G^*(u, \lambda))$.

The reduction up to this point can be summarized as follows.

Lemma 5.1 *Let $(x, \sigma, \lambda) \in (X - \{0\}) \times \mathbb{R} \times \Lambda$ be sufficiently close to $(0, 0, \lambda_0)$. Then $F(x, \sigma, \lambda) = 0$ if and only if $(x, \sigma) = (x^*(u, \lambda), \sigma^*(u, \lambda))$ for some $u \in U - \{0\}$ close to zero and such that $G^*(u, \lambda) = 0$.*

Now we go back to the variational structure of (5.7), as expressed by (5.10). We set

$$\mathcal{S}^*(u, \lambda) := \mathcal{S}(x^*(u, \lambda)),\ \mathcal{H}^*(u, \lambda) := \mathcal{H}(x^*(u, \lambda), \lambda) \tag{5.27}$$

and

$$\mathcal{F}^*(u, \sigma, \lambda) := -(1 + \sigma)\mathcal{S}^*(u, \lambda) + \mathcal{H}^*(u, \lambda). \tag{5.28}$$

(Observe that we do not replace σ by $\sigma^*(u, \lambda)$). From (**7**) we see that if $y \in X \cap Y$ and $u \in U$ then

$$\omega_Z(\zeta(u), \dot{y}) = -\omega_Z(\tfrac{d}{dt}\zeta(u), y) = -\omega_Z(\zeta(A_s u), y) = 0.$$

In combination with (5.25), (5.12), (**33**) and (**35**) this implies:

40. $\mathcal{S}^*(u, \lambda) = \widetilde{S}(u) + O(\|u\|^{2k+2})$ as $u \to 0$; hence $\mathcal{S}^*$ is of class C^{2k+1} near $u = 0$, and C^∞ for $u \neq 0$.

41. $\mathcal{H}^*(u, \lambda) = H_{NF}(u, \lambda) + O(\|u\|^{k+2})$, such that $\mathcal{H}^*$ is of class C^{k+1} near $u = 0$ and C^∞ for $u \neq 0$.

42. $\mathcal{S}^*(e^{A_s\varphi}u, \lambda) = \mathcal{S}^*(u, \lambda)$, $\mathcal{H}^*(e^{A_s\varphi}u, \lambda) = \mathcal{H}^*(u, \lambda)$ and $\mathcal{F}^*(e^{A_s\varphi}u, \sigma, \lambda) = \mathcal{F}^*(u, \sigma, \lambda)$, for all $\varphi \in S^1$.

The next calculation forms a crucial step which will allow us to reformulate the characterization of the solutions of (5.7) as given by Lemma 5.1 in terms of the functional $\mathcal{F}^*$. Let $(u, \sigma, \lambda) \in (U - \{0\}) \times \mathbb{R} \times \Lambda$ be such that $(u, \sigma^*(u, \lambda), \lambda)$ is sufficiently close to $(0, 0, \lambda_0)$. Using (5.28) and the properties (**7**) and (**39**) one finds then for all $\tilde{u} \in U$ that

$$\begin{aligned}
&D_u\mathcal{F}^*(u, \sigma, \lambda) \cdot \tilde{u} \\
&= D_x\mathcal{F}(x^*(u, \lambda), \sigma, \lambda) \cdot D_u x^*(u, \lambda) \cdot \tilde{u} \\
&= [D_x\mathcal{F}(x^*(u, \lambda), \sigma^*(u, \lambda), \lambda) + (\sigma^*(u, \lambda) - \sigma)D_x\mathcal{S}(x^*(u, \lambda))] \cdot D_u x^*(u, \lambda) \cdot \tilde{u} \\
&= \omega_Z(F(x^*(u, \lambda), \sigma^*(u, \lambda), \lambda)\,,\, D_u x^*(u, \lambda) \cdot \tilde{u}) + (\sigma^*(u, \lambda) - \sigma)D_u\mathcal{S}^*(u, \lambda) \cdot \tilde{u} \\
&= \omega_Z(\zeta(G^*(u, \lambda)), \zeta(\tilde{u}) + D_u y^*(u, \lambda) \cdot \tilde{u}) + (\sigma^*(u, \lambda) - \sigma)\tilde{\omega}(X_{\mathcal{S}^*}(u, \lambda), \tilde{u}) \\
&= \tilde{\omega}(G^*(u, \lambda) + (\sigma^*(u, \lambda) - \sigma)X_{\mathcal{S}^*}(u, \lambda)\,,\, \tilde{u}). \qquad (5.29)
\end{aligned}$$

Lemma 5.2 *Let $(x, \sigma, \lambda) \in (X - \{0\}) \times \mathbb{R} \times \Lambda$ be sufficiently close to $(0, 0, \lambda_0)$. Then we have that $F(x, \sigma, \lambda) = 0$ if and only if $x = x^*(u, \lambda)$ for some sufficiently small $u \in U - \{0\}$ such that*

$$D_u\mathcal{F}^*(u, \sigma, \lambda) = 0. \qquad (5.30)$$

Proof By Lemma 5.1 it is sufficient to show that for all $(u, \sigma, \lambda) \in (U - \{0\}) \times \mathbb{R} \times \Lambda$ close to $(0, 0, \lambda_0)$ (5.30) is equivalent to

$$G^*(u, \lambda) = 0 \quad \text{and} \quad \sigma = \sigma^*(u, \lambda). \qquad (5.31)$$

From (5.29) we see immediately that (5.31) implies (5.30); we can use (5.29) because (5.31) implies that $\sigma^*(u, \lambda) = \sigma$ is small.

Conversely, suppose that (5.30) holds; then $\sigma^*(u, \lambda)$ is small, and we can again use (5.29). Indeed, (5.30) implies

$$X_{\mathcal{F}^*}(u, \sigma, \lambda) = 0, \qquad (5.32)$$

where $X_{\mathcal{F}^*}$ is the Hamiltonian vectorfield corresponding to $\mathcal{F}^*$ via the symplectic form $\tilde{\omega}$ on U. Properties (**40**) and (**41**) show that $D_u X_{\mathcal{F}^*}(0, 0, \lambda_0) \cdot \tilde{u} = A_n \tilde{u}$. A straightforward Liapunov-Schmidt reduction on (5.32) then shows that u must be close to $U \cap \ker A_n$, which by (**31**) implies that $\sigma^*(u, \lambda)$ is small. From (5.29) and (**38**) we obtain then

$$G^*(u, \lambda) = (\sigma - \sigma^*(u, \lambda))X_{\mathcal{S}^*}(u, \lambda) \qquad (5.33)$$

and

$$(\sigma - \sigma^*(u, \lambda))(X_{\mathcal{S}^*}(u, \lambda)\,,\, A_s u) = 0. \qquad (5.34)$$

From (**40**) we see that $X_{\mathcal{S}^*}(u, \lambda) = A_s u + O(\|u\|^{2k+1})$, such that $(X_{\mathcal{S}^*}(u, \lambda), A_s u) \neq 0$ for $u \neq 0$ and small. Therefore (5.34) implies $\sigma = \sigma^*(u, \lambda)$, and (5.33) gives then $G^*(u, \lambda) = 0$. We conclude that (5.30) implies (5.31), which proves the lemma. □

Comparing (5.28) with (5.19) and using Lemma 5.2 we see that we are almost finished: to conclude the proof of Theorem 3.2 it is sufficient to find a way to transform the term $\mathcal{S}^*(u, \lambda)$ in (5.28) into the term $\widetilde{S}(u)$ appearing in (5.19); this must be done in a way which respects the S^1-invariance given by (**42**). We will use the S^1-equivariant version of the Morse lemma proven in the appendix. This result shows that there exists a mapping $\Phi : U \times \Lambda \to U$ with the following properties:

43. $\Phi(u,\lambda) = u + O(\|u\|^{2k+1})$.

44. Φ is of class C^{2k}, and C^∞ for $u \neq 0$.

45. $\Phi(e^{A_s\varphi}u,\lambda) = e^{A_s\varphi}\Phi(u,\lambda)$, $\forall \varphi \in S^1$.

46. $\mathcal{S}^*(\Phi(u,\lambda),\lambda) = \widetilde{S}(u)$ for (u,λ) near $(0,\lambda_0)$.

It follows that $\Phi(\cdot,\lambda)$ is, for λ near λ_0, a diffeomorphism on a neighborhood of $u = 0$. Replacing u by $\Phi(u,\lambda)$ in the foregoing we see that $\mathcal{F}_0(u,\sigma,\lambda) := \mathcal{F}^*(\Phi(u,\lambda),\sigma,\lambda)$ has the form (5.19), with $\widetilde{H}(u,\lambda) := \mathcal{H}^*(\Phi(u,\lambda),\lambda)$; the equation $D_u\mathcal{F}^*(\Phi(u,\lambda),\sigma,\lambda) = 0$ is then equivalent to (5.18) or to (3.13). Finally, if we set $\Psi(u,\lambda) := x^*(\Phi(u,\lambda),\lambda)(0)$ then (**35**) and (**45**) imply that

$$x^*(\Phi(u,\lambda),\lambda)(t) = (\Gamma_t\, x^*(\Phi(u,\lambda),\lambda)(0) = x^*(\Phi(e^{A_st}u,\lambda),\lambda)(0) = \Psi(e^{A_st}u,\lambda). \tag{5.35}$$

We can then reformulate Lemma 5.2 as follows.

Lemma 5.3 *Let $(x,\sigma,\lambda) \in (X - \{0\}) \times \mathbb{R} \times \Lambda$ be sufficiently close to $(0,0,\sigma_0)$. Then we have $F(x,\sigma,\lambda) = 0$ if and only if $x = x^*(\Phi(u,\lambda),\lambda)$ or equivalently $x(t) = \Psi(e^{A_st}u,\lambda)$ for some sufficiently small $u \in U - \{O\}$ satisfying (3.13).* □

We have now all the elements to formally conclude the proof of Theorem 3.2. We first assume that H is in normal form up to order $(k+1)$, i.e. we have the set-up assumed at the beginning of this section. It is then easily seen that the mappings $\widetilde{H}$ and Ψ introduced above satisfy the properties (i), (ii), (iii) and (vii) of Theorem 3.2; property (iv) follows from Lemma 2.4 and the fact that $\widetilde{H}(e^{A_s\varphi}u,\lambda) = \widetilde{H}(u,\lambda)$ for all $\varphi \in S^1$. It remains to prove (v) and (vi). Now $\tilde{x}(t)$ is a nontrivial $T = (1+\sigma)^{-1}T_0$-periodic solution of $(3.1)_\lambda$ if and only if $\hat{x}(t) := \tilde{x}((1+\sigma)^{-1}t)$ is a T_0-periodic solution of $(5.4)_\lambda$, which in turn is equivalent to $F(\hat{x},\sigma,\lambda) = 0$. By Lemma 5.3 and for $(\hat{x},\sigma,\lambda)$ sufficiently close to $(0,0,\lambda_0)$ this is equivalent to

$$\hat{x}(t) = \Psi(e^{A_st}u,\lambda), \tag{5.36}$$

with (u,σ,λ) satisfying (3.13). Now (5.36) is equivalent to

$$\tilde{x}(t) = \Psi(\tilde{u}(t),\lambda), \text{ with } \tilde{u}(t) := e^{(1+\sigma)A_st}u, \tag{5.37}$$

while from our observation (**b**) above we know that $\tilde{u}(t)$ is a T-periodic solution of (3.8) if and only if $\tilde{u}(t)$ has the form given in (5.37), with (u,σ,λ) satisfying (3.13). This concludes the proof of Theorem 3.2.

6 Appendix: The equivariant Morse lemma

In this appendix we prove a parameter-dependent version of the equivariant Morse lemma (see Arnold [1976]). Let U be a finite-dimensional real vectorspace, and Γ a compact Lie group acting linearly on U; let $< \cdot,\cdot >$ be an inner product on U for which the action of Γ is orthogonal, i.e. such that $< \gamma u_1, \gamma u_2 > = < u_1, u_2 >$, $\forall \gamma \in \Gamma$. If $f : U \to \mathbb{R}$ is differentiable we define the gradient $\nabla f : U \to U$ by the relation $Df(u)\cdot\tilde{u} = < \nabla f(u), \tilde{u} >$; if f is Γ-invariant (i.e. $f(\gamma u) = f(u)$, $\forall \gamma \in \Gamma$), then ∇f is Γ-equivariant: $\nabla f(\gamma u) = \gamma \nabla f(u)$, $\forall \gamma \in \Gamma$. As in the paper Λ denotes a Banach space of parameters.

Theorem 6.1 *Let $S^* : U \times \Lambda \to \mathbb{R}$ have the form $S^*(u, \lambda) = S(u) + R(u, \lambda)$, with $S : U \to \mathbb{R}$ a nondegenerate quadratic form, and $R : U \times \Lambda \to \mathbb{R}$ a smooth function such that $D_u^j R(0, \lambda) = 0$ for $0 \leq j \leq m-1$ $(m \geq 3)$ and for λ near some $\lambda_0 \in \Lambda$. Assume that S and R are Γ-invariant. Then there exists a smooth Γ-equivariant mapping $\Phi : U \times \Lambda \to U$, with $\Phi(u, \lambda) = u + O(\|u\|^{m-1})$ as $u \to 0$ and uniformly for λ near λ_0, such that*

$$S^*(\Phi(u, \lambda), \lambda) = S(u), \quad \forall (u, \lambda) \in U \times \Lambda \text{ near } (0, \lambda_0). \tag{6.1}$$

Proof The idea of the proof is to construct a smooth Γ-equivariant mapping $\eta : \mathbb{R} \times U \times \Gamma \to U$ such that

$$S(\varphi(t, u, \lambda)) + tR(\varphi(t, u, \lambda), \lambda) = S(u), \tag{6.2}$$

where $\varphi(t, u, \lambda)$ denotes the flow of the (non-autonomous) equation

$$\dot{u} = -\eta(t, u, \lambda); \tag{6.3}$$

then (6.1) follows immediately by taking $\Phi(u, \lambda) := \varphi(1; u, \lambda)$.

Differentiating (6.2) we see that (6.2) will hold if

$$R(u, \lambda) =< \nabla S(u) + t\,\nabla_u R(u, \lambda),\ \eta(t, u, \lambda) > . \tag{6.4}$$

Let $\psi(t, u, \lambda) := \nabla S(u) + t\,\nabla_u R(u, \lambda)$; ψ is smooth, Γ-equivariant, $\psi(t, 0, \lambda) = 0$ and $D_u\psi(t, 0, \lambda)$ is invertible, by the nondegeneracy condition on $S(u)$. It follows that the mapping $(t, u, \lambda) \mapsto (t, \psi(t, u, \lambda), \lambda)$ is a diffeomorphism on a neighborhood of $[0, 1] \times \{0\} \times \{\lambda_0\}$, with an inverse of the form $(t, u, \lambda) \mapsto (t, \widetilde{\psi}(t, u, \lambda), \lambda)$; clearly $\widetilde{\psi}$ is smooth, Γ-equivariant, and $\widetilde{\psi}(t, 0, \lambda) = 0$. It follows that we can write

$$\widetilde{R}(t, u, \lambda) := R(\widetilde{\psi}(t, u, \lambda), \lambda) =< u, \tilde{\eta}(t, u, \lambda) >, \tag{6.5}$$

with

$$\tilde{\eta}(t, u, \lambda) = \int_0^1 \nabla_u \widetilde{R}(t, su, \lambda)\, ds; \tag{6.6}$$

$\tilde{\eta} : \mathbb{R} \times U \times \Lambda \to U$ is smooth, Γ-equivariant, and $D_u^j \tilde{\eta}(t, 0, \lambda) = 0$ for $0 \leq j \leq m-2$. Replacing u by $\psi(t, u, \lambda)$ in (6.5) shows that (6.4) holds with

$$\eta(t, u, \lambda) := \tilde{\eta}(t, \psi(t, u, \lambda), \lambda) \tag{6.7}$$

The mapping $\eta : \mathbb{R} \times U \times \Lambda \to U$ is smooth, Γ-equivariant, and satisfies $D_u^j \eta(t, 0, \lambda) = 0$ for $0 \leq j \leq m-2$. The result follows then immediately. □

Remark If, as in the paper, $R(u, \lambda)$ is only of class C^{m-1} at $u = 0$, but C^∞ outside of $u = 0$, then one can easily check that ψ, $\widetilde{\psi}$, $\tilde{\eta}$, η and Φ are of class C^{m-2} at $u = 0$, and C^∞ outside of $u = 0$.

References

Abraham, R. and Marsden, J. [1978] *Foundations of Mechanics* (Second Edition), Benjamin/Cummings Publ. Co, Reading, Mass.

Arnold, V. [1976] *Wave front evolution and equivariant Morse lemma*, Comm. Pure Appl. Math., **29**, 557–581.

Cushman, R. and Sanders, J. [1986] *Nilpotent normal forms and representation theory of* $\mathfrak{sl}(2,\mathbb{R})$, Contemp. Math., **56**, A.M.S., Providence, 31–51.

Duistermaat, J.J. [1984] *Bifurcation of periodic solutions near equilibrium in Hamiltonian systems*, Lect. Notes in Math., **1057**, Springer-Verlag, Berlin, 57–105.

Golubitsky, M., Stewart, I. and Schaeffer, D. [1988] *Singularities and Groups in Bifurcation Theory*, Vol. II. Springer-Verlag, New York.

Humphreys, J. [1972] *Introduction to Lie Algebras and Representation Theory*, Springer-Verlag, New York.

Iooss, G. and Adelmeyer, M. [1992] *Topics in Bifurcation Theory and Applications*, Advanced Series in Nonlinear Dynamics, Vol. 3, World Scientific, Singapore.

Knobloch, J. [1992] *Reduction of Hopf bifurcation problems with symmetries*, preprint.

Knobloch, J. and Vanderbauwhede, A. *Bifurcation of periodic solutions at multiple resonances in reversible systems*, in preparation.

Marsden, J. [1992] *Lectures on Mechanics*, L.M.S. Lect. Note Series 174, Cambridge Univ. Press.

Moser, J. [1976] *Periodic solutions near an equilibrium and a theorem by Alan Weinstein*, Comm. Pure Appl. Math., **29**, 727–747.

Rabinowitz, P. [1986] *Minimax Methods in Critical Point Theory with Applications to Differential Equations*, Reg. Conf. Series in Math., **65**, A.M.S., Providence.

Sanders, J., Vanderbauwhede, A. and van der Meer, J.-C. *Hamiltonian Hopf Bifurcation*, in preparation.

Vanderbauwhede, A. [1982] *Local Bifurcation and Symmetry.* Res. Notes in Math., **75**, Pitman, London.

Vanderbauwhede, A. [1989] *Centre Manifolds, Normal Forms and Elementary Bifurcations*, Dynamics Reported, **2**, 89–169.

van der Meer, J.-C. [1982] *Nonsemisimple 1:1 resonance at an equilibrium*, Cel. Mech., **7**, 131–149.

van der Meer, J.-C. [1986] *The Hamiltonian Hopf Bifurcation*, Lect. Notes in Math., **1160**, Springer-Verlag, Berlin.

van der Meer, J.-C. *Degenerate Hamiltonian Hopf bifurcations*, to appear in the Fields Institute Communications volume, Conservative Systems and Quantum Chaos, L. Bates and D. Rod, editors.

Weinstein, A. [1973] *Normal modes for nonlinear Hamiltonian systems*, Inv. Math., **20** 47–57.